Geology

and

Plant Life

Geology and Plant Life

The Effects of Landforms and Rock Types on Plants

ARTHUR R. KRUCKEBERG

University of Washington Press

Seattle and London

Printed in the United States of America
Design by Veronica Seyd

Library of Congress Cataloging-in-Publication Data

Kruckeberg, Arthur R.

Geology and plant life : the effects of landforms and rock types on plants / Arthur R. Kruckeberg.

p. cm.

Includes bibliographical references (p.).

ISBN 0-295-98203-9 (alk. paper)

1. Phytogeography. 2. Plant ecology. 3. Geology. I. Title.

QK101 .K847 2002

581.7—dc21 2002016597

The paper used in this publication is acid-free and recycled from 10 percent post-consumer and at least 50 percent pre-consumer waste. It meets the minimum requirements of American National Standard for Information Sciences—Permanence of Paper for Printed Library Materials, ANSI Z39.48-1984.

Frontispiece: Austere, windswept landscape on limestone (karstic) pavement, Yorkshire Dales National Park, near Whernside, England. Photo by A. L. Kruckeberg.

To my three most inspiring mentors who led me to see
the functionally reciprocal linkages between geology and botany:

HANS JENNY, soil scientist

HERBERT MASON, systematist and philosopher

G. LEDYARD STEBBINS, evolutionist

Contents

Preface

"The plant world exists by geological consent, subject to change without notice." It was historian Will Durant who coined the original truism, only using "civilization" rather than "plant world." It will be my task to demonstrate the truth of this epigram. Before all other influences began to fashion life and its lavish diversity, geological events (processes and their products) created the initial environments—both physical (terrain) and chemical—for the evolutionary drama that followed. It is the terrestrial floras of the planet that are most subject to the influences of landforms and lithology. Plants first must locate themselves in particular places on the land. Where they find a "home" is first fixed by the lay of the land: mountains, valleys, plains. All such surface heterogeneities are mainly the products of geology.

Then particular habitats become even more finely subdivided by differences in rock types and their derived soils. Most regional climates, commonly thought to be the prime conditioners of regional vegetation, are products of geology, especially in the making of mountainous terrain. This reciprocity (geology ←→ plants), so often overlooked by ecologists bent on studying interactions between organisms, is the central theme of this book.

For over forty years, I have straddled the academically artificial boundary between botany and geology. It was the stark and unique world of serpentine floras that led me into this borderland science of geobotany—or as I have called it, geoedaphics. Serpentines make the clearest statement on the power of geology to influence plant life. So the linked themes of lithology and landforms as they can mold floras and vegetation make up the substance of this book.

Hence, in somewhat the sense of C. P. Snow's "two cultures," which deplores the gaps between disciplines, I offer the reader an attempt at a synthesis, probing numerous linkages between the two natural sciences—geology and botany.

A. R. Kruckeberg
May 2002

Acknowledgments

To attempt a synthesis between geology and the plant world, I have been challenged to reach beyond my area of competence, ecological and evolutionary botany. If the challenge has been met, it has been achieved with the generous and wise counsel of other scientists. Three geologists—Robert Coleman of Stanford University, and Steven Porter and Joseph Vance of the University of Washington—responded to my pestering and often naive queries on many occasions. Sections of the book dealing with serpentine ecology and metallophytes benefited from consultations with Richard B. Walker of the University of Washington and two New Zealand scientists, Roger Reeves and Robert Brooks. Dwight Billings of Duke University led me to his discovery of the effects of drastic contrasts in soils on plant distribution.

I am especially grateful to Robert Ornduff of the University of California at Berkeley, who, among the several reviewers of the manuscript, gave it a most thorough scrutiny. His painstaking critiques led to major changes in the text.

My home Department of Botany (University of Washington) has facilitated in many ways the genesis of the book. I am deeply indebted to M. Kay Suiter, expert word processing operator, who faithfully and wisely produced the several drafts of the text.

I was fortunate in having University of Washington Press staff give generously of their talents once again; special thanks to editor Leila Charbonneau and to Naomi Pascal, Marilyn Trueblood, and Veronica Seyd.

To all these individuals, and to many others too numerous to name, my deepest gratitude. Support from the National Science Foundation is also very gratefully acknowledged.

Geology

and

Plant Life

1 The Geology-Plant Interface

"Had I just landed on the far side of the moon? All rock and no life here? No . . . I had reached the high barren backbone of the Mayacamas Mountains of California. I looked down on the vineyards of the fertile Napa Valley and off toward the rugged summit of Mount St. Helena." Thus I might have written in my field notes many years ago, standing in the midst of this moonscape of a serpentine barren, where only a few widely spaced endemic crucifers (*Streptanthus brachiatus*) confirmed that the place was indeed on our planet. Here, as in many other spots around the world where ferromagnesium-rich mantle rock surfaces, the plant response to the serpentine rock and soil is so striking. Serpentines eloquently illustrate the impact of bedrock geology on plants. Sparse vegetation is often drastically different in appearance from that on adjacent nonserpentine geologies, and is usually accompanied by a unique and highly endemic flora. Besides exemplifying the influences of rock type on plant life, serpentines can also demonstrate the effects of landform on plant cover. Steepness of slope, compass direction of exposure, the presence of rock outcrops, talus, or alluvial soil make for a marked difference in serpentine vegetation and flora. The serpentine story is told in several of the forthcoming chapters.

Fields of Geology and Botany Concerned with the Plant-Geology Interface

Plants, more than most animals, are captives of their inanimate environments. All terrestrial higher plants are tethered to some kind of underpinning: soil, rock, water, or other plants. In turn, the anchoring media are the complex products of physical and biological processes and materials. A major component of the origin and character of particular habitats is geological. Geological processes

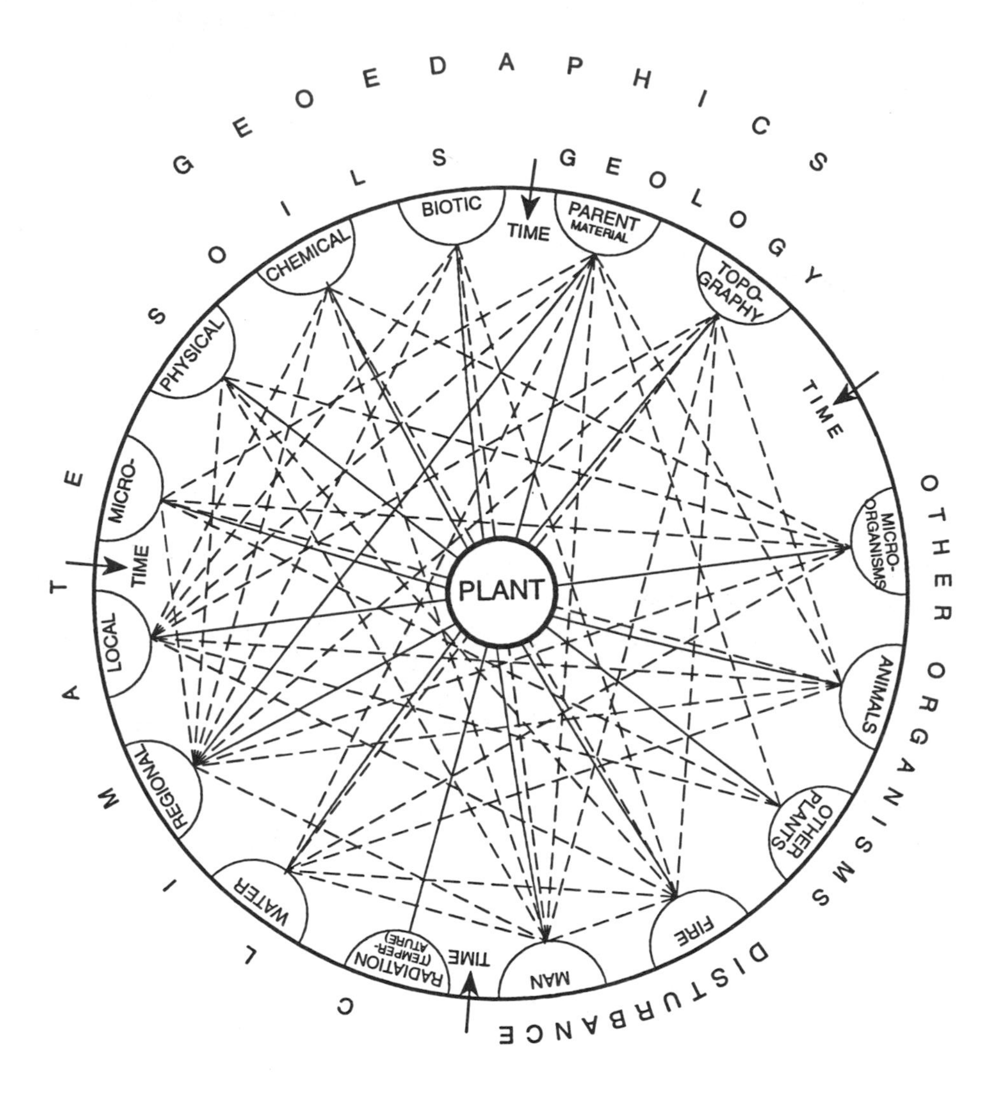

1.1 The plant-geology interface in relation to the total environmental complex. Adapted from Billings 1952.

and products lead to the formation of a particular setting—substrate or landform. It is this interplay of geology and plant life that is the focus of this book. Many facets of earth sciences bear upon the daily and evolutionary lives of plants. The geology-plant interface can be visually represented, borrowing the model devised by plant ecologist Dwight Billings years ago (1952). Figure 1.1 displays that part of the interdependent (holocoenotic) environment addressed here. Those facets of geology that significantly influence plant life would read like the chapter headings of a geology textbook. First, the processes and products that give shape to the Earth's surface—global and regional tectonics—operate all the way from shaping and positioning continents to regional mountain-building, volcanism, faulting, subduction, and kindred processes. They provide the subject matter of physical geology and geomorphology. Though transitory, the material products of the Earth's crustal movements, when exposed to the biosphere, become significant for plant life. Different parent materials (bedrock) and their particular mineral compositions can yield, by weathering, unique soil types for plants.

Changes in landforms and their lithologies are incessant. Weathering is a function of quality of parent materials, climate, and actions of organisms. The kind and amount of weathering yield a secondary geological product, the regolith—rock fragments, soils, and sediments. Erosional geology (sedimentary and glacial) then becomes an innate partner with lithology in creating substrates for plants. While pedology (soil science) is traditionally, and conveniently, a separate earth science discipline, it is really a part of geology. Soils, the weathered products of surface materials (igneous, metamorphic, and sedimentary rocks, as well as organics), are the domain of the soil scientist and the usual materials that support plant life. But soils do have an intimate geological connection, in time and in substance.

The component of time must be folded into all these linked geological processes and products. Spans of time, whether measured by a leisurely geological time clock or a more rapid method, serve to frame the geology-plant interface in years or millennia, marking beginnings and terminations. Historical geology and paleobiology thus can address time-bounded ways that geology and plants interact.

Just as the many branches of geology are functionally related to terrestrial plant life, many of the disciplines in botanical science also deal with the domain where "plants meet geology." A case can be made for a geological connection to all of plant sciences. "In nature, everything is connected to everything else," to paraphrase John Muir. Yet it is more to the point to single out those branches of botany where the influences of geology on plants are most significant.

The field of higher plant systematics recognizes the ways that geology influences plants; taxonomic status is given to plant populations affected by geologic processes and products. Gaps in the spatial display of geology yield discontinuities in plant habitats. The result—speciation—fosters a rich diversity in kinds of plant life. The taxonomist recognizes the discontinuous arrays of plants with formal names in the taxonomic hierarchy. I will come back to this fundamental attribute of nature—discontinuity—later.

Systematics, in the broadest sense, attempts to discover and describe the attributes and kinships of organisms. The best systematic treatment of a region's biota portrays the diagnostic features of living things. In so doing, it draws upon ecology, morphology, biogeography, paleobiology, and physiology for definitive judgments on the evolution of a group of plants. As a discipline of synthesis, systematics must also draw upon the influences of geologic factors. Surely systematists should include geological science in their training and in the appraisal of biological relationships.

Other botanical disciplines come under the influence of geology. Shape and structure of plants can be molded by earth-related stimuli: Differences in plant morphology and anatomy can be evolutionarily (or ecologically) conditioned by a plant's location in a geologically heterogeneous landscape. Linked to form (morphology and anatomy) is function (physiology); geology promotes particular physiological consequences. Mineral nutrition, water relations, rates and types of photosynthesis and respiration, tropisms, growth rates—all such physiological processes can be conditioned by substrates (lithology and soils) and topography (landforms, altitude, etc.). Plant responses to geology in form and function are well illustrated by the serpentine "syndrome." Serpentine geology brings about adaptive responses in structure (like dwarfism, reduction in leaf size, etc.) and in function (like tolerance to high magnesium and nickel content, drought, etc.). This linkage of plant structure and physiology will be given detailed treatment in later chapters.

But the most relevant discipline that witnesses geological influences on plants is plant ecology. Environments for plants have both biotic and abiotic (or physical) components. The physical environment of plants is created by climate and geology. Edaphic ecology focuses on the latter abiotic factor. Conventionally, for the plant ecologist the edaphic factor is the domain of soil-plant interactions. But I have contended (Kruckeberg 1986) that this is too narrow a construct. Besides soils, other geological influences (lithology, topography, etc.) shape plant life. The term I use for this broader scope for edaphics is *geoedaphics* (Kruckeberg 1986, p. 456). The particular ecologies of individual plant species and of plant communities inevitably have a strong geoedaphic component. Parent materials, slope, and exposure (lay of the land), continuous or discontinuous patterns of landforms, and particular lithologies—all critical features of plant environments—are the results of geology. Thus soils, the edaphic factor, are shaped primarily by one component (lithology, or parent materials) of the broader geoedaphic matrix. The domain of ecology that I call *geoedaphics* (see Glossary) has also been called *geoecology* (Huggett 1995).

The ecology of single plants, populations, and species (autecology) and the ecology of groupings of species into communities or associations (synecology) are under the influence of geology. The conceptual setting for all environmental interfaces—organisms with geology, organisms with climates, and organisms with other organisms—is the ecosystem. Yet no ecologist takes on the task of analyzing the interactive totality of abiotic and biotic controls of an ecosystem. Rather, the ecologist chooses to study some particular piece of the ecosystem fabric. Choice of ecological study is dictated by any number of channeling influences: personal bias and training, peer pressure and fashion, and most any other of the biopolitical stimuli that bias the choice of scientific study. Any of these can determine the type of ecological problem to be investigated. Witness the late twentieth-century preoccupation with biotic interactions in ecology. The findings from this organism-to-organism preoccupation have been impressive. The output of biotic ecology ranges from reproductive ecology (pollination syndromes, breeding systems, and population dynamics) to studies of herbivory and other symbioses (coevolutionary systems), as well as antagonisms (competition, inhibition, and parasitism). All such pursuits have yielded a wealth of understanding of how organisms deal with each other. A similar preoccupation with biotic interactions emanates from community studies. The synecologist has developed a rich encyclopedia of concepts and generalizations on the development of communities (succession) and their perpetuation (climax), their classification (phytosociology), and the many ways that the species within a community interact to create a physiognomy unique to that vegetation type.

All well and good. But where in these studies of biotic interactions do we find any meaningful descriptions of their physical (abiotic) environments? I contend that habitats provide the essential setting in which organism-to-organism interactions take place. Accounts of the physical setting are generally either "givens"—assumptions underlying a field of study—or overlooked in the zealous pursuit of a more fashionable scientific problem. The current preoccupation of most ecologists with biotic interactions lessens not a whit the need to know the abiotic setting. The components of the physical environment require but two words: climate and geology; or even one word, geoecology, in the sense of Troll (see Gerrard 1990 and Huggett 1995). Climate and geology each has its own hierarchy of environmental factors. For climate we recognize global, regional, local, and even microclimate as determinants of where plants grow. Other aspects of climate that influence the kinds and numbers in a biota are: intensity and quality of light, seasonal and daily temperature, and moisture. We need not pursue here any of these essential "givens" in the formulation Environment + Plants = Flora/Vegetation. Later we look at how geology shapes climate and how climate shapes geology (Chapters 3, 4, 5). However, the other abiotic factor, geology, is the major subject of the rest of this book. A cautionary confession should be made here: I am as guilty as any other single-purposed ecologist in selecting one "subroutine" in the ecosystem network. Hence a further "given" is in order: I fully appreciate the seamlessness of ecosystems—their animate and inanimate elements are interconnected. Only ecologists take on the ecosystem piece by piece. The naturalist knows better.

Botanical "Laws" Shaped by Geology

The presumption that there are universal and inviolate laws governing the relationship between plants and substrate is difficult to defend. "Laws" in general are in short supply in biological sciences. And they are fewer in number and less inviolate as one scales up through levels of complexity, especially beyond the individual organism to the ecosystem and biome. Yet I discern at least causal relationships, generalizations, and even testable hypotheses that have some predictive value in the realm of geology's connections with plant life.

The ideas that follow, then, may fall short of inflexible laws, but may serve as guiding principles for the geology-plant context. I am reminded of a maxim coined by G. Ledyard Stebbins (1982): "The only absolute rule in biology is that there are exceptions to every rule!"

The first set of "laws" are those broad, all-encompassing concepts that transcend particular branches of botany.

1. *Discontinuities in geological phenomena cause discontinuities in the distribution of organisms.*

I have argued elsewhere (Kruckeberg 1969b): "Discontinuity, the isolation of constituent elements of a system, manifests itself at all levels of biological systems. From macromolecule to ecosystem, gaps between elements of a system isolate the structural and functional integrity and uniqueness of the elements." Systematists, ecologists, and evolutionists recognize the significance of discontinuity for the genesis and perpetuation of the world's diverse biota. Discontinuity in time and place fosters isolation, the prime determiner of discrete, distinctive biological units, from populations and species to higher categories and from plant associations to biomes. Isolation, with its attendant generation of biotic diversity, results from the interplay between variant populations of organisms and their world of discontinuous environments. Though reinforced by internal and biological barriers to gene exchange, this isolation is ultimately achieved by discontinuities in the expression of climatic and physiographic-geological conditions, on micro to macro scales. The apparent preference of populations and species for habitats circumscribed by particular climatic and geologic features is the universal result of environmental discontinuity. The creation of discontinuities by geology is seen everywhere on our planet. Landscapes, topography, and lithology are heterogeneous and inevitably have boundaries—they are discontinuously arrayed. Imagine a physical world without the rich tapestry of landforms and rock types—a monotony of one crustal surface stretching from pole to pole. Such a landscape would foster only insensible, gradual shifts in a homogeneous biota, where climate would be the sole determiner of clinal shifts in the variation of one organism. Chapters 4 to 7 of this book deal with the effects of geologically induced discontinuity.

2. Environments are shaped by geology.

Habitats—the places where organisms live—are primarily fashioned by geologic events and materials. The four great world biomes on land (desert, grassland, savanna, and forest) originally developed as a consequence of particular geological processes. Although climate is the direct determiner of these four major vegetation types (biochores of Dansereau 1957), climates are often influenced by such geological events as mountain-building, as well as by size, shapes, and locations of landmasses. And at lesser dimensions of landscapes, the effects of geology are often paramount. Chapters 3, 4, and 5 explore the implications of the environment-geology connection.

3. Evolution is shaped by geology.

It is in the realm of the origins of species that geology provides some of the ingredients of discontinuity and isolation—the prime conditioners for speciation. The regional to local geological differences multiply varied environments to create unique evolutionary lineages. Of course, not all the species of a regional biota are the direct product of geologically induced discontinuity. Biotic interactions can further elaborate the diversity of a particular biota. But the primary fashioner of a flora (and fauna) will have been geology. Chapter 6 is devoted to the evolution-geology interface.

4. Distributions of organisms are shaped by geology.

Biogeography makes obvious connections with geology. The creation and distribution of landforms and particular substrates (lithologies and soils), when cast in a geographic framework, are the main determinants of species distribution. The influence of geology on plant distribution is the subject of Chapter 7.

Geobotanical "Laws" According to Disciplines

Some generalizations or concepts that link botany and geology are best cataloged according to a particular botanical discipline.

PHYSIOLOGICAL CONCEPTS

1. Ranges of plants are limited by tolerances to geoedaphic factors (adapted from Cain 1944). Genetically fixed tolerance ranges are the universal physiological constraints on where plants may grow. While climate is a major external determinant, it is evident that geoedaphic factors of elevation and exposure in mountains, "rain-shadow effects," parent material, and soils have modified local to micro climates to elicit specific tolerance ranges. Of course, tolerance ranges are genetically fixed responses to climatic and geoedaphic stimuli.

2. Parent materials are primary determinants of the mineral status of substrates. Bedrocks of different mineral compositions yield, upon weathering, soils with correspondingly different mineral compositions. Besides serving as inorganic nutrients, mineral elements from bedrock may sometimes be toxic, or play no known role in plant metabolism. Accumulation or exclusion of certain elements by plants is common on soils derived from parent materials with unusual mineral constituents. This linkage between lithology and elemental uptake by plants is a major subject of Chapter 5.

3. The water balance of plants can have a geologic component. There are both direct and indirect influences of geology on the water economy of plants. Landforms contain features that foster or impair optimal water relations. These are most apparent in mountainous terrain, where slopes may influence runoff or retention of water, and exposures determine amounts of precipitation; change in elevation in mountains is the arbiter of rain versus snow as the form of water storage. Water economy and its link to geology are treated in Chapter 5 (especially under "Wetlands").

4. Geoedaphic factors can induce physiological stress in plants. Habitats where landforms or lithology create unfavorable conditions for growth may induce physiological stress, either by physical or chemical means. Stresses are both real and imagined. Whenever environments appear to limit plant growth, it is often claimed that the plants are encountering stress. But I contend that only when such stress factors occur with some degree of unpredictability or newness does the plant exhibit some stress response. In contrast, when the stress factors have been accommodated to by evolved adaptation (genetically fixed tolerances) in a species, then it is no longer under stress; it is suited to that environment. Only when it was evolving tolerance did it have to pass or fail the test of physiological stress.

5. Many other genotype-phenotype responses are reactions of the geoedaphic

environment and may be expressed as particular physiological attributes. Though some such traits may be once removed from a direct geoedaphic cause, they are linked with geology nonetheless. Primary and secondary metabolism, growth phenomena, with their particular phenotypic expressions, are in part the evolved responses to the abiotic environments fashioned by geology. Even at the biochemical level, geology exerts influences. Should we not expect the "central dogma" (DNA → proteins → phenotype) in some of its specific manifestations to be under the selectional control of geological processes? This would mean that many metabolites play roles that are induced by geoedaphic stimuli. An example: The mechanism for the accumulation of nickel by some species on serpentines is likely to be explained at the molecular level.

CONCEPTS RELATING PLANT MORPHOLOGY TO GEOLOGY

Function makes form and form makes function. The ties between morphology-anatomy and physiology are close. Yet morphological and anatomical features can be phenotypic manifestations of habitat. Plant form and geology, just as with physiology, have evident linkages.

1. Particular manifestations of plant form, especially in the vegetative phases of life histories, can result from particular geologies. These are best seen where unusual substrates, such as serpentine, gypsum beds, and limestone, evoke unique morphological responses. The "serpentinomorphoses" such as those described by European geobotanists (Ritter-Studnika 1968; Pichi-Sermolli 1948) are well-known morphological indicators of exceptional substrates.

2. Life-form, while primarily under the control of climate, can be conditioned by geoedaphic factors. The extreme case of life-form response—on serpentine substrates—is expressed as a shift in life-form spectra. Whittaker (1975) notes that this shift is a worldwide phenomenon. For example, in Oregon, Douglas fir forest is replaced by an open pine woodland on serpentine, and in the tropics, the shift is dramatic, from tropical forest to savanna or scrub.

3. Anatomical traits may be linked to geoedaphics. If there are morphological responses to geology, it follows that the cells and tissues that form the structural and ultrastructural bases of external form are also linked to geology. For example, the various serpentinomorphoses have anatomical foundations. Nanism, glaucescence, and sclerophylly are examples of anatomical changes induced by life on serpentines. Similar responses in anatomy are found on other specialized substrates. Geomorphological settings may also induce anatomical changes: elevation, slope, rock substrates (crevices, talus, etc.), and exposure have their effects on plant structure.

CONCEPTS IN TAXONOMY AND EVOLUTION

Systematics, taken broadly to include taxonomy and evolution, is a fertile milieu in which to examine the influences of geology. The connections are diverse, ranging from speciation and infraspecific differentiations to regional biotic diversity, as well as kindred evolutionary phenomena. The richness of regional floras often is a measure of geological diversity—diverse landforms, lithology, and kindred geoedaphic influences. It is in the realm of systematics that the concept of discontinuity (see previous discussion) plays a crucial role. The following dicta will be discussed in detail in Chapters 4, 5, and 6.

1. Within a regional climate, the amount of floristic richness of the area is largely a manifestation of geologic diversity. The contrast in species diversity between homogeneous landscapes (may wet tropics be the exception?) and highly heterogeneous terrains can be striking. The rich and highly endemic flora of California has, as its prime genesis, the rich array of geological processes and products that have operated over time in the California Floristic Province (Raven and Axelrod 1978; Stebbins and Major 1965; Kruckeberg 1985, 1986).

2. The events of racial differentiation and speciation are fostered in frequency and in sheer numbers by the environmental diversity, mainly created by geology. This dictum naturally follows from the previous one on species diversity. Diversity manifests itself as variable populations, racial (ecotypic) diversification, and speciation. The discontinuity and the environmental challenges fashioned out of geomorphic and lithologic variety provide the greatest opportunity for development of diversified biota, especially vascular plant floras.

3. Geologic diversity challenges the genetic resources of populations to produce a rich array of adaptive peaks. Adaptation to particular habitats results from inherited consequences of natural selection. Given the emphasis here on abiotic environments, particular geoedaphically created habitats will promote particular adaptive responses. Whether or not all species-specific responses to geology are the direct products of natural selection has to be answered on a trait-by-trait basis. For instance, tolerance to heavy metals and other nutritional imbalances on serpentine soils of *Calochortus* species is judged to be exaptations—"side effects" of the selection process—not direct adaptive responses to cope with high levels of heavy metals (Fiedler 1985). Yet I have no hesitancy in asserting that unique geological stimuli can evoke adaptive responses.

CONCEPTS IN PALEOBIOLOGY

If the present is a key to the past, then all the concepts and generalizations in this section should apply to past biota, especially to vascular land plants. Indeed the uniformitarian principle should be most germane to the interplay of geology and plants in times past.

1. As soon as land plants began to proliferate in the middle Paleozoic, encounters with contrasting landforms and lithology will have elicited adaptive responses.

2. Changes in the past wrought by successive geological events—mountain-building, continental drift, exposures to new and distinct parent materials—brought about change in the composition and location of floras. Chapters 6 and 7 amplify the geoedaphic influences on the paleobiology of plants.

CONCEPTS IN ECOLOGY

By its very nature in dealing with the environments of organisms, ecology should be the prime conceptual framework for the interaction between geology and plant life. Despite the current preoccupation of ecologists with biotic interactions, plant ecology still fruitfully studies the edaphic milieu of higher plants. Since much ecological research borders on, and depends on, systematics, evolution, and physiology, it is reasonable to expect that the dicta presented above for each of these disciplines will have relevance for ecologists. There are also concepts that reside most comfortably in plant ecology, some autecological and others with a synecological focus:

1. Some plant species have strong geoedaphic indicator status, with a high fidelity for either particular substrates or for geomorphically unique sites. Well known are indicators for azonal substrates like serpentines, limestone and dolomite, shale, and gypsum. Whittaker (1954a) pointed out that there is a hierarchy of plant indicator status. Besides the population with ecotypic indicator status for a geoedaphic habitat as well as the substrate-dependent edaphic indicators at the species level, there is the community level of plant indicator ranking. Particular plant communities may characterize a geologically determined habitat. Serpentine chaparral and serpentine grassland are good examples of indicator communities (Kruckeberg 1985).

2. Geoedaphic factors can determine the principal attributes of habitats and niches for species in a community. Nearly every geological factor, from parent material to topography, especially when discontinuously arrayed in space, can fashion particular habitats. The determination of a particular niche for higher plants is less certain. Since nearly all plant species share photosynthesis as their resource base, the green plant niche is less clear-cut than is the niche for animals. But when the habitat has some chemical or physical attribute that singularly makes demands on plants, then geoedaphically circumscribed niches seem real.

3. Azonal habitats whose attributes are the result of geoedaphic factors most

often harbor edaphic climax vegetation, though successional stages on edaphic sites may also occur. Plant communities set apart from a region's zonal vegetation by some geoedaphic peculiarity have been appraised as seral stages, destined to take on the character of the surrounding climatic climax in time. Whittaker (1960) took issue with this judgment; he contended that the edaphically constrained community is a self-perpetuating one, its own climax vegetation created by geoedaphic conditions. However, I suggest that successional stands on edaphically unique habitats will occur most often when the geoedaphic conditions are of recent origin or recurrent (e.g., volcanism, new landform changes like mudflows, stream meanders, landslides, and the like).

How geology conditions ecological responses is dealt with mainly in Chapters 4 and 5, though the other chapters are also germane to the ecological theme.

BIOGEOGRAPHY AND GEOLOGY

Geology and the distribution of organisms are intimately linked. From major geological consequences of continental drift and plate tectonics to regional and local effects of geology, the stamp of the Earth's crustal changes is unmistakable on the biogeography of higher plants. This pervasive connection between geology and the distribution of plant life gets major attention in Chapter 7.

It was in the classic *Foundations of Plant Geography* by Stanley Cain (1944) that concepts for plant distribution were codified and evaluated. Cain prefaced his list of principles (pp. 10–11, 1944; Table 7.1 of the present work) and their amplification (chapter 2 and elsewhere in Cain's book) by a cogent caveat: "A perfect set of principles of plant geography has never been written and probably never will be, if for no other reason than the practical impossibility of defining the exact content of the science." This reservation is appropriate not only for the present résumé of biogeographic "laws" but for all other such "laws" offered here. Yet we do seek guiding dicta.

Of Cain's thirteen principles, two (climate control is primary, edaphic control is secondary) are best combined into one for the purpose of the geoedaphic emphasis of this book.

1. Though "climatic control is primary," within a given climatic region geoedaphic control is a paramount determinant of plant distribution. Cain's fourth principle concerning the physical environment relegating edaphic control to a secondary position is true only in the global or continental context. Floristic and vegetational diversity within a regional climatic zone reflects the nature of the terrain in all its geological intricacies and discontinuities.

2. Geology affects—and effects—climates. Regional climates are often fashioned

by physiographic features, favoring or constraining the prevailing climate. It is mostly mountainous terrain that exerts decisive influences on climate, by virtue of leeward/windward effects (rain-shadow phenomena), by elevation, and by the nature of the mountains' topography (sharp, serrated mountains with deep canyons and valleys versus gentler topography). And at a smaller scale—mesoclimates to microclimates—landforms and even lithologies can cause particular local climates. "It is well known that the interactions between land surfaces and the atmosphere, and the resulting exchanges in water and energy, have a tremendous effect on climate" (Wood 1991, p. vii). This opening sentence of a recent book, *Land Surface—Atmosphere Interactions for Climate Modeling,* clearly supports the concept just espoused.

3. Discontinuities in landforms and lithologies promote "insular" patterns of plant and animal distribution. In their 1967 classic on island biogeography, MacArthur and Wilson state succinctly this concept: "Insularity . . . is a universal feature of biogeography." I have elaborated on this theme—the geoedaphic influence on insularity—to test the application of the island biogeography theory for mainland "islands" fashioned out of discontinuities in geology (Kruckeberg 1991b).

4. Dispersal and establishment of vascular plants depend on suitable geoedaphic conditions. Beijerinck's law is appropriate here: "Everything is everywhere, but the environment selects" (quoted in Sauer 1988). One thinks of the "cloud" of airborne spores present everywhere in the Earth's atmosphere. But where they land to reestablish their kind is a matter of environmental suitability. The spores of the serpentine endemic fern *Polystichum lemmonii* presumably are "everywhere" but may establish only on serpentine substrates of northwestern North America.

ANTHROPOGENIC HABITATS AND PLANTS IN RELATION TO GEOLOGY

Topography and lithology have been modified by humans from ancient to modern times. Changes in the physical features of the Earth's crust include alteration of landforms, mining, agriculture and forestry, urbanization, and any other of the multifarious human activities that impact physical environments.

1. Human disturbance of landforms and substrates most often encourages the invasion of exotic species; weedy floras result (Baker and Stebbins 1965; Baker 1995; Faber 1998). This subject is dealt with in Chapters 7 and 8.

2. Plants often accommodate genotypically to human-altered substrates. Tolerance to the heavy metals of mine spoils is the best known of such responses. The most recent of the extensive literature reviews on heavy metal tolerance are the collections edited by J. Shaw (1989) and Brooks (1998).

3. The geoedaphics of sites disturbed by past human cultures can be used in archeological studies. A recent monograph, *Phytoarchaeology* by Brooks and Johannes (1990), gives a thorough account of this aspect of the geology-plant connection to cultural history.

4. Disturbance by human activity of geoedaphically unique habitats puts their unique biota at risk. Plants endemic to many azonal soils are often threatened or endangered by habitat alteration (mining, timber or crop harvest, recreation, etc.).

Reflections on Geoedaphic "Laws" and the Environmental Complex

Much of what follows in this book expands upon the generalizations just presented. Though my bias is intentionally on the side of environments created by geology, there is no escaping the underlying limitations that such a bias engenders. The entire complex of environmental factors needs envisioning as an interactive (holocoenotic) system in many vectors and relationships. Billings (1952) eloquently portrayed this concept of wholeness and continued to reaffirm it (Strain and Billings 1974); see Figure 1.2.

One way to compensate for the bias inherent in singling out one set of factors (here, the geoedaphic ones) is to couch those that one chooses to study within a broader analytical framework. Various versions of the functional-factorial formulation, originated by Hans Jenny (1941), can serve as models. Jenny's first formulation was for soil: $S = f(cl, o, r, p, t)$. The five independent variables are climate, organisms, topography, parent material, and time. The Jenny formulation was modified by Major (1951) to portray the factors forming vegetation: $V = f(cl, o, r, p, t)$. The most recent adaptation of Jenny's formula (Kruckeberg 1986) accounts for biological diversity (*B.D.*) in the functional equation $B.D. = f(cl, o, r, p, t)$. In each version of the Jenny formula, the geoedaphic components are two independent variables, topography (r) and parent materials (p). Thus when r or p are variables, and the other factors (cl, o, t) are kept constant, distinct soils or vegetation or displays of biotic diversity will result. This approach thus preserves the holocoenotic view of environment while allowing study of a particular factor.

I see one defect in this conceptual scheme. As I have maintained for certain of the dicta presented in this section, all the Jenny variables but time are not wholly independent of each other. For example, regional climate can be a major consequence of physiography (see Figs. 1.1 and 1.2). Hence the more accurate rendition of the holocoenotic view of the environmental complex is the classic Billings

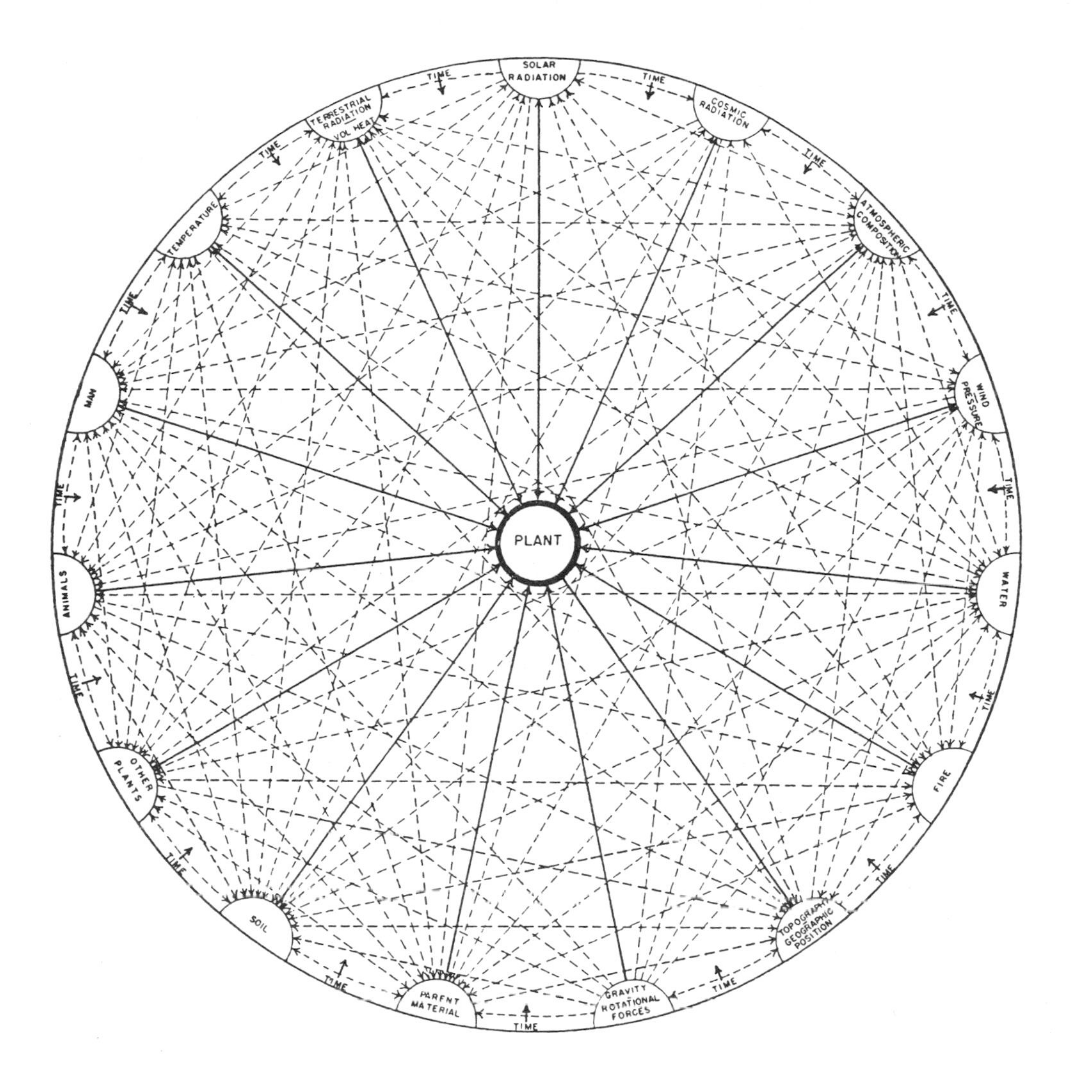

(1952) diagram. It portrays the multitude of connections between plants and their environments (Fig. 1.2).

1.2 The mutually interacting (holocoenotic) factors of the environment as they relate to the plant. Solid lines: relations between factor and plant; dashed lines: relations between factors; reciprocal effects: arrows at both ends of line; time: short inwardly pointing arrows at inner edge of circle. From Billings 1952.

A heuristic version of the Jenny "clorpt" formula recently appeared in a book that elegantly complements the substance of the present work. Huggett (1995) calls his version the "brash" equation. It consists of five interdependent terrestrial spheres: b = the biosphere, r = the toposphere, a = the atmosphere, s = the pedosphere (soils), and h = the hydrosphere. I recommend Huggett's *Geoecology* text as companion reading with the present book.

In Billings's portrayal of "a holocoenotic environmental complex," four of the environmental factors are geoedaphic: soils, parent material, gravitational and rotational forces, and topography/geographic position. Yet their effects on plant life are interwoven with all the other environmental factors. Indeed, this is the key message in the Billings scheme: Plant life responds holocoenotically. It should be pointed out that he realized that "the only group of factors not affecting the plant directly is that of topography/geographic position. These affect the plant only through other factors." This notion is portrayed with a nondirectional line between Plant and Topography, the only such linkage in the diagram.

Jenny (1980), in his discussion of serpentine soils, was aware of the linkages and feedback loops of the whole environmental complex. He called this totality of interactive environmental factors the "serpentine syndrome," realizing that not only geoedaphic factors but other physical and biotic factors in combination with geoedaphics yield a particular manifestation of the syndrome. Jenny's syndrome idea was portrayed diagrammatically for ultramafic substrates in Kruckeberg (1985, p. 25). It should work equally well for other geoedaphic syndromes, and serve to put the geologic component in its broader context—the complex of environmental factors (Fig. 1.1).

The holocoenotic web merits a bit more embellishment. If all the factors of the complex were of equal potency in molding vegetation and flora, then a rather homogeneous biological response might result. But there are degrees of potency of single factors in the environmental complex. Some biotic, climatic, or geologic factors may have an overriding effect, such as to shift the responses of plants or vegetation types. The venerable principle of "limiting factors" needs invoking here. In fact, Billings (1952) contended that the two principles—holocoenotic environments and limiting factors—can be joined. In the context of geoedaphic factors, ecologists recognize that an azonal substrate, like serpentinite or shale, can become a limiting factor. Any geoedaphically induced factor that falls to the minimum required—or exceeds the maximum tolerance—for a species, may arbitrate the survival of that species. A corollary of the limiting factor concept is the "trigger" factor. Billings (1952) invokes this notion for those instances where the limiting factor may undergo rapid departure from its limiting status. It is most often invoked for anthropogenic changes in an ecosystem—the introduction of an alien species or a change in the physical habitat wrought by human activity. Can the notion of the trigger factor be applied to changes in some attributes of the geoedaphic complex? It seems worth a try.

Two types of geoedaphically induced trigger factors are possible, and in fact do occur—natural and human-induced shifts in a geoedaphic attribute. Sudden changes in landforms brought on by vulcanism, rock avalanches and landslides, floods, and advances or retreats of glaciers come easily to mind. Sudden changes in lithology are less likely, though vulcanism creates new parent materials as well as changes in landform. Geoedaphic "triggers" pulled by human hands range from alterations of landforms to new exposures of parent materials (mine tailings, ballast, etc.). So the John Muir epigram takes on an added meaning: "When we try to pick out anything by itself, we find it hitched to everything else in the universe." And some are hitched together more strongly than others.

2 Geobotany: Its Historical Roots

No written record exists on how primitive humans sensed the influence of landforms and substrates in molding their environment and governing their way of life. Yet man, "the thinking reed," must have mentally stored, cataloged, and reacted to impressions of his surroundings. Indeed, the very survival of early humans depended on this store of sensory impressions.

It is to the naturalists of the Western world that we turn for glimmerings and insights relevant to our theme. This is not to say that other early cultures took no note of their physical surroundings. To be thorough, a search of literature on natural history and science produced by ancients in China, Japan, India, and the Near East could be rewarding. But the mainstream of environmental writings flows from the Western world.

Early Influences

The classical period yields little for our plants-and-geology theme. From the thorough account of ancient views on nature by Alexander von Humboldt (1849), one infers an apathy on the part of Greek and Roman writers to record their impressions of the natural world in the Mediterranean basin. In volume 2 of his great work *Cosmos,* von Humboldt gives a masterful and insightful account of ancient descriptions of nature. His subtitle, "The difference of feelings excited by the contemplation of nature at different epochs, and among different races of man," reveals the scope of his essay.

Greek and Roman natural historians were singularly oblivious to the impress of geology on their ancient landscapes. What of the Greek view of inanimate nature?

Von Humboldt paraphrases a German poet-dramatist-historian: "If we bear in mind the beautiful scenery with which the Greeks were surrounded, and remember the opportunities possessed by a people living in so genial a climate, of entering into the free enjoyment of the contemplation of nature, and observe how conformable were their mode of thought, the bent of their imaginations, and the habits of their lives to the simplicity of nature, which was so faithfully reflected in their poetic works, we cannot fail to remark with surprise, how few traces are to be met amongst them of the sentimental interest with which, we, in modern times, attach ourselves to the individual characteristics of natural scenery." One searches through the writings of Aristotle, for example, only to find that his preoccupations in the field of biology were with man and other animals. But dismissing him as a chronicler of plants and their environments may be unfair, for according to Nordenskiold (1935), Aristotle's botanical writings appear to have been lost.

Also, at least one of his disciples, Theophrastos, can be credited with a geobotanical observation. He described the unique *phrygana* vegetation on limestone in Greece 2000 years ago (p. 460, in Dierschke 1975).

Gleanings on geoedaphics from Roman literature are hardly more fruitful. Of this dearth, von Humboldt says: "That which we miss in the works of the Greeks . . . is still more rarely to be met with amongst the Romans. A nation which . . . evinced a decided predilection for agriculture and other rural pursuits, might have justified other expectations; but with all their disposition to practical activity, the Romans, with the cold severity and practical understanding of their national character, were less susceptible of impressions of the senses than the Greeks, and were more devoted to every day reality than to the idealising poetic contemplation of nature" (p. 382). Should not the encyclopedic natural history of Pliny the Elder yield morsels of geobotany? Von Humboldt found that despite Pliny's passion to give an account of the natural world, the result was "unequal in style, an indiscriminate and irregular accumulation of facts . . . deficient in individual delineations of nature." Not to mention its frequent enumeration of superstition and fable as fact.

Yet that Mediterranean cradle of Western civilization possessed ample wild nature, richly embellished and diversified by landforms and substrates in variety. Limestones and serpentine in both Greece and Italy displayed their singular, eye-catching landscapes. But perhaps the ancients looked with fear upon such settings—"Nature's pudenda" in the words of an anonymous sixteenth-century English writer who felt intimidated by wild places.

Romans were preoccupied with the human domain—politics, wars, intrigue, as well as the arts and other cultivated pleasures. Some descriptions of nature are found in the works of the great military campaigners, of whom von Humboldt says: "It is only in the writings of the great historians, Julius Caesar, Livy, and Tacitus that we meet some examples . . . , where they are compelled to describe battle fields, the crossing of rivers or difficult mountain passes, in their narrations of the struggle of man against natural obstacles" (p. 388). But he went on to note: "No description has been transmitted to us from antiquity of the eternal snow of the Alps, reddened by the evening glow or the morning dawn, of the beauty of blue ice of the glaciers, or of the sublimity of Swiss natural scenery, although statesmen and generals, with men of letters in their retinue, continually passed through Helvetia on their road to Gaul. All these travellers think only of complaining of the wretchedness of the roads, and never appear to have paid any attention to the romantic beauty of the scenery through which they passed. It is even known that Julius Caesar, when he was returning to his legions in Gaul, employed his time whilst he was passing over the Alps in preparing a grammatical work, entitled *De Analogia*."

From classical times down to the Linnaean era, preoccupation with human activities—spiritual and secular—precluded much in the way of descriptions of wild nature, much less of geological influences on landscapes. So we now embark on the eighteenth-and nineteenth-century journeys of natural historians who portrayed and sought to explain plant life in the contexts of the physical world. It was in these fecund centuries of the Enlightenment and beyond that floras were written, the harvests of collectors working globally were systematized, and a science of plant geography emerged. In particular, early floras of those centuries took note of the discontinuous distribution of plants; thus did they sense the influence of climate and geology on habitats. My guide to the roots of plant geography is chapter 2 of *Historical Plant Geography* (1943), by the Russian geobotanist E. V. Wulff.

1800–1850

While there were glimmerings of ideas on plant distribution in the eighteenth century (Willdenow 1792; Stromeyer 1800), it was the landmark work by Alexander von Humboldt, *Ideen zu einer Geographie der Pflanzen* (1807), that delineated the scope of the new science of plant geography. This vigorous, handsome German man of letters, science, and leisure was an all-around naturalist and creative natural scientist—perhaps, apart from Darwin, the most productive and imaginative scientist of his century. A person of comfortably independent means, von Humboldt was able to travel widely. His most famous expedition was to South America, where he came to know that continent's vast and endemic vegetation, from the wet tropics of the Amazon to the higher reaches of the Andes. His fascination with mountain summits (he made elevational readings of major peaks) gave him a special insight on the distribution of vegetation in relation to altitude, slope, and exposure. Substrate and lithology and their influences on plant life were only briefly noted by von Humboldt. He anticipated Franz Unger's (1836) recognition of the vicariant distribution of two *Rhododendron* species on different substrates in the Alps of Europe. Further, in his extensive measurements of mountain elevations in South America, he also noted the contrasts in lithology. But I fail to detect his making any major connections between rock types and flora. See Figure 2.1 for portraits of Humboldt and other notable geobotanists.

It is to Franz Unger (1836) that we turn for a full-blown conceptualization of the geology-plant connection. Unger (1800–1870) had a long and productive career in plant science; he made influential contributions to the growing science of botany in areas of cell biology, anatomy and morphology, paleobotany, and plant pathology. But his earliest adventures with plant life were ecological. In that great nineteenth-century encyclopedic work, *The Natural History of Plants,* Anton Kerner von Marilaun offered a vivid picture of the patterning of vegetation by substrate that set Unger to developing his chemical concept of plant distribution. I quote at length from the English version of this picturesque account (Kerner and Oliver 1902, 2: 495–496):

> The little town of Kitzbuhel, in the Northeast Tyrol, has a very remarkable position. On the north rises the Wild or Vorder Kaiser, a limestone chain of mountains with steep, pale, furrowed sides, and on the south the Rettenstein group, a chain of dark slate mountains whose slopes are clothed far up with a green covering. The contrast presented by the landscape in its main features is also to be seen in the vegetation of these two mountain chains. On the limestone may be seen patches of turf composed of low stiff Sedges, Saxifrages whose formal rosettes and cushions overgrow the ledges and steps of the rugged limestone, the yellow-flowered Rhododendron, and white-flowered Cinquefoil adorning the gullies, dark groups of Mountain Pines bordered with bushes of Alpine Rose; and opposed to these on the slate mountains are carpets of thick turf composed of the Mat-grasses sprinkled with Bell-flowers, *Arnica montana* and other Composites, groups of Alpine Alder and bushes of the rust-colored Alpine Rose—these are the contrasts in the plant-covering which would strike even a cursory observer, and would lead a naturalist to ask what could have been the cause. No wonder that the enthusiastic Botanist, Franz Unger, was fascinated by this remarkable phenomenon in the vegetable world. In his thirtieth year, furnished with a

Alexander von Humboldt (1769-1859)

Franz Unger (1800-1870)

Anton Kerner von Marilaun (1821-1898)

Herbert Mason (1896-1994)

Hans Jenny (1899-1992)

Josias Braun-Blanquet (1889-1992)

Dwight Billings (1910-1997)

2.1 Notable scientists who have dealt with the geology-plant interface. Their contributions to geoedaphics are discussed in the text. Courtesy of Hunt Institute for Botanical Documentation, Carnegie-Mellon University, Pittsburgh.

comprehensive scientific training, he came as a doctor to Kitzbuhel, and with youthful ardour he used every hour of leisure from his professional duties in the investigation of the geological, climatic and botanical conditions of his new locality, devoting his fullest attention to the relations between the plants and the rocks forming their substratum. The result of his study was his work, published in 1836, *On the Influence of Soil on the Distribution of Plants as shown in the Vegetation of the Northeast Tyrol,* which marked an epoch in questions of this sort. The terminology introduced in the book found rapid entrance into the Botanical works of the time. Unger divided the plants of the district accordingly to their occurrence on one or other of the substratums—in which lime and silica respectively predominated—into (1) those which grow and flourish on limestone only; (2) those which prefer limestone, but which will grow on other soils; (3) those which grow and flourish on silica only; and (4) those which, whilst preferring silica, will grow on other soils.

The essence of Unger's view—that mineral content of rocks and soils is the major edaphic influence on substrate-specific plant distribution—has been substantiated over and over again in modern times. This contention will be amply justified in later chapters. Unger's attempt to quantify the mineral nature of the substrate differences was to carry out analyses of ashed plant parts. As Kerner pointed out, this approach failed, but the key concept of mineral differences remains viable. Kerner became a disciple of Unger's ecological ideas and carried out transplant and pot test studies on species from limestone and siliceous rock habitats. He explained the plant responses as follows (Kerner and Oliver 1903, p. 498):

> The difference in the vegetation on the closely adjoining limestone and slate mountains . . . can be accounted for most satisfactorily in the following way. Plant-species which demand or prefer a siliceous soil are absent from limestone mountains wherever their roots would be exposed to more free lime than is beneficial; if present they would be weakened, and thus vanquished in the straggle with their fellows, to whom the larger quantity of lime is harmless, and they would eventually perish. These plants flourish luxuriantly, however, on slate mountains, because there the soil does not contain an injurious amount of lime. The absence of species, demanding or preferring lime, from slate mountains can be explained in the same way.

It is curious to note that Unger's earliest work, cited above, is not considered by botanical historians as his chief contribution. Besides being the co-author (with

Endlicher) of a popular botany text, Unger made major contributions in paleobotany and historical plant geography, as well as in plant anatomy, plant pathology, and cell theory. His stand in debate with Schleiden on cell division has been upheld and his views on evolution, though attacked by the clergy of the day, presaged some of Darwin's ideas. Interestingly, Unger was a teacher of Gregor Mendel in Vienna. "Unger's involvement in the working out of the cell theory and its application to the fertilization process may well have played a crucial role in equipping Mendel for the cytological interpretation of his breeding experiments" (*Dictionary of Scientific Biography,* 1978, p. 542).

Unger appears to have pioneered the Chemical Soil Theory (Braun-Blanquet 1932), which asserts that the inorganic constituents of the parent rock and derived soil strongly influence the response of plants. It was not unexpected, then, that a contrasting Physical Soil Theory would emerge. It belittled the chemical effects and emphasized the importance of physical properties (texture, particle size, porosity, etc.) in determining the nature of plant responses. The first acclaimed proponent of the physical theory was Jules Thurmann in his *Essai de Phytostatique* (1849). Thurmann stressed both textural differences (psammitic or coarse-textured soils versus pelitic or fine-grained clayey soils) and the weathering capacity of parent rocks (eugeogenous rocks, high in silica that weather readily, versus dysgeogenous rocks like limestone and chert that weather slowly). And so a lively debate was joined; the two "hostile camps," so described by Braun-Blanquet (1932), kept the contrasting theories in the air for the remainder of the nineteenth century. The debate continued on in the early twentieth century, until soil science acquired its major breakthrough—the discovery in the 1920s of the colloidal fraction and its role in cation exchange in soils. With this salient discovery and other influences, adherence to either the chemical or physical theory dissipated.

Like evolution, geoedaphics "was in the air," as detected in the writings of Victorian scientists. British botanists in the early nineteenth century embellished their local floras with arguments for the causes of local plant diversity. The edaphic factor (soils and their geologic environments) was either given prime status or rejected as a significant cause of local floristic makeup. One writer, H. C. Watson (1833), took issue with this single-factor approach as much too simplistic. Gorham (1954) has pointed out that Watson's paper was an early, if not initial, recognition that plant distribution has multiple causes. Watson was explicit in setting forth the holocoenotic view of the environment, later to be championed by most ecologists (e.g.,Billings 1952; Strain and Billings 1974) and by plant geographers (Cain 1944). Yet Watson's 1833 paper does carry the geoedaphic element in the environmental complex to a lucid and fruitful level. His seems to have been the only early nineteenth-century British contribution to our geoedaphic theme. One wonders if Charles Darwin may have read the Watson paper. By 1833 he was halfway round the world on the HMS *Beagle,* tacking up and down the South American coast. Young Darwin, then and upon his return home, was deeply committed to the science of geology. Did he read the Watson paper when he got back to England? I can find no specific reference or clue in Darwin's later writings that he was moved by the geology-plant linkage; though he did acknowledge the help of Watson's counsel on other botanical matters (Darwin 1887).

By the mid-nineteenth century, floristics, plant geography, and plant ecology were flourishing disciplines with many proponents, mostly writing descriptive accounts of plants and habitats. Nearly every major monographic essay or book recognized the significance of the physical environment for the discontinuous distribution of plants. Most works emphasized climate as the primary determinant; the role of geology was usually cast in a historical context—the change in landforms, and floras, over geologic time.

So prodigious became the outpouring of literature in the broad area of geoedaphics, from the mid-nineteenth century to the present, that my decision here will be to pick and choose. I continue the history of geobotany by citing the major milestones rather than producing an encyclopedic review. Major conceptual landmarks in a discipline are often recorded in reviews, monographic works of a period, and in textbooks, as well as in major revolutionary original works. I now sample, in two chronological sections, major contributions to geoedaphics from the 1850s to the mid-twentieth century.

1850–1900

Perhaps the most influential biologist of this period, and indeed of all times since, is Charles Darwin. So extensive and diverse are his contributions to biology and geology that anyone preparing a historical account of a relevant subject is bound to ask, "What did Darwin say about it?" So I have attempted to probe his vast writings for geobotanical contributions. That he was a plant ecologist is easily asserted. Of this central theme in Darwiniana, John Harper (1967) reminds us: "The theory of evolution by natural selection is an ecological theory—founded on ecological observation by perhaps the greatest of all ecologists." Given Harper's bias toward demographic problems in ecology, his eulogy of Darwin is not surprising: Darwin was preoccupied with numbers. This is Darwinian plant ecology.

Yet I find no evidence in his best known writings that Darwin observed, heard about, or contemplated the effects of geology on plant distribution or evolution.

His apparent omission of the geoedaphic influence on plant life is most curious, especially since he was familiar with the floras of nearby chalk and associated non-limestone substrates in Britain, or through a vicarious familiarity, via correspondence, with other floras of the world. His well-known studies of orchid pollination were done at the Orchid Bank, a chalk habitat in north Kent (Lousley 1950, p. 62).

However, the vast Darwiniana does not fail us! It is axiomatic that Darwin, in his long productive life, had something to say about most every subject in natural history. So it was not too surprising to find in one of his lesser-known works, *Journal of Researches* (1839), an observation on the influence of differences in lithology on plant diversity and the luxuriance of vegetation. It was during one of his forays into the interior of Australia, while on the *Beagle* voyage, that he observed a marked change in vegetation. He had been trekking for miles in the Blue Mountains between Sydney and Bathhurst across sedimentary rock (mostly sandstone) that supported a rather uninteresting vegetation. Then abruptly his path crossed into an exposure of granite. Of this encounter he wrote: "With the change in rock, the vegetation improved; the trees were both finer, and stood further apart; and the pasture between them was a little greener, and more plentiful" (Darwin 1839). It was Richard Beidelmann (pers. comm.), writing on the travels of western explorer John C. Fremont, who cited the above geoedaphic gem by Darwin. For Fremont recorded in his journal a similar geology–plant life discontinuity, encountered when his party left the basalt of southern Idaho (1843): "[We] found ourselves suddenly in granite country. Here the character of the vegetation was very much changed; the *Artemisia* disappeared almost entirely . . . and was replaced by *Purshia tridentata,* with flowering shrubs and small fields of *Dieteria divaricata* (*Machaeranthera canescens,* Hoary Aster), which gave bloom and gayety to the hills. These were every where covered with a fresh and green short grass, like that of early spring." As Beidelmann notes, the discerning Fremont was at the southwestern edge of the great granitic Idaho batholith.

Darwin recognized the significance for biotas of geologic discontinuities—in time and in space. In the later chapters of *Origin of Species* (1859), he enlarges upon the succession of organisms through geologic time and devotes a whole chapter to the incompleteness of the geologic (fossil) record. In his chapters 12 and 13, on geographic distributions, he stresses the importance of barriers to dispersal, especially of oceans separating continents and islands from one another, as well as land masses separating bodies of fresh water. These considerations of barriers to dispersal created by geologic events are indeed a part of my thesis: landform discontinuity is one aspect of geoedaphically induced isolation. One wonders why Darwin did not include the isolating effects of mountains or of lithologies in these chapters. That he was aware of the effects of soil differences on plant responses is revealed by his simple pot experiments with cultivated plants on altered soil conditions (chap. 23 in *The Variation of Animals and Plants under Domestication,* Darwin [1868]).

Darwin's two botanical friends and constant correspondents, Asa Gray and Sir Joseph Dalton Hooker, gave much thought to the problems of plant distribution. Both men furnished Darwin with useful information on plant geography and both wrote extensively on the origin and dispersal of floras. They, like Darwin, were principally preoccupied with the history of floras as conditioned by changes in landforms and the effects of those changes on climate. Gray's major essays on the relationships between floras of eastern North America and eastern Asia (Gray 1846, 1860) allude to the geologic cause of the floristic affinities—the existence of an early Tertiary land bridge between the two continents. In a similar vein, Hooker accounted for discontinuities in related floristic elements by noting changes in landmass, long-distance dispersal, and climate change. Hooker's insightful essays on distribution of floras were introductory pieces to his several floras dealing with the Galapagos, Australia and Tasmania, and the Antarctica. Though Darwin, Gray, and Hooker all recognized the *historical* changes induced by geologic events, I find no hint of any awareness of local to regional distributions, endemism, vicariant species, and so forth, that are caused by discontinuities in geological attributes. In fact, all three believed that present differences in physical features of habitat cannot explain plant distribution. Rather, they thought that past changes in climate and in the distribution of landmasses caused discontinuities in plant distribution. I will contend, however, that present gaps in physiography and lithology can effectively isolate populations and species, thus contributing to their ongoing evolutionary divergence.

Plant ecologists acknowledge the writings of Anton Kerner von Marilaun as the parental ideas of the science (Conard 1951). While Kerner's monumental *Natural History of Plants* (German version, 1888–1891; English version, 1902–1903) is his best-known work and does give a full exposure to the edaphic factor in ecology, it is his earlier book (1863), *Plant Life of the Danube Basin,* that Conard calls "the immediate and direct parent of all later works on Plant Ecology" (Conard 1951, p. vii, foreword to the English version of Kerner's 1863 book). Repeatedly in this work Kerner notes the sharp vegetational contrasts of limestone versus crystalline (silicic) rocks. His dramatic prose style captures the sharp contrasts: "Even the ordinary tourist, who is not at all concerned with the geological structure or the plant life of the lands he tours, must notice the contrast presented by the longitudinal valleys of the Alps between the calcareous mountains on the one side and the crystalline mountains on the other" (Kerner, in Conard 1951, p. 167). Later on, Kerner proposes that the vegetational contrasts of limestone with crystalline substrates result not so much from the chemical differences in parent material as from differences in rates of weathering. This in turn causes differences in soil depth and texture. With this view, Kerner sides with the Physical Soil theorists, dating from Thurmann (1849). Kerner also observed the now well-known occurrence of altitudinal displacement of floras: alpine species and communities displaced downward in elevation on limestone (p. 202). This phenomenon was later observed by others, as on serpentines at low elevations harboring plants from higher elevations (Kruckeberg 1969b; Whittaker 1975).

Kerner's encyclopedic *Natural History of Plants* devotes several pages to the substrate–vegetation/flora syndrome. In volume 2 (pp. 495–499, Kerner 1903), he gives a thorough account of the contrasts between limestone and crystalline rocks. Following a delightful portrayal of Franz Unger's landmark observations at Kitzbuhel (quoted earlier in this chapter), Kerner records results of his own experiments with calcicole and calcifuge plants grown reciprocally on limestone and siliceous (acidic) soils. He concludes that the reciprocal culturing did not change the phenotype and that the two substrates were toxic to their nonnative populations.

A new synthesis in plant ecology emerged at the end of the nineteenth century. It was the physiological interpretations of plant adaptation that Schimper stressed in his landmark work, *Plant Geography upon a Physiological Basis* (1903, English version). In that massive compendium, soils are given special treatment, showing Schimper to be aware of local differences in flora caused by differences in substrate. His recognition of discontinuities in geoedaphic factors led him to make the following generalization: "The inequality of the powers of resistance [I would translate as *tolerance*—A.K.] of different species is to a great extent responsible for the differences in the floras of substrata that differ chemically from one another." This concept became Schimper's Second Law,* in the view of the American ecologist W. B. McDougall (1949): "Plant distribution within limited areas is determined almost entirely by edaphic factors. This fact is well expressed by what we may call Schimper's Second Law which states that 'The local distribution of plants and of plant communities is determined chiefly by the nature of the soils, either directly, or in its relation to other factors'" (p. 129). This concept is the primary thesis of the present book. It would appear that I have illustrious company

---

* Schimper's First Law states: "The type of flora in so far as it depends on existing factors is determined primarily by heat" (McDougall 1949, p. 110).

in carrying this particular ecological torch. The idea that geoedaphics shape floras and vegetation within a climatic region has appeared in one form or another ever since Unger's early observations in the Tyrol (1836). A recent restatement of the concept comes from Hans Jenny, noted soil scientist: "The growth of vegetation is mainly determined by the character of the parent material, whether limestone, igneous rock, sand deposit or clayey shale" (Jenny 1941, p. 65, in writing about immature soils).

But back to Schimper! In a fascinating account of nineteenth-century German ecology, Cittadino (1990) evaluates Schimper not so much as an innovator but as a synthesizer of scattered observations, packaged in the physiological frame of reference: "Schimper's 1889 work introduced no new methodological principles or points of view; rather, it drew attention to a body of research that had been in existence for many years. . . . Schimper and the other German botanists who brought physiology outdoors in the 1880s and 1890s were not interested in founding a new discipline; they were interested mainly in broadening the scope of scientific botany" (Cittadino 1990, p. 115).

1900 to the Present

If in North America, geoedaphic consciousness did not surface perceptibly in the nineteenth century (as it did in Europe), then by the turn of the century some glimmerings of geobotanical awareness were seen. Most notable were the early writings of Merritt Lyndon Fernald, keeper of the Gray Herbarium at Harvard. His 1907 paper on the soil preferences of northeastern North American high montane plants is a remarkable contribution. He put its most salient theme in these words: "When we examine the lithological character of the regions in which these plants occur we find a very striking coincidence between the soil-forming rocks of these mountains and cliffs and the distribution of plants which cover them." The habitats of subalpine and alpine plants of northeastern North America abound in lithological diversity, from acid igneous granites and gneisses to limestone and serpentine. Fernald made this connection most clear when he wrote: "the alpine plants are much more dependent upon the chemical constituents of the soil than has been generally supposed."

Two other eastern U.S. botanists, Pennell and Wherry, noted the fidelity of plants to rocks, notably on the eastern North American serpentines (see Dann 1988 for a summary of their findings). Besides his notable geobotanical studies of serpentine in Maryland and Pennsylvania, Wherry did pioneering work on plant distribution associated with pH—his "indicator" attribute of variations in substrates.

By the early twentieth century the study of plant ecology was in full swing, especially in Europe and North America. The science's most influential protagonists gave soils, with their rich diversity, a fair treatment as significant partners in determining "why plants grown where they do." Yet there was scarcely any recognition of the all-embracing influence of geology on the distribution of plants. Most authors leaned heavily on their nineteenth-century predecessors—Schimper, Kerner, Thurmann, and Unger—for their early recognition of the part played by geological phenomena in plant distribution. I have reached this assessment after a fair sampling of early to mid-twentieth-century texts in plant ecology: Braun-Blanquet (1932), Weaver and Clements (1938), Tansley and Chipp (1926), and Warming (1906). Their books were the major influences molding plant ecology early in the century. This admittedly limited search may have overlooked accounts that might well harbor some significant geoedaphic insights. I would welcome hearing about any such scattered writings.

By sheer chance, I had just such a random encounter with a little-known early twentieth-century contribution to the link between geology and plant life. In sifting through the reprints of my former ecological colleague George B. Rigg (1872–1961), I found a gold mine of early papers on plant ecology. Rigg had treasured papers by his major professor, Henry Chandler Cowles of the University of Chicago, and one such memento caught my eye. Cowles is best known as a community ecologist who turned the dynamics of the Chicago area dunes into major concepts for synecology. But here was a side of Cowles less well known. In 1901 he wrote "The Influence of Underlying Rocks on the Character of Vegetation," published in a journal not usually consulted by plant ecologists. The stimulus for Cowles's pioneering (or prescient) contribution came also from the Great Lakes region, but on shoreline vegetation confined to a rich array of solid bedrock types. Cowles began his essay with these perceptive words: "It is a matter of common observation that different soils have different plants. Everyone expects to see a change in the natural forest covering as he passes from one soil to another. . . . So true is this that in many places a bird's-eye view of the forest is sufficient to indicate the nature of the soil. One may go even farther; in many places it is possible to tell the nature of the rock by means of the trees that grow above it." While Cowles did give lithology its due importance in this paper, he treated the effect of parent materials as a transient one. Bare rock with its xeric plant cover gives way in time to mesophytic vegetation. Only with exceptional substrates (serpentinite, limestone, etc.) did Cowles concede that rock type is primal in determining plant cover.

The early twentieth-century tome by Eugene Warming (*Oecology of Plants,* 1906) commands our attention. While the book, like those of other early plant ecologists, is mostly devoted to descriptions of plant community types, Warming also recognized the significance of soil diversity in the creation of plant communities. Two quotations support that statement. The first, "The nature of the nutrient substratus, or edaphic conditions, largely determines the habitats of plants and their topographical distribution" (p. 40);† thus begins his lengthy chapter on the properties of soils and their significance for plant distribution. A few pages later, the subject of the edaphic factor in the origin of species is raised: "Distinctions in soils have probably led to the separation of new species" (p. 56). Warming then refers to the three well-known substrates that support local endemics: zinc soils (with their "calamine" species), serpentine endemics, and limestone/dolomite taxa. One notes that he recognized the intimate relations of certain plants to what he called "rock soils" (currently called lithosols or entisols): "[I]t is the nature of the rock that determines what vegetation can develop upon it." Chapter 17 ponders the question, "Are the chemical or physical characters of soil the more important?" Like others just before him, Warming expressed no preference for one theory or the other. Not until Braun-Blanquet (1932) delineated how both the chemical and the physical properties of soils are crucial in defining substrate preferences for plants was the debate resolved.

Warming's chapter on soils gives a clear account of the role of competition in modifying the plant response to soils: "Plants are evidently . . . tolerably impartial as regards soil, . . . *so long as they have no competitors*" (Warming's italics, p. 71). A well-known example of the competition effect was reported by Tansley (1917), who did pot tests with two *Galium* species on contrasting limestone and normal soils. And in recent times, the importance of competition was stressed by Gankin and Major (1964) in their study of edaphic endemics in California.

Plant ecology in North America during the early part of the twentieth century was dominated by the pioneering efforts of Cowles (1901a, b), Clements (1928), Gleason (1939), and Harshberger (1911). All recognized in varying degrees the importance of the geoedaphic factors in determining local to regional distributions of species and communities. During that period, the works of Clements (1928) and Weaver and Clements (1938) exerted a major influence on ecological studies in North America. The Clementsian school of plant ecology became the prevailing force in ecological studies of the time. But to Clements, his colleagues, and adherents the role of geology in determining the distribution of plants was minimal. In the most influential Clementsian work, the textbook *Plant Ecology*

† Warming reminds us that the word *edaphic* (for soil) was coined by Schimper.

(Weaver and Clements 1938), the emphasis is on phytosociology. It stresses the role of succession and climax in creating community types. Hardly recognized are geoedaphic factors that can influence community dynamics. In a chapter on soils, the authors give a brief account of azonal soils, in which local geoedaphic features override climatic influences.

However, in the earlier, encyclopedic work by Clements, *Plant Succession and Indicators* (1928), we find a significant recognition of geological influences. Both topography and parent material are given prominent exposure in the two sections of the book. Clements (pp. 36–55) ascribed a dominant role to the effects of topography in initiating succession: "All forces which mold land surfaces . . . add to the land or take away from it" (p. 36) and thus become initial causes in succession. He then described the various topographic initiators of succession. What he focused on there is one of the most critical elements of what I would call the "geoedaphic syndrome"; hence physiographic processes and consequences will receive detailed accounting in Chapter 4.

In a later section of Clements's 1928 book, dealing with plant indicators, he once again pointed to the role of topography and soils (p. 295). Each of these geoedaphic factors has its singular plant indicators—species or communities that signalize a geoedaphic effect. We shall return later to this concept of plant indicators, so thoroughly elaborated by Clements.

It is curious that in neither of the major works by Clements cited above is there any recognition of the linkage between diverse lithologies (parent materials) and plant life. Surely in his wide travels across North America, he must have seen the dramatic effects of rock types (ultramafic, shale, limestone, etc.) on plant distribution. Perhaps his midwestern prairie locus conditioned his conceptual focus. The relatively homogeneous but subtly changing landforms and substrates would not provoke a clear perception of substrate effects.

The geologic diversity abounding in North America should have inspired ecologists and plant geographers of the early twentieth century to think geoedaphically. Besides Clements, one looks to the writings of Cowles, Gleason, and Harshberger for a recognition of the connection between geology and plant life. Cowles, fascinated with the dunes vegetation around the Great Lakes, did reveal his appreciation for the effects of local topography on plant distribution. Indeed, as I related earlier, Cowles saw beyond the dune vegetation of Lake Michigan. His 1901 paper, along with Fernald's (1907) seminal contribution, made known the geology–plant-life connection in the most explicit terms. Gleason and Cook (1927) had the good fortune to witness geoedaphic diversity in Puerto Rico, and recorded their impressions of the role of physiography and lithology in giving local, often abrupt discontinuities in floristics and vegetation types. The field experiences in Puerto Rico and elsewhere must have been critical stimuli in evolving two major Gleasonian concepts: the principle of discontinuous distribution (Gleason and Cronquist 1964) and the individualistic concept for plant communities (Gleason 1939). Both ideas reflect the geoedaphic bases of unique and discontinuously arrayed habitats. From Harshberger's (1911) encyclopedic account of North American vegetation one expects an awareness of the influence of geology on plants. Harshberger was Germanically thorough in describing North American plant life, but he did not speculate on why plants grow where they grow. So it remains for Harshberger's readers to extract from the voluminous accounts of rich native plant diversity, inferences on what might be the causes. At the very least, the Harshberger atlas provides a telling impression of the influence of topographic complexity on plant distribution.

If the Clementsian school, of plant ecology held sway in North America for decades, its phytosociological counterpart in Europe has had an even longer life. Braun-Blanquet, founder of the Zurich-Montpellier school, is best known in America for his book *Plant Sociology* (1932, the revised English version by Fuller and Conard). The fundamental social unit, as perceived by the Braun-Blanquet school, the plant association, derives its characteristics from both species composition and the interplay of a particular set of biotic and abiotic factors. It is not surprising that the edaphic factor gets prominence with proponents of the Zurich-Montpellier plant sociologists. After all, they and their ancestral mentor, Anton Kerner, carried out their ecological studies in the rich geoedaphic landscapes of the Alps of central Europe. The phytosociological consequences of a highly diverse physiography and a range of parent materials has been the recognition of a superabundance of plant associations. But what is germane to our geoedaphic theme is the thorough treatment of geoedaphic influences by Braun-Blanquet, and indeed all others of his predisposition. The 1932 text devotes 145 pages to soils in the section "Synecology or Community Economics." The physical and chemical attributes of both zonal and azonal soils get full attention. All well and good for the edaphic ingredient; less satisfactory is the treatment of the physiographic influences. A scant eight pages are devoted to "orographic" factors (altitude, slope, and exposure).

The Braun-Blanquet heritage of plant sociology continues to the present day. Under the aegis of the Rübel Institute in Zurich, notable work in geobotany has appeared over the years. Students and colleagues of Drs. Ellenberg, Ludi, and Rübel have acknowledged the importance of the geoedaphic factor in molding plant associations.

A distinctly different product of the Zurich phytosociological milieu was personified by the famous soil scientist Hans Jenny. Although trained in soil physics in Zurich (biographical sketch by J. Olsen, pp. viii–ix in Jenny 1980), Jenny made use of his early soils studies in collaboration with Braun-Blanquet. Their notable contribution (Braun-Blanquet and Jenny 1926) linked soil acidification with plant succession. It was as a young pedologist at Berkeley that Jenny created a new way of looking at soil genesis. In his landmark book, *Factors of Soil Formation* (1941), Jenny stated the fundamental paradigm of soil formation in the elegantly simple factor-function formula: $S = f(cl, o, r, p, t)$. His r (topography) and p (parent materials) are the substance of the present book's treatment of geoedaphic factors. The application of the formula to any soil formation process underlines the interplay of *all* factors. Though my message is to emphasize geoedaphic factors, I fully adhere to the interactive roles of cl (climate), o (organisms), and t (time). In Jenny's chapter "Parent Material as a Soil-forming Factor" there is a passage that could serve well as an epigraph for the present book: "Within a given climatic region, the growth of vegetation is mainly determined by the character of the parent material, whether limestone, igneous rock, sand deposit, or clayey shale" (p. 64, Jenny 1941). Although Jenny is speaking about "very immature soils," the essence of the idea can be applied to many azonal soil types.

This account of the historical roots of geoedaphics would not be complete without mention of mid-twentieth-century contributors to the subject. Herbert Mason's two landmark papers on "The Edaphic Factor in Narrow Endemism" (1946a, b) stimulated a flurry of research on the nature of plant restriction to azonal soils in California (e.g., Kruckeberg 1951, 1954; Gankin and Major 1964; McMillan 1956; Walker 1954; Walker, Walker, and Ashworth 1955; Vlamis and Jenny 1948). Mason's own words make a clear declaration of the significance of the geoedaphic factor: "[T]he edaphic factor occurs spatially in a manner that is most apt to be related to highly restricted patterns of distribution among plants" (Mason 1946a, p. 218). He makes clear the connection between floristic diversity and lithological variety: "Much of the diversity of the California flora results from the superposition of lithological features across the areas of special climatic conditions, thus creating local habitats that are occupied by special populations of plants" (1946b, p. 245). In retrospect, it seems inevitable that California would have inspired a full-bodied, vigorous geoedaphic outlook. In the absence of any obliteration of flora by continental glaciation, California's plant life could continue to flourish right through the Pleistocene. The region's great geological richness was to be matched by an unsurpassed floristic diversity ever since the Late Tertiary (Raven and Axelrod 1978). Yet the geoedaphic El Dorado of the California Floristic Province had to await Herbert Mason of the mid-twentieth century to put its natural treasure in a geologic mode. As I have written elsewhere (Kruckeberg 1985, pp. 3, 4), early California botanists like Brewer, Greene, Jepson, and Abrams were too preoccupied with the naming of California's floral richness to reflect on why the state was so rich in endemics.

At about the same time in Europe, a similar recognition of geoedaphic influences on Mediterranean-type flora emerged. W. B. Turrill's monumental work on the *Plant Life of the Balkan Peninsula* (1929) offers parallels to the California story. From northwestern Yugoslavia to southeastern Greece, the Balkan terrain abounds in physiographic and lithological diversity—truly, an "old-world California." Turrill began his encyclopedic work with these prophetic words: "Geologically it [the Balkan Peninsula] has had a varied history which has resulted in a wide range of rocks and soils. . . . [I]t is especially the interaction of the geographical, geological, and climatic factors which makes the Balkan Peninsula so fascinating for students of plant distribution" (p. xiii). In a lengthy chapter devoted to geology and soils, Turrill discussed the significance of geological structure, lithology, and geologic history for the distribution of plants throughout the peninsula. Endemism within the peninsula gets deserved attention in a later chapter: regional and local endemics total 1754 species out of a total flora of 6530. Turrill made no effort to account for the causes of the high endemism in ecological terms. One must infer that edaphic endemism is a major outcome of the discontinuous nature of landscapes and substrates. Yet nowhere did Turrill satisfactorily explain the probable causes of the endemism. We are left to conjecture that much of the endemism can be attributed to geoedaphics.

By midcentury, geoedaphics had been restored to the pantheon of environmental factors that influence the distribution of plants. What Mason did for Pacific Coast plant ecology with his 1946 papers, Hans Jenny and Harold Lutz did for soil science—cementing the link between soils, lithology, and vegetation. I will return repeatedly to the seminal contributions of Jenny (1941 and 1980), with his brilliant synthesis of soil-forming factors and his recognition of parent material as a crucial soil-forming factor. Less well known is the contribution of Harold J. Lutz to the geology-soil-plant triad. Lutz's special field, forest soils, may have kept his contributions from being more widely recognized in plant ecology. But his 1958 paper, "Geology and Soil in Relation to Forest Vegetation," captures the essence of geoedaphics—the substance of the present book. Lutz gives us chapter and verse on the significance of diverse lithologies for forest vegetation. It is truly a landmark paper; perhaps only made less accessible to mainstream plant ecologists of the day by its specialized appearance in the proceedings of a forest soils conference (Lutz 1958).

I have made the case in this chapter that awareness of geoedaphic influences on plant life has had precedents—ancient and modern. Yet during the formative years of plant ecology, from the mid-nineteenth century to World War II, geoedaphics assumed an inordinately modest place on the ecogeograhic stage. In the chapters to come I attempt to justify its being given a more central position. Before we embark on that challenge, it is well to draw attention to some notable geoedaphic references that review certain facets of the subject. As reviews compiled in the mid-twentieth century, they open doors to an extensive, though scattered literature on geoedaphics. I will draw upon these reviews and their abundant literature citations throughout the book.

GENERAL REVIEWS AS SOURCES

These cover a wide range of substrate-plant phenomena. The proceedings of a conference on substrate and vegetation (Dierschke 1975) was devoted exclusively to geoedaphic matters. Two major reviews by Helmut Kinzel (1982, 1983) focus on the ecophysiology of substrate-plant interactions. Stålfelt's text, *Plant Ecology,* is mainly devoted to the soil environment (Jarvis and Jarvis 1972, English edition of Stålfelt 1960). The recent book by Richard Huggett (*Geoecology,* 1995) is an essential companion work to the present one. See also Anderson et al. (1999).

WORKS ON HEAVY METAL HABITATS

Heavy metals (copper, zinc, lead, nickel, etc.), whether in naturally occurring substrates or in disturbed, artificial sites (mine tailings, garbage dumps, etc.), have a pronounced inhibitory effect on most vegetation, and yet have elicited remarkable evolutionary responses through genetic tolerance. Ever since the classic work of Bradshaw and co-workers in the 1960s on the tolerance of plants to heavy metals, the subject has evoked a steady output of published work. This literature has been reviewed recently (Shaw 1989).

WRITINGS ON ULTRAMAFIC (SERPENTINE, ETC.) SUBSTRATES

The worldwide occurrences of serpentine and other ultramafic rocks and soils, with their attendant unique floras, has fascinated geobotanists for years. Major reviews, beginning with Krause (1958) and Proctor and Woodell (1975), continue up to the present (Kruckeberg 1984; Brooks 1987, 1998; Roberts and Proctor 1992; Baker, Proctor, and Reeves 1992; Jaffré et al. 1997). The serpentine story will be told in a variety of contexts (floristics, biogeographic, taxonomic, and evolutionary) throughout this volume.

3 Geoedaphics and Other Environmental Influences: Their Reciprocal Relationships

The deep truth of interconnectedness underlies all of science, certainly biology. Realization of the complex network of the organism-environment interplay has taken a variety of expressions. Poets, philosophers, and naturalists—all have sensed the reticulate oneness of the living world. In the words of John Muir (1911): "When we try to pick out anything by itself, we find it hitched to everything else in the universe." The idea took on scientific expression in the mid-twentieth century when Friederichs (1927) and then Allee and Park (1939) introduced the term "holocoenotic" to stand for the whole of the environmental complex, in which "no one factor can change without affecting all others" (Strain and Billings 1974). Dwight Billings (1952) fleshed out the concept for vegetation and gave us his now familiar diagram of the environmental complex (Fig. 1.2). The same idea was expressed in mathematical form, with the genesis of soils set in a functional-factorial equation, $S = f(cl, o, r, p, t)$ (Jenny 1941). Jenny's five independent variables—climate, organisms, topography, parent materials, and time—act in concert to yield a given soil. He saw the equation as providing a tool to study soil genesis; by holding all other factors constant, the effect of, say, parent material as a variable could be tested. The same functional-factorial equation was modified by

Major (1951) to analyze the genesis of a given vegetation, and a further modification can be used to put the origin of biological diversity (*B.D.*) in a holocoenotic frame: $B.D. = f(cl, o, r, p, t)$ (Kruckeberg 1986). The most recent variant of the Jenny functional-factorial equation, by Huggett (1995), is discussed above in Chapter 1.

Admittedly, there is a single-mindedness in taking the geoedaphic factor complex out of the larger environmental context. It is the message of this chapter to deny that I take a single-factor approach in explaining the distribution of plants. But beyond the acceptance of the holocoenotic principle, I want to examine the possible, even inevitable, reciprocal action of the several independent and dependent environmental factors. In particular I will argue that geoedaphic factors exert influence on other environmental factors, just as geoedaphic factors are under the influence of other features of the environment.

What are those "other features" that temper geoedaphic effects on land-dwelling biota? The most obvious are climate, other organisms, and time. Further, the ways in which bodies of water interface with the land create unique conditions for terrestrial biota, especially plants. Other influences are less tangible. Opportunities (or the lack thereof) created by chance, and even capriciousness of place and process, can have significant evolutionary and distributional outcomes.

But climate takes the center of the biospheric stage as the overriding determinant of plant distribution. At least that is the conventional wisdom. Plant ecologists and phytogeographers from the early nineteenth century to the present have named regional climate as the primary control of plant life. An early formal statement by the plant geographer Good (1931) cast the role of climate thusly: "Plant distribution is primarily controlled by the distribution of climatic conditions." The same principle, tersely put, leads the list in Stanley Cain's set of principles for plant geography: "Climatic control is primary" (see Table 7.1). In his hierarchy of controls of vegetation, the geoedaphic factor comes fourth: "Edaphic control is secondary" (Cain 1944). But casting the more encompassing geoedaphic factor in the role of a secondary control is misleading; it is no less significant than climate. Within a regional climate, geoedaphic controls are primary. This idea was framed succinctly by Jenny (1941): "Within a given climatic region, the growth of vegetation is mainly determined by the character of the parent material." To which I would add: "and topography." All well and good; but the prevailing view of environmental controls is still that in any major biotic region climate is primary and edaphic controls are secondary. Let me try to develop a countervailing idea.

In saying geology influences climate, or geology exerts controls on climate, one must ask "how?" The answers stem from two very different aspects of climate—regional and local. Both are affected by geologic phenomena. If I am able to make a plausible case for this notion, it should follow that the Good-Cain principles of plant geography need revision. I will state my own versions of certain principles: (1) geologic control of the biosphere is primary; (2) control by regional climate is primary only when it can be shown that geology does not influence regional climate; and (3) edaphic controls (soils, *sensu stricto*) are secondary for a given regional climate, but are primary within that region. Support for these principles is summarized below.

1. *Geological control is primary.*

By this principle, I am emboldened to say that vegetation is controlled primarily by geological processes and products. Further, the prevailing regional climate that elicits a particular vegetation depends on the geology of the region. The most salient geological influence in this context is topography. Physiography, the heterogeneity of land surfaces of all kinds, has its genesis in geology. The effects of surface heterogeneity (Fig. 3.1) can shape climates and in turn can foster vegetation of a particular type. Thus climate and vegetation are dependent variables; surface heterogeneity, resulting from geomorphic effects, is the only independent variable in the environment-plant matrix. At least this is true at time zero for a given landform. The surface heterogeneity then may be changed by climate and vegetation, but without a particular landform these subsequent influences take on a different pattern. The most evident examples of this vector (surface heterogeneity—climate—vegetation) are mountains, especially major mountain ranges. The two cordilleran axes of western North America, extending from the far north in Canada to Mexico, are the major arbiters of regional climate, both west and east of the Rocky Mountain and Sierran-Cascadian chains. Similar examples of the connection between surface heterogeneity and climate are found on all other continents—the Andes of South America, the Alps of Europe, the Himalayas of Asia, to name the most commanding massifs. Surface heterogeneity, most telling in mountain systems, is, at time zero, a product of tectonic geology. Regional climate and resultant vegetation all follow as consequences of the mountain-building phenomena. In his introduction to *Land Surface—Atmosphere Interactions for Climate Modeling,* Wood (1991, p. vii) states: "The interactions between land surfaces and the atmosphere, and the resulting changes in water and energy have a tremendous affect [*sic*] on climate." This idea is further elaborated by Sellers in one of the articles in Wood's book: "The important interactions between atmosphere and land surface involve the exchanges of radiation, sensible heat, latent heat and momentum. . . . [T]hese exchanges have direct impact on the fields of wind vector and precipitation for the atmosphere and surface temperature and soil moisture for the sur-

3.1 On level terrain, etched limestone creates habitable crevices. Yorkshire, England. Photo by A. L. Kruckeberg.

face" (p. 85). In sum, surface heterogeneity on both regional and local scales exerts a dominant influence on climate.

It is in mountain environments that the effects of surface heterogeneity are most telling—not only within the mountain terrain but in adjacent lowlands. Gerrard's (1990) critical examination of the mountain environment (its physical geography) adds support to my claim for the primacy of geology in the equation linking surface heterogeneity and climate: "Mountains are extremely important components of the Earth's surface [since] about 36 per cent of the land area of the world is composed of mountains, highlands and hill country." To this I would add that intervening lowland terrain is also influenced by the delimiting mountains of a region and in turn creates conforming climates. Maritime versus continental climates, and major rain-shadow effects, have regional consequences as the result of the extreme surface heterogeneity of mountains.

3.2 Irregular surfaces are spectacularly displayed in montane terrain, here in the North Cascades of Washington State. Photo by Mary Randlett.

2. Control by regional climate is primary only if geological influences are absent.

Ultimately all physical attributes of the Earth—and of extraterrestrial landscapes—are shaped by geology. The genesis of our planet and its metamorphosis are the outcome of geophysical phenomena. Given that causation—planet Earth as a geophysical entity—then all climates are geophysical in origin. Earth's shape and relationship to the sun yield variations in global atmosphere. Were there no regional surface heterogeneity, there would at least be a minimum of climates as a function of latitude: polar, temperate, and tropical. In this context, climatic control of vegetation is primary. For instance, there is no immediate geological cause of major climate systems that operate in the tropics. Climate is primary in the eastern sector of Amazonia. But the Andean massif deflects tropical-rainy climate and thus the mountain range does determine regional climate in its vicinity.

3. Edaphic controls (soils, in the strict sense), though they may be conditioned by a regional climate, are primary within that region.

Edaphic controls are a subset of geoedaphic influences on vegetation. Soils, the edaphic factor, develop under a particular regime of regional climate. Yet within a region's prevailing climate, substrate and landform heterogeneity vastly elaborate the diversity of the edaphic environment. To paraphrase Jenny (1941, pp. 64–65): Within a given climatic region, the growth of vegetation is mainly determined by the character of the topography and the parent materials. I contend that geoedaphic controls—topography and lithology—are primary within a regional climate.

Microclimate is defined as "climate near the ground." Such ultralocal climates vary in character with the terrain; surface heterogeneity on a confined scale yields a variety of local microclimates. From this observation, we may derive a corollary to principle 3: Geoedaphic influences foster microclimates. I will elaborate on this idea in Chapter 4.

Where land and water meet, the habitats of plants and animals reveal a variety of distinctive features. The qualities of any land adjacent to water are derived in part from the geological events and processes that created the land-water junction. The consequences may be local, regional, or even global. Local climates of lake and stream borders are the most notable of these consequences. In addition, special terrain irregularities may result where land meets fresh or salt water: beaches of all kinds, terraces, cliffs, canyons, and valleys are surely geoedaphic consequences of the land-meets-water link. Seasonal climates of the Northern and Southern Hemispheres differ because of differences in distribution of land and marine waters in the two hemispheres. "Northern Hemisphere summers are warmer and winters are colder than those of south of the Equator. The reason is the rapid seasonal heating and cooling of the great continental land masses dominating north latitudes" (Akin 1991, p. 13). Also, solar radiation responds differently to land than to water surfaces and masses. Albedo, specific heat, translucence, evaporation, and mobility are all properties of solar radiation that can differ dramatically in this regard (Akin, p. 16). Global atmospheric circulation with its westerlies, cyclonic, and anticyclonic storm tracks, pressure cells, and so forth, can be both independent of landmasses and influenced by them. So we would have to add the possibility of geoedaphic modification of global weather patterns.

The deep truth, that life modifies its contexts, has relevance everywhere in nature and throughout this book. Geological processes and products are altered by organisms, just as the geology may select for particular kinds of life. Life as a geological force, though not a new idea, is a concept gaining wider acceptance. The simple expression "geoedaphics—climate—organisms" must be viewed not only as reversible linearly but also as cyclical. At the very least, organisms in a particular climate effect changes in landforms and substrates. Life as a geologic force is exemplified in the genesis of organic substrates (e.g., limestone, corals, humic soils), the organically induced weathering of rocks, and the creation of certain landscapes (slope retention by vegetation, Mima mounds, atolls, sand dunes, etc.). This reciprocity between organisms and the inorganic environment has taken expression in recent years as the Gaia principle: Planet Earth lives! (see Rambler, Margulis, and Fester 1989; Westbroek 1991).

Each of the following three variables must be reckoned with in evaluating the genesis, development, and character of a particular ecosystem.

1. *Time.* Jenny's functional-factorial equation makes time an independent variable. The age of a landform can be crucial in determining the composition of its biota; and the time span of occupancy of a place by its organisms surely can affect the quality of the landform. A newly formed Krakatau, or a just-congealed lava flow, is a very different physical setting for plants than an older land surface like the lavas of the Columbia Plateau.

2. *Area.* Size, shape, and contiguity of landform can influence the quality of biota in a variety of ways. Isolation, by physical discontinuity, has the most telling effects on organisms. A lone, isolated volcanic peak has a very different colonizational potential than does a coherent chain of peaks.

3. *Evolutionary Opportunity.* Variables in time and area conspire to influence evolutionary events. These can be independent of other environmental controls. A newly created oceanic island (such as Surtsey) provides an evolutionary opportunity whose course is unpredictable. Random occurrences of waif biota, colonists, and migrants are chance samplings that bias the evolutionary outcome of an initial colonization. So we must add to climate and other organisms the roles of chance and even capricious happenings in modifying evolutionary outcomes, regardless of the geoedaphic constraints.

The message of this chapter is that not only can other environmental factors influence the expression of geologic phenomena, but geology also influences the expression of other environmental factors (climate, organisms, etc.). Climate is tempered by geology, and geology is tempered by climate. This reciprocity operates globally, regionally, and locally to create the milieus for richly diverse kinds of plant life. So back to basics: The environment is holocoenotic, even as I make a case for the primacy of geoedaphic influences.

4 Landforms (Geomorphology) and Plant Life

Imagine a terrestrial world with no mountains, no hills, no valleys. Suppose that monotonously flat stretches of land provided the setting where life first appeared and began to evolve. On the other hand, would life in any degree of diversity have evolved on a global billiard ball? The surface of the real world throughout most of geologic time has had a wealth of texture. The land has been thrown into all sorts of configurations, inevitably transient, changing over geologic history. But for some spans of time, a particular topographic diversity existed at any place on the globe to foster an evolving biological diversity.

Both landforms and lithology are included in the "geo" component of the term *geoedaphic* as used throughout this book (see Glossary). Surface heterogeneity—irregularity in land surfaces—is a preeminent attribute of the geoedaphic factor and can decisively control the nature of vegetation and flora, and thereby the distribution of animals and other biota. Irregular land surfaces range from the undulating ground and crevices of level limestone karst to the major variations in slope and exposure of mountains. Examples of the many ways surface heterogeneity can influence plant life make up the substance of this chapter.

Before looking at this linkage between landforms and plants, some useful concepts merit discussion. Topography takes on special significance when viewed as an independent variable in soil formation and in promoting variations in flora. Jenny (1941) demonstrated its role as a soil-forming factor, emphasizing the effect of uneven land surfaces on soil profile development, water table, and erosion. Major (1951) adapted the Jenny formula, using topographic variation as an independent variable in the formation of vegetation; he illustrated the effects of landform on vegetation as "toposequences," especially in arid mountains: "In the Santa Catalina Mountains of southern Arizona at several topographic levels vegetation is strikingly different on north and south slopes of ridges" (p. 404). Kruckeberg (1986) extended the use of Jenny's topography variable (*r*) to signify its role in the genesis of biological diversity: $B.D. = f(cl, o, r, p, t)$. When all other factors are constant, variations in topography can promote variant climates, soils, vegetation, and flora.

The holocoenotic view of the biosphere is crucial to understanding the interface of landform and vegetation. Each variant factor in the environments of organisms operates as a two-way system or even multidirectionally. Topography shapes climate and climate shapes topography. In turn, topography shapes vegetation, and, more subtly, vegetation can shape topography. These assertions will be fleshed out with examples later in this chapter. The term *phytogeomorphology* embraces these reciprocal relationships. Phytogeomorphology, the study of the interdependence of vegetation and landforms, is the central theme of a recent book by Howard and Mitchell (1985). These authors link the factors of climate, landforms, and soils with vegetation. They define geomorphology as "the product of climate acting upon geology." Viewed as a one-way process, it means that landforms are products of climate. This seems to be the prevailing definition of geomorphology, rather than its unbiased, descriptive dictionary definition: "The study of the characteristics, origin, and development of land forms" (*Random House Dictionary*, 1967). While not denying the influence of climate on topography, I prefer to view them in a bidirectional fashion: topography $\longleftrightarrow$ climate. In fact, Howard and Mitchell (1985) make that very point: "World-wide vegetation distributions clearly mirror . . . variations of climate-induced topography and topographically-induced climate" (p. 3).

Types of Landforms Affecting Plant Life

All landforms that are exposed at the Earth's surface can be expected to influence flora and vegetation in diverse ways. Hence any classification scheme of geomorphological features has relevance for the explanation of plant distributions. The scheme of J. Tricart (1965; in Pitty 1971), reproduced here as Table 4.1, classifies landforms hierarchically by size of area. Rankings range from continents and ocean basins, regional and local relief, down to the microscopic. Moreover, this graded scale of magnitude is put in a time frame, from 10^9 years for the largest geomorphic units to 10^2 years for microscale surface heterogeneity. Further, the Tricart-Pitty system has a causal component: for each level, the interplay of climate and relief as well as basic mechanisms controlling relief are given.

Table 4.1. Classification of geomorphological features (after Tricart 1965, in Pitty 1971)

Order	*Units of Earth's surface in km²*	*Characteristics of units, with examples*	*Equivalent climatic units*	*Basic mechanisms controlling the relief*	*Time span of persistence*	*Plant life* (vegetation, flora)*
I	10^7	Continents, ocean basins	Large zonal systems controlled by astronomical factors	Differentiation of Earth's crust between sial† and sima†	10^9 years	Continental endemism in many vegetation types (e.g., Australia, North America, Eurasia)
II	10^6	Large structural entities (Scandinavian Shield, Tethys, Congo basin)	Broad climatic types (influence of geographical factors on astronomical factors)	Crustal movements,as in the formation of geosynclines.† Climatic influence on dissection	10^8 years	Subcontinental endemism and regional vegetation types (e.g., taiga of Scandinavia tropical forests of Congo, desert biota of western North America)
III	10^4	Main structural units (Paris basin, Jura)	Subdivisions of the broad climatic types, but with little significance for erosion	Tectonic units having a link with paleogeography; erosion rates influenced by lithology	10^7 years	Regional endemism and indicator vegetation types within regional topographic units (e.g., forests, alpine vegetation of central Europe, tepuis of Venezuela, etc.)
IV	10^2	Basic tectonic units; mountain massifs, horsts, fault troughs	Regional climates influenced predominantly by geographical factors, especially in mountainous areas	Influenced predominantly by tectonic factors; secondarily by lithology	10^7 years	Vegetation of a major mountain range, with regional endemics (e.g., Sierra Nevada and Rocky Mountains of America)

Table 4.1. *(continued)*

Order	*Units of Earth's surface in km²*	*Characteristics of units, with examples*	*Equivalent climatic units*	*Basic mechanisms controlling the relief*	*Time span of persistence*	*Plant life* (vegetation, flora)*
			Limit of isostatic adjustments			
V	10	Tectonic irregularities, anticlines, synclines, hills, valleys	Local climate influenced by pattern of relief (adret, ubac),† altitudinal effects, slope exposure	Predominance of lithology and static aspects of structure	10^6–10^7 years	Local diversification of plant life (incl. narrow endemics) linked with variations in relief and lithology (e.g., serpentine, granitic, lava exposures in western North America)
VI	10^{-2}	Landforms; ridges, terraces, cirques, moraines, debris, etc.	Mesoclimate, directly linked to the landform (e.g., nivation hollow)	Predominance of processes, influenced by lithology	10^4 years	Nearly same as V (e.g., vegetation on ridge tops vs. north or south slopes; flora of limestone or serpentine outcrops)
VII	10^{-6}	Microforms; solifluction lobes, polygonal soils, nebka,† badland gullies	Microclimate, directly linked with the form (e.g., lapies [karren]†)	Predominance of processes, influenced by lithology	10^2 years	Ultralocal contrasts in plant life influenced by differences in mircorelief (and microclimate) and lithology (e.g., flora of rock outcrop vs. talus)
VIII	10^{-8}	Microscopic (e.g., details of solution and polishing)	Microenvironment	Related to processes and to rock texture		Mostly cryptogamic flora (lichens, algae, fungi) on minute surfaces of substrate (rock, sediments, soil surfaces)

* This column added by the present author.
† See Glossary.

A case can be made that each level of the graded array of landforms (relief) influences plant life. Thus the grossest level of continent formation and position isolates as well as connects world floras. And at the other end of the scale, microforms of rock and soil heterogeneity (slope, exposure, texture, color, etc.) condition the success of individual plants (i.e., the microsite of Harper 1977). Rather than construct an exhaustive catalog of all the plant-topography connections that geomorphic features foster, I will single out certain ones as illustrations of the landform-plant interface, especially those that profoundly influence the distribution and evolution of land plants. After all, it should not be too difficult for any geobotanically inclined reader to see in any landform the potential for the selective inclusion or exclusion of plants.

MOUNTAIN ENVIRONMENTS: LANDFORM VARIATIONS

"Extremely important components of the Earth's surface," and perhaps the most topographically complex of landforms on our planet, mountains comprise about 36 percent of the world's land area (Gerrard 1990); see Figure 4.1. And mountains come in all sizes and shapes, from low to high ones, and from sharply serrated, steep-sloped terrain to gently rounded summits. Price (1981) defined a mountain as "an elevated land form of high local relief, e.g., 300 m (1000 ft), with much of its surface in steep slopes, usually displaying distinct variations in climate and associated biological phenomena from its base to its summit." Mountains are geologic in origin, but may be altered in height, slope, and so forth, by other environmental factors. And such terrain plays host to a wide array of biological phenomena—ecophysiological, speciational, biogeographical, and evolutionary. Mountains are singular and important nurseries for the birth and development of distinctive floras. Size, elevation, topographic conformation (structure), orientation, degree of isolation, effects on climate (and mountains affected by climate), as well as lithology—all have important influences on montane floras and faunas. We next examine some of these variables that can effect biological responses.

Spatial Continuity or Discontinuity

Western North America is an ideal locale for studying a wide range of geoecological effects. Three major mountain systems, the Coast Ranges, the Sierra Nevada–Cascade Range axis, and the Rocky Mountains, exhibit spatial coherence. These three cordilleran systems, by their internal integrity and north-to-south orientation, are inhabited by species that often occur widely in one or the other of the mountain systems. Examples abound: *Abies magnifica* spp. *magnifica* in the Sierra Nevada, closed-cone pines (*Pinus muricata, P. radiata, P. attenuata*) of the Coast

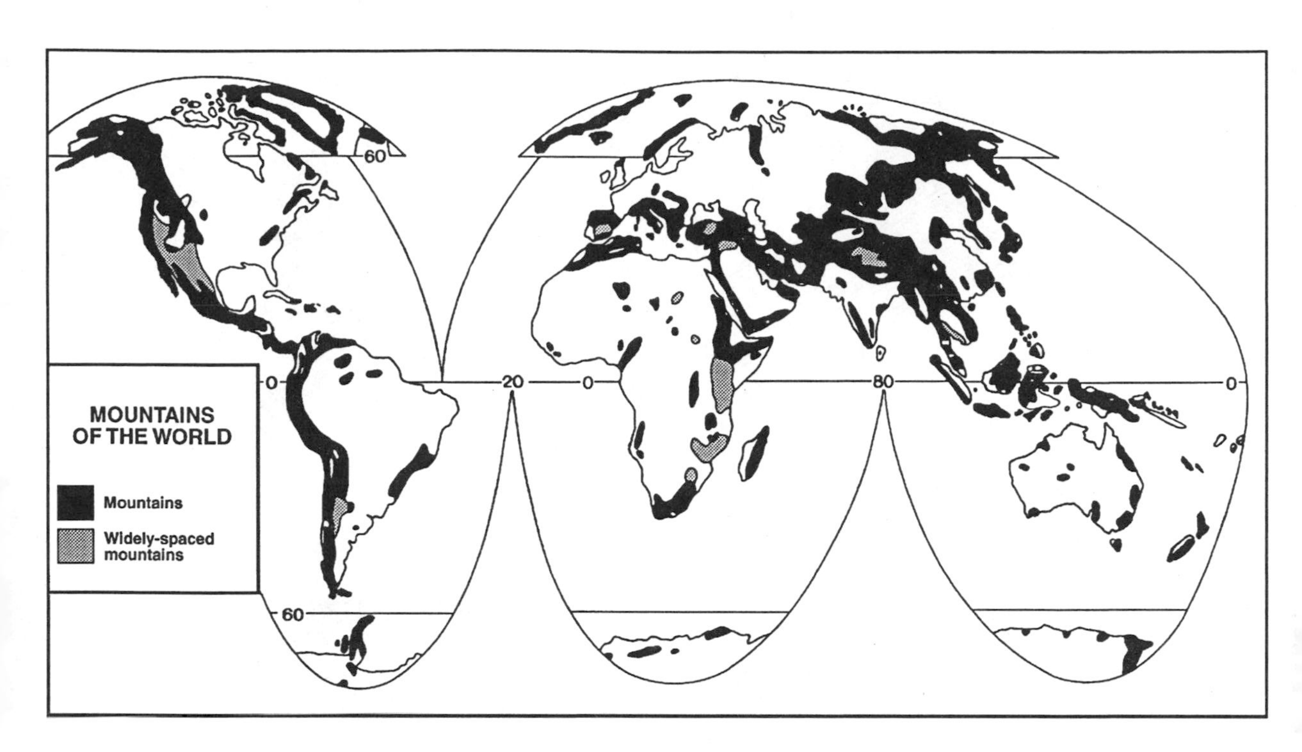

4.1 Global distribution of mountains. About 36 percent of the world's land surface is mountainous. Submarine mountains omitted. From Gerrard 1990.

Ranges, and *Picea pungens* restricted to the Rocky Mountains. The cordilleran endemism of these conifers is mirrored by that of many shrub and herbaceous species that are confined to each mountain system: *Chamaebatia foliolosa* of the Sierras, *Ceanothus thyrsiflorus* in the Coast Ranges, and *Eritrichium nanum* of the Rocky Mountains are examples.

Separation of mountains one from another (either as isolated ranges or as single remote peaks) has another biological significance. This spatial discontinuity is well represented in western North America. Montane, largely timbered, "high islands" in a "sea" of sagebrush and bunchgrass plains are beautifully portrayed (Fig. 4.2) in the basin and range country of Nevada and Utah (Billings 1978 and Trimble 1989) and in the isolated peaks and local mountain ranges in Montana, east of the Rocky Mountain Continental Divide. Both kinds of mountains, continuous (cordilleran) and discontinuous (isolated peaks and ranges), have been highly significant staging environments for fostering ecotypic (racial) differentiation of wide-ranging species as well as local speciation; they are cradles of evolutionary diversification. Both kinds of mountain systems can be found elsewhere in the world, such as East Africa, southern South America, and southern Europe (Fig. 4.1, from map 1.1 in Gerrard 1990). While endemics spawned by mountains

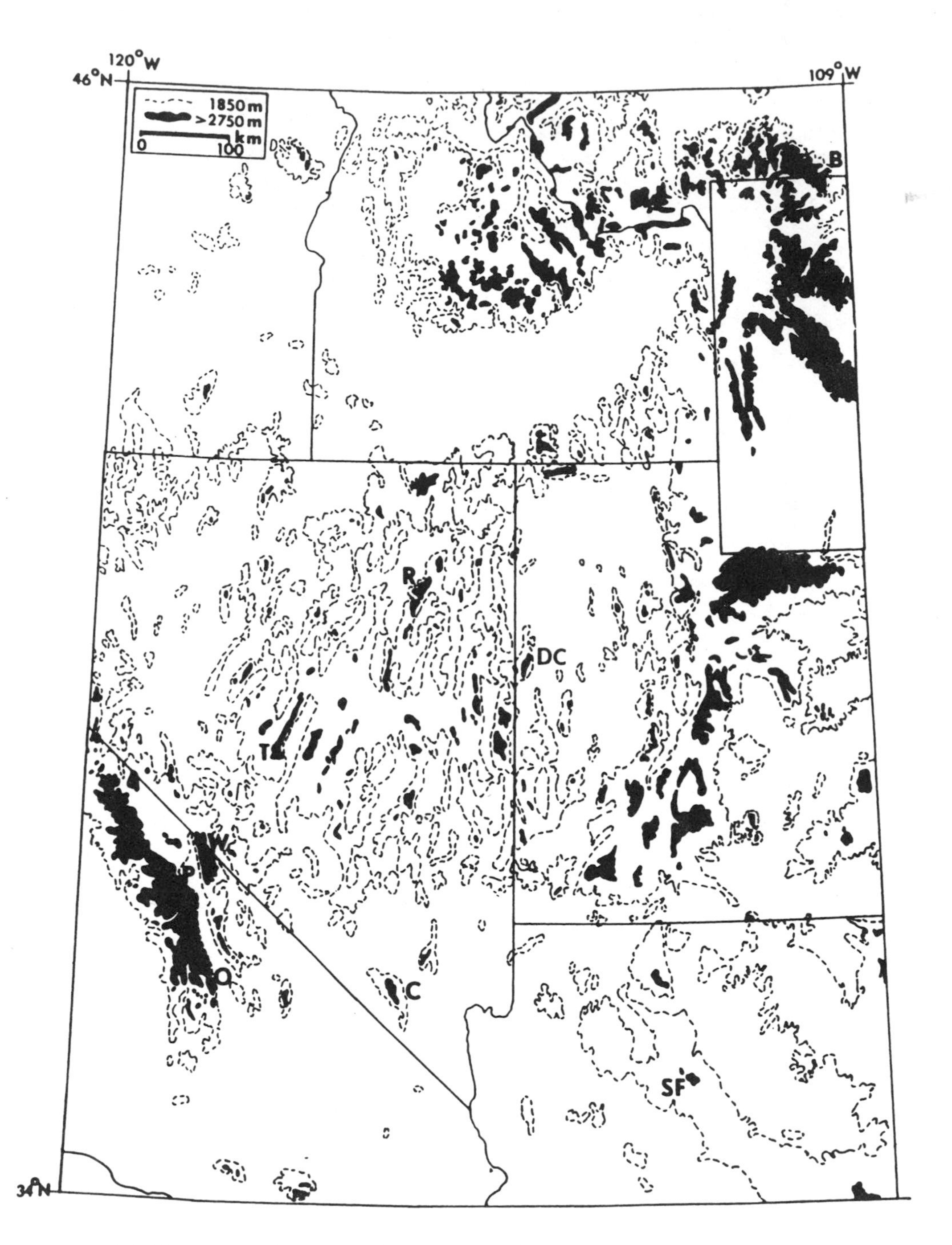

4.2 The summits of mountains in the Great Basin of western United States form a discontinuous pattern of isolated ranges. This insularity influences plant distribution. Black areas are alpine summits. Dashed lines (1850-meter contours) estimate lower limits of alpine vegetation at full glacial. From Billings 1978.

4.3 Spectacular isolated landform, Shiprock, New Mexico. Photo by B. van Diver.

capture the attention of the biogeographer, there is another biogeographic feature of cordilleran and isolated mountains. It is the widespread appearance of many species, mostly of the subalpine and alpine habitats, that occur on more than one massif, and may stretch their distributions intercontinentally solely on mountains. Taxa that spring to mind are *Silene acaulis, Saxifraga oppositifolia, Potentilla fruticosa,* and *Empetrum nigrum;* all four occur in high montane habitats in both North America and Eurasia, and *E. nigrum* has a close relative, var. *rubrum,* in South America.

Sedimentary Rock Formations as Landforms

Some of the most spectacular landforms are often fashioned out of "layer-cake" geology—horizontal beds of sedimentary rocks, dissected into plateaus, mesas, canyons, and other sculptured natural art forms (Fig. 4.4). Witness the grand display of dissected topography in the Canyonlands of western North America and the Grand Canyon of the Colorado River (Fig. 4.5). Then there is the massive Drakensberg Mountain complex of South Africa.

Grandest of all, though, are the *tepuis* of Venezuela, massive and discrete sedimentary mesas arising out of the Guyana Shield. The *Lost World* of Conan Doyle has been found again by twentieth-century scientists, naturalists, and adventurers.

4.4 Effects of alpine glaciation leaving tarns and irregularly planed surfaces. Rampart Ridge, Cascade Range, Washington State. Photo by Bob and Ira Spring.

The tepui (tepuy) topography occupies 500,000 square kilometers of the ancient sediments of the Guyana Shield. Tepui is a local Indian word for mountain, and what mountains they are! The more than a hundred vertical-sided tepuis are lofty sandstone mesas rising above the lowland like gigantic fortresses cut off from each other by intervening low-elevation savanna and tropical forest. One tepui, Mount Roraima, soars up to 2772 meters (9,094 ft). It and others of the Roraima group and the Chimantá massif have sheer vertical sides

4.5 Vertical walls and talus fans caused by erosion of layered sedimentary rock. Oxbow on San Juan River, Utah. Photo by R. Kiser.

and tower over 5,000 feet above the lowlands (George 1989; Huber 1992). See Figures 4.8 and 4.9.

The remarkable tepui formations have nurtured a rich biota, fashioned out of the many elements of the singular topography. Foremost is the element of isolation: each tepui is separated from its nearest neighbor by a vertical drop to the lowland forest. Then high elevations of tepuis consort with equatorial climate to promote tropical cloud forest conditions. Summit surfaces of most tepuis can vary dramatically in relief to impose great contrasts in habitat types. Though they look flat from afar, tepui summits may be dissected into an amazing kaleidoscope of landforms. Sandstone pillars sculpted into bizarre shapes, or simply deeply pockmarked surfaces, create a variety of habitats and microsites. The rock itself adds another variant to the tepui syndrome. The quartzitic sandstone readily weathers, but yields a soil low in nutrients. Intermittent intrusions of igneous diabase yield a more nutrient-rich soil to promote a richer floristic diversity (Huber 1992).

The consequences of isolation, elevation, landform, climate, and surface heterogeneity are a rich and highly endemic tepui flora. For just the Chimantá massif alone, there are over 400 vascular plant species in 55 families and 160 genera.

4.6 Contrast of vertical sandstone rock surface with plants only in crevices and nearby talus with plant cover. Mount Lilian, Kittitas County, Washington State. Photo by author.

4.7 Montane surface heterogeneity on a grand scale: shaded-relief image derived by computer from 1:24,000 scale topographic maps of the Chelan quadrangle, Cascade Range, Washington State. Image realized by R. Haugerud.

4.8. The sheer vertical walls of Roraima tepui, one of the complex massive sandstone plateaus of Venezuela, rises nearly 5,000 feet above the surrounding forest. Photo by G. Picon.

4.9 The irregular summit surface of Roraima tepui, Venezuela. Photo by G. Picon.

Twenty-nine of the species are endemic to just Chimantá. There are even endemic genera: *Chimantea* and *Achnopogon* (Compositae), *Mallophyton* (Melastomataceae), *Adenanthe* (Ochnaceae), *Wurdackia* (Eriocaulaceae), and *Acopanea* (Theaceae). The largest families are Compositae with 14 genera and 29 species, Ericaceae with 11 genera and 22 species, Orchidaceae with 12 genera and 21 species, and the Melastomaceae with 10 genera and 13 species.

The vegetation of tepuis contrasts strikingly with that of the lowland tropical forest. Each of the richly diverse—rococco!—continuously rain-drenched surfaces promotes a different vegetation type. One finds cloud "forests" of shrubs, nearly barren surfaces with no soil, scattered clumps of herbs (orchids, sundews, etc.), and even swamps. Short of a most arduous visit to encounter the surrealist world of the tepuis, I commend two references (George 1989 and Huber 1992) with lavishly illustrated text. The tepui phenomenon truly makes a bold statement for the role of landform in molding floras and vegetation.

Another montane world where sedimentary landforms promote floristic diversity is in the Drakensberg Mountains of South Africa. Like the tepuis, these mountains are elevated and sharply dissected mesas, though with a different combination of lithologies. Their imposing montane landscape is partitioned into two topographically distinct lithologies. The Low Berg is composed of sandstone and rises gently from the surrounding lowlands; it is topped by a massive vertical-walled basalt cap, the Main or High Berg. The slopes and summits of both Low and High Bergs support a richly endemic flora (Pearse 1978). For the Cathedral Peak area (1200–2000 m), a total of 419 genera and 907 species have been recorded. The value for the entire Drakensberg range is much higher (Pearse 1978). The flora is rich in monocots: grasses (Poaceae) and members of the Liliaceae, Iridaceae, and Amaryllidaceae. Notable dicot families are the Proteaceae and the Ericaceae.

The origins of the remarkable temperate and tropical sedimentary landforms known as *karst* have intrigued geomorphologists for decades (see Jennings 1985 and White 1988). The key attribute of karstic landforms is the differential dissolution of sedimentary rock, mostly limestone and dolomite (Fig. 4.10). The resulting dissection of surfaces takes many forms, some of which appear as mountainous terrain: for example, tower and cone karst in the tropics (Fig. 4.11), mogotes of Cuba (steep-sided hills), and the classic cockpit karst of Jamaica (innumerable hillocks, from the air looking like gigantic Mima mounds); see Figures 4.12, 4.13, 5.19 to 5.23, and Table 4.2. Other karstic landforms appear as dissected flat surfaces; the limestone pavements of Yorkshire, England, and The Burren of western Ireland are notable examples (Figs. 5.10 to 5.13). Since karst topography is closely linked to sedimentary lithology (especially carbonate rocks), a fuller account of this set of

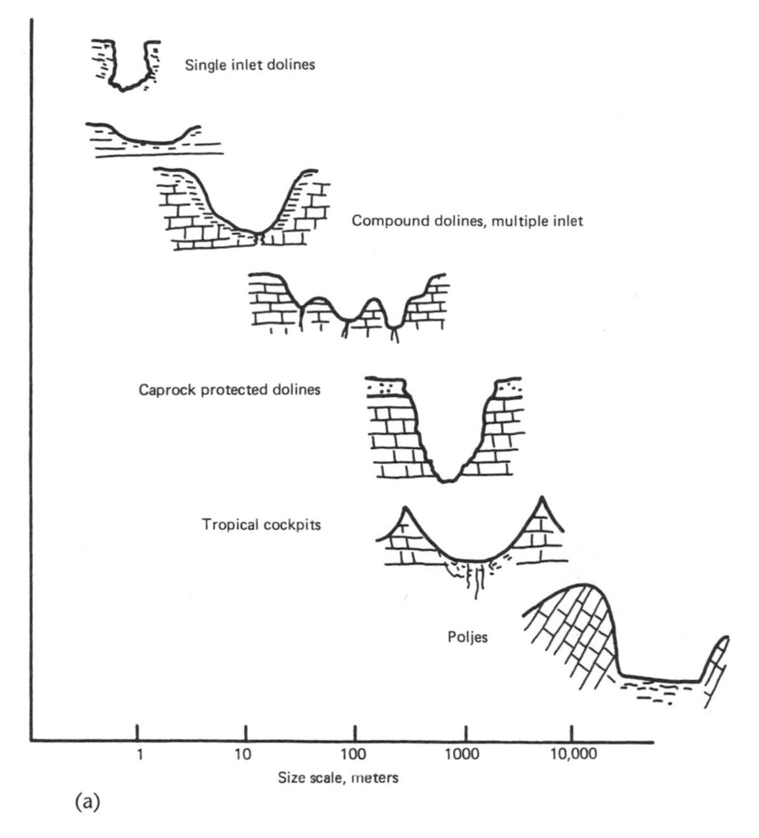

(a)

4.10 Types of karst formation (diagrams from White 1988):

(a) Profile diagrams showing size scale and types of karstic closed depression features.

(b) Two cross-sections of Adriatic karst showing the stairstepping of polje (closed basin) levels from the montane interior to the coast.

(c) Some profiles of cone and tower karst.

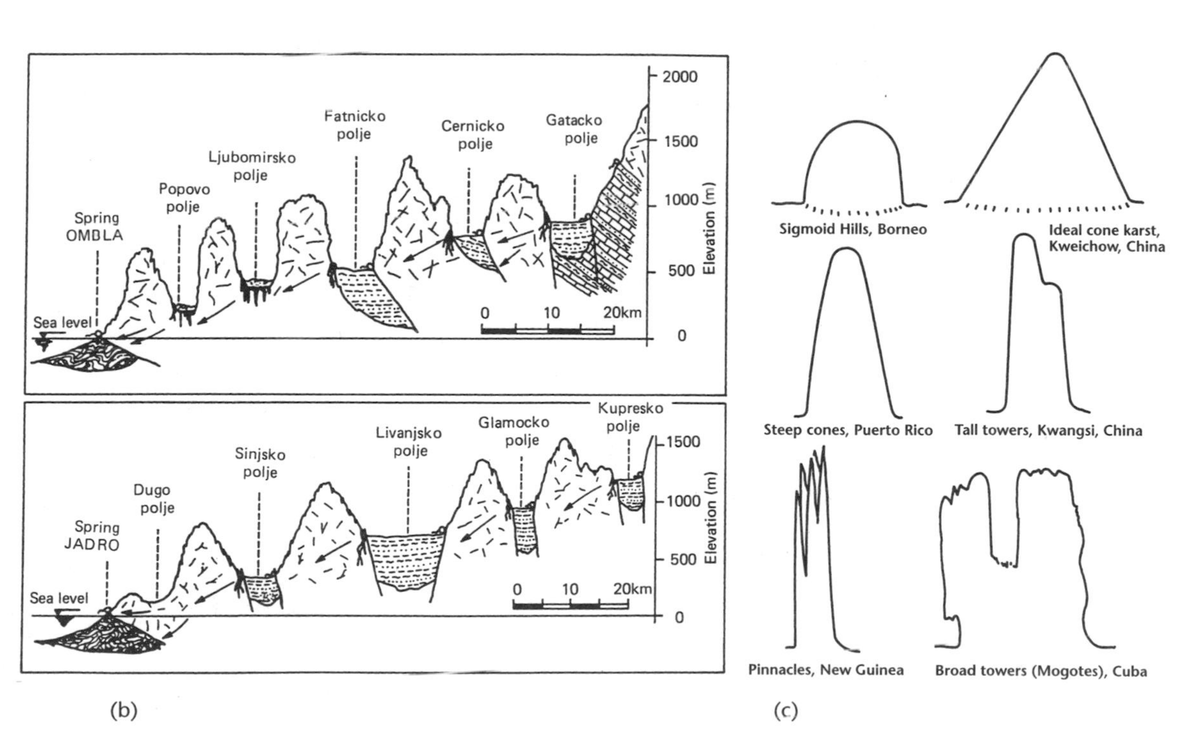

(b)

(c)

4.11 Tower karst (limestone) along Guilin Li River, China, the classic example of this tropical karstic formation. Photo by B. van Diver.

opposite:

4.12 Pinnacle karren (karst) near sea level, southwest of Mayaguez, Puerto Rico. Height of pinnacles, 2 to 3 meters. Photo from White 1988.

4.13 Temperate sedimentary relief in Olympic Mountains, Washington State. Folded rocks of eastern Olympics. Photo by R. Tabor.

Table 4.2. Surface landforms associated with karst.

Landform	*Salient features*	*Biotic potential*
Gorges	Deep and narrow with steep sides. France: Grands Causses, Verdun Gorge. UK: Kardale Scar, Yorkshire. Greece: Vicos Canyon. Features: natural bridges, blind valleys, steep heads.	Endemics on sheer walls and ledges, riparian species in canyon bottoms.
Dolines	Closed depressions (sink- or swallow-holes) of oval to circular shapes and sizes. Yugoslavia and northeastern Italy (Istria). USA: Driftless area, Iowa; Mitchell Plain, Indiana.	Sides of dolines can harbor unique flora or plant associations. Some broad flat doline basins used for crops (poljes).
Cockpit Karst (cone karst)	Closed depressions in tropical humid karst, similar to dolines but with sides lobed convexly inward, making them star shaped. Jamaica, Java.	In Jamaica, xeric tops and slopes with little or no soil; rich endemic flora (Proctor 1986); cockpit depressions with fertile soil and large trees (or cropland).
Poljes	Large closed depressions with flat floors (polje = field) and steep-sided limestone walls. Yugoslavia, Cuba, New Guinea, Morocco, Italy, Malaysia. Karst-margin plain, a similar feature.	Unique flora (?) mostly confined to slopes and vertical walls arising from floor of polje; floor often lacustrine or farmed.
Uvalas	Closed depressions with uneven floors, often formed by intersecting dolines (e.g., a complex of dolines). Yugoslavia, Tasmania, New Zealand, Morocco. Karst valleys: closed depressions, often "riddled with dolines," are similar to poljes and uvalas; also called blind valleys, gulfs, or karst windows.	Possibly unique plant life on slopes, though often used for pasturage.
Tower karst	Tall, nearly vertical stacks of limestone amid flat plains. Mostly in humid tropics: southeastern China, Malaysia, Cuba, Puerto Rico. Cone karst and mogotes are similar landforms.	Slopes and tops of towers likely sites for unique flora.
Mogote karst	Steep-sided residual hills scattered over nearly level terrain. Cuba, Puerto Rico.	Endemic flora and unique vegetation on slopes and summits in Cuba (Borhidi 1991).
Dogtooth karst	Akin to pavement karst: level terrain but jagged, corrugated (toothlike) limestone. Caribbean area.	Some endemics recorded for Cuba and Cayman Islands.
Pavement karst	Parallel crevices created by dissolution along jointing of platform limestone. Regolith removed by repeated glaciation. Northern England, The Burren of western Ireland, Spitzbergen, Greenland.	Often distinctive arctic or high latitude flora in crevices.
Karst caves	Subterranean spaces (passages, caverns, canals, tubes, rooms, etc.) either dry or with water; entrances from vertical to horizontal. Worldwide, but best known in Europe (France, Balkans) and North America (Mammoth Cave, Kentucky; Carlsbad Caverns, New Mexico).	Entrances may support unique flora; possibly endemic ferns and bryophytes. Well known for unique animals.

SOURCES: Borhidi (1991), Gerrard (1988), Jennings (1985), Skinner and Porter (1992), White (1988).

spectacular landforms and the unique plant associations and endemism associated with karst is taken up in Chapter 5.

Volcanoes

The case can be made that volcanoes possess unique attributes in addition to many of the features of other mountains (Gerrard 1990). The characteristics of shape, isolated position, and recurrent volcanic activity put them in a special class of mountain types. The relevance of volcanic landscapes, their formation, and their instability to the colonization, evolution, and persistence of biota clearly calls for dealing with them as distinct from other mountains. The physical geology of volcanoes and volcanism is treated thoroughly in Gerrard (1990); he devotes an entire chapter to "Volcanoes as Mountains."

Shape, size, and elevation of volcanoes significantly affect plant life. Shield volcanoes (lava shields such as those in Hawaii) have gentle slopes (less than 7 degrees) while the spectacular cones of stratovolcanoes, such as Fujiyama and the Cascade high peaks in western North America, are often as steep as 45 degrees. Steepness of slope, in combination with the nature of the volcanic substrate, exerts telling influences on plants. Steep slopes with loose volcanic material, like pumice, evoke characteristic responses—especially morphological, anatomical, and spacing—in plants. For instance, one of the most characteristic flowering plants on steep pumice slopes of the stratovolcano Mount St. Helens (Washington State) was (and still is, after the 1980 eruption) *Eriogonum pyrolaefolium*. Its woody roots can extend several meters upslope from the locus of the leaf and inflorescence rosette; the upslope rootstock is almost completely buried in pumice. This is apparently an adaptation to survival on these steep, unstable slopes of loose pumice.

Elevations and Shapes of Volcanoes and Plant Life. Volcanoes often emerge well above the surrounding mountain terrain. This is illustrated by the chain of volcanoes in the Cascade Range of western North America. Each of its volcanoes, from Mount Lassen in northern California to Mount Garibaldi in southern British Columbia, projects well above its contiguous montane relief. In the southern Cascades, volcanoes or their remnants range from 8000 to 10,000 feet (2700–3300 m) in elevation, while nearby terrain hovers around the 3000 to 5000 foot level (1000–1700 m). In the North Cascades of Washington State (Figs. 4.14–4.16), the contrast is similar but with higher nonvolcanic peaks. Mount Baker, a spectacular stratovolcano in northern Washington, is 10,731 feet (3285 m), while the peaks of the rugged Picket Range just to the east, a suite of *Gipfelfluren* (accordant) summits (Thompson 1962, 1990), average around 9000 feet (2900 m). Moreover, the contrast in relief is dramatic: Mount Baker has less precipitous slopes than do the

4.14 Mount Rainier, highest (4,392 meters) of Washington's Cascade volcanos. Photo by Bob and Ira Spring.

4.15 *Hulsea algida* (Asteraceae) on scoria, Mount Adams, Washington. Photo by author.

steep flanks of the serrated metamorphic peaks of the Picket Range. The higher elevations of the Cascadian volcanoes permit habitats just below permanent snow and ice to support a truly alpine flora, less likely to appear on the surrounding montane terrain at lower elevations. For instance, only on the higher volcanic peaks of the Cascades does one encounter the alpine tundra life-forms and habitat types. They are amply displayed on the flat, pumice surface of the summit of Burroughs Mountain (7300 feet, 2400 m), a companion summit on the north flank of the volcano Mount Rainier (Franklin and Dyrness 1973). These alpine tundra habitats, though infrequent in the Cascades, have characteristic features: low ericaceous shrubs and cushion plants as well as a cryogenically induced micro-

4.16 Eroded remnant of volcanic plug, Wenatchee Mountains, Washington; local endemics in rock crevices and on scree. Photo by author.

topography (stone stripes, solifluciton surfaces, etc.). These elevational disparities between volcanoes and adjacent montane relief are found elsewhere (Japan, Alaska, South America, East Africa, etc.). They, too, portray the same kind of floristic and vegetational contrasts caused by differences in elevation and substrates.

Spacing of Volcanoes and Floristic Isolation. The high degree of isolation of individual volcanic peaks, or peak systems, is often dramatic and undoubtedly contributes to local biotic endemism and other biological consequences of discontinuity. The degree of isolation varies all the way from the completely isolated, lone volcanic systems like Mount Cameroun in West Africa or Mount Ararat in Asia Minor, to the north-south-trending chain of semi-isolated volcanoes in the Cascade Range of western North America. A spectacular example of a mainland island-archipelago of volcanoes is the well-studied group in East Africa. The great volcanoes of Mount Kilimanjaro, Mount Kenya, and the Ruwenzori Mountains support unique floras, rich in endemics, some exhibiting bizarre life-forms, such as the arborescent herbs in *Senecio* and *Lobelia* (Hedberg 1970; Lind and Morrison 1974). In the temperate zone, archipelagos of mainland volcanoes lavishly populate mountain systems in Japan and western North America. I have examined the floristic consequences of the "island chain" of volcanoes in the Cascade Range (Kruckeberg 1987a), in connection with the recovery of plant life on posteruption Mount St. Helens. Emerging along the south-to-north transit of the Cascadian cordillera, there are 14 cones and calderas, from Mount Lassen and Mount Shasta in northern California to Mount Garibaldi in British Columbia. The subalpine-to-alpine summits of all 14 are separated from one another by midelevation montane forest ecosystems. Though the distance between any two cones may not be great (average is 58 miles or 93 km),* they are indeed floristic "islands." Some Cascadian volcanoes harbor local endemics (e.g., *Pedicularis rainierensis* and *Castilleja cryptantha* are found only on Mount Rainier). Other high volcanic cones have endemics (e.g., *Botrychium pumicola* on Mount Mazama, Oregon; and *Arnica viscosa* from Mount Shasta, California) to the nearest volcanoes to the north in southern Oregon. (See Table 2.1 in Kruckeberg 1987a for lists of plants on the Cascade volcanoes.) I return to the intriguing biogeographical aspects of volcanoes as islands in Chapter 7.

Soils of Volcanoes. Volcanic soils are derived from a variety of parent materials: acidic to basic igneous rocks, like andesite and basalt, to ejecta like pumice and tephra. Basic volcanic rocks weather more rapidly than do acidic rocks. A range in soil fertilities exists; basic rocks yield more fertile soils than acidic rocks. Flora and vegetation take on particular attributes derived in part from particular volcanic soils. Thus endemism, life-form, and community structure take on special characteristics due to soil type differences.

Recurrence of Volcanic Activity and the Biota. In contrast with other mountain systems of different origins and a greater stability of landscapes, some volcanoes are notoriously restless. Plant life may be wiped out instantly and massively via eruptions, yet often makes a rapid recovery. The establishment of vegetation follow-

---

* Range in miles for distances separating pairs of isolated cones: Mount St. Helens ⟷ Mount Adams = 38.5 miles to Mt. Rainier ⟷ Glacier Peak = 104 miles.

ing eruptions is well documented. The classic cases of Krakatau (Simkin and Fiske 1983) and Surtsey (Fridriksson 1975) are extremes; the emergence of completely barren island landscapes following eruption witnessed plant colonization and establishment over short time spans. Colonists on these island volcanoes were recruited from elsewhere by long-distance dispersal. In other instances, the restive volcanoes reside in places where posteruption immigrants grow just beyond the perimeters of the volcanic activity. Two well-documented case histories (Mount St. Helens and Hawaii's Kilauèa Iki) illustrate the remarkable resilience and opportunism of the pioneering biota in locally active volcanic landscapes. The 1959 eruption of Kilauea on the island of Hawaii gave ecologists an ideal opportunity for observing, over a period of several years, the inexorable return of plant life (Smathers and Mueller-Dombois 1974). Various new substrates were created: massive pahoehoe lava, spatter, and pumice, which came to serve as time-zero substrates for a new wave of plant life. About 500 hectares of montane rain forest and seasonal forest were devastated. Algae, ferns, and flowering plants were the initial colonists. Recruitment on the new surfaces was aided by two consequences of the preexisting vegetation. First, recolonization was accelerated in sites where organic remains of the devastated areas were near the surface. Second, vegetation was hastened by a "holdover" effect: plants that survived within the eruption landscape in sheltered microsites provided propagules for recolonization.

Mount St. Helens, in southwestern Washington State, was, for a short time, the "Fujiyama of North America." Its symmetrical cone was destroyed on May 18, 1980, following the latest of frequent eruptions of a volcano "in a hurry" (Harris 1980). At least two earlier eruptions are known in historical times. Between 2500 B.C. and A.D. 1800, fifteen eruptions were dated by geologists (Harris 1980, fig. 14-2). This most restless of the Cascadian volcanoes barely allowed time for vegetation to reestablish during dormant intervals of 100 to 500 years. Compared with other volcanoes of the Cascade Range, the pre-1980 flora of Mount St. Helens was depauperate; it had no local endemics, and many of the species found on nearby volcanoes were absent (Fig. 4.17). The parkland and timberline community types were underrepresented, and there was an atypical blend of low and high elevation conifers (Kruckeberg 1987a). The massive, explosive eruption of May 18 appeared to destroy much of the plant life on the slopes of the volcano. The upheaval took a variety of forms: huge avalanches of rock and snow; rock, superheated steam, and gases blasted out of the north slope; repeated pyroclastic flows; tephra fallout; and mudflow (Fig. 4.18). All this material inundated slopes of the former cone, and airborne ashfall (tephra) was delivered far beyond the mountain. These varied manifestations of the eruption each had a different effect on the vegetation.

4.17 Pre-eruption view of Mount St. Helens, Cascade Range, Washington, the quiescent "Mount Fuji of North America." This 1954 photograph shows the pre-1980 volcano's mostly forested slopes to the snowline. Kalama River valley in foreground. U.S. Forest Service photo.

The directed lateral blast laid waste to forests in its path (downed timber of the blast zone); pyroclastic and mudflows buried upper montane vegetation to depths of 15 to 100 meters, and tephra deposits covered ground vegetation at depths of 5 to 20 centimeters. What was the prospect for recovery of the vegetation following this latest eruptive episode? This question prompted a major research effort by scientists of diverse callings. The devastation of plant life offered an unparalleled opportunity for plant and animal biologists to observe the rebirth of life from ground zero at time zero. Studies on colonization, plant succession, population ecology, and comparative ecophysiology have gone on ever since the 1980 eruption. Critical to recovery of plant life has been the survival of herbaceous plants that served as foci for recolonization (del Moral and Wood 1988). Several years after the eruption, devastated areas with no nearby sources of "infection" remained largely barren, while devastated sites adjacent to two pockets of living vegetation showed good recovery. Variations in depth of tephra deposits have a direct bearing on the recovery of perennial plants. In a series of papers, Antos and Zobel (1985a, b, c; Zobel and Antos 1986, 1987)

4.18 The truncated cone of Mount St. Helens and devastated forests about three months after the May 18, 1980, eruption. Photo by author.

reported a range of accommodation by species to the tephra deposits. Some woody species developed adventitious buds or rhizomes within the tephra (e.g., *Spiraea, Salix, Alnus, Ribes,* etc.); and some herbaceous species sent perennating buds into the tephra and thus have survived the burial. Some forest-dwelling members of the Liliaceae were unable to send perennating buds into the tephra and thus eventually died. However, one liliaceous plant, false hellebore (*Veratrum viride*), a tall robust perennial, proved an exception. Plants buried under 12 to 18 centimeters of tephra emerged without difficulty, to flower (Zobel and Antos 1987). Moreover, when intentionally buried with an additional 40 centimeter increment of tephra, the false hellebore shoots still emerged, but only if they had already begun elongating prior to the experimental burial. Though this species can cope with being buried under substantial depths of tephra, depths of 40 centimeters can prevent shoot emergence and eventually kill the plants.

Ashfall (airborne tephra) from the 1980 eruption settled on agricultural land and native vegetation many miles leeward of the volcano. Ecological effects of the new encrustation were significant. One positive consequence was a decrease in water loss by a kind of mulching effect (Chapin and Bliss 1988). Probably a similar water retention resulted from the change in albedo (light reflectance) at soil surfaces. Most preeruption substrates and habitats had dark soil surfaces; the "top dressing" of tephra increased the albedo. For example, subalpine heather meadows in the Goat Rocks, about 30 air miles east of Mount St. Helens, changed from dark to light soil surfaces, following tephra fallout. Seen in 1983, these plant communities, mantled with their light-colored top dressing, appeared to be in excellent condition, even surpassing their usual luxuriance (Kruckeberg and Tanaka 1983). The deposition of airborne tephra reached eastern Washington to fall on both agricultural lands and native steppe desert vegetation. The effects on crops, reported by Cook et al. (1980), were both positive and negative. Alfalfa hay crops, already in full growth by May 18, suffered lodging (flattening). However, yields of wheat, potatoes, and apples were expected to be above normal for 1980. No significant contribution to the nutrient status of the soil was expected from the tephra depositions. Substantial short-term mechanical damage to herbaceous perennials (acaulescent and prostrate dicot species) was observed later in the 1980 growing season (Mack 1981). Yet the damage was not so severe as to impede the eventual recovery of the plants. Mack also noted the change in albedo which could reduce water loss, thus affecting the water regimes of native soils. He suggested that the new soil surface created by tephra crusts could become seedbeds and "safe-sites" for next-generation progeny of both native and weed species. At this writing (2001) there are tephra crusts still persisting in sagebrush-bunchgrass habitats east of the Columbia River.

Research on vegetation recover on Mount St. Helens continues. Seedling establishment has differed, depending on microsite; flat surfaces have lower numbers of seedlings than do sites of uneven terrain (Shiro, Titus, and del Moral 1997; Titus and del Moral 1998). Succession from bare volcanic debris to at least a pioneer, early successional plant cover has been rapid, with perennial herbs, often weedy species, predominating. Woody species have been slower to recolonize; willows and red alder (*Alnus rubra*) assume low but increasing prevalence, while the reestablishment of a typical Pacific Northwest conifer forest is still in the distant future (Franklin et al. 1985; Dale 1991).

Volcanic substrates (both raw parent materials and soils) can provide unique environments for plants; they can be studied from both ecological and evolutionary biases. In sum, volcanoes as a special kind of surface heterogeneity have physical and temporal properties that can have profound effects on floras and vegetation.

Effects of Size and Elevation of Mountains on Plant Life

Most of the Earth's mountains are arrayed in systems of massive continuity. The Himalayas, Rocky Mountains, and Andes loom large not only in areal extent but in bulk. The continuous mountain systems foster internal consistency in biota, arrayed both in elevation and in latitude-longitudinal extent. Less common are orogenies that yielded single, isolated mountains or small, isolated mountain ranges. Examples of the latter are the isolated ranges of Australia (the MacDonnell Range and the Snowy Range), the semi-isolated granite domes of western Australia and southeastern United States, and the smaller mountain ranges east of the Rocky Mountain Continental Divide in Montana (Meloy 1986). The most spectacular of wholly isolated single peaks are of volcanic origin, exemplified by the oceanic volcanoes of the western Pacific and the isolated continental volcanoes like Mount Cameroun and Mount Ararat. These single peaks or small mountain ranges are ideal evolutionary laboratories, where endemism and racial (ecotypic) differentiation can flourish (Kruckeberg 1991b).

For the massive mountain systems, their very internal integrity may provoke a different kind of evolutionary response. We expect to find kinship among the taxa. Endemism in such large massif systems can involve many variant species on a generic theme, such as *Primula* and *Meconopsis* in the Himalayas and *Schizostylis, Tecophilea,* and *Francoa* in the Andes. The biogeographic consequences of size and contiguity of mountain systems are taken up in Chapter 7.

Change in elevation in the ascent of a mountain or mountain system has a profound effect in shaping the floristic composition and the physiognomy of vegetation along the sloping gradient. Indeed it is the most obvious ecologic manifestation of *Hochgebirge* (high mountains). It was Alexander von Humboldt (1807) who first formalized the phenomenon of altitudinal zonation, in his classic studies on plant geography. His many traverses of Andean inclines inspired the notion of zonation of vegetation. From sea level to the upper limit of vegetation, imposed by permanent ice and snow, the quality of plant life tracks the change in altitude. Life-form, species composition, and community structure are all influenced by altitude. The most direct cause of the gradient or zonal display of vegetation is, of course, the way mountains control mesoclimates, especially in causing increases in rainfall and decreases in temperature with increase in altitude. I will examine in detail the relation of mountains to climates shortly below. Changes in vegetation with elevation have elicited a variety of zonation schemes by geoecologists. In North America, the zonation concept of T. Hart Merriam held sway for years. It was an elaboration on the simpler zonal sequence of Schimper (basal, montane, and alpine zones; Schimper 1903). Merriam's Life Zone concept (1895, 1898) owed its conceptual origins to his experiences in the mountains of western North America. Merriam contended that for each species along a gradient, there are certain *fixed climatic limits* that prevent the spread of the species beyond its limits of tolerance to other altitudes (Jones 1936). Species with coincident or overlapping tolerance ranges would then congregate in a "life-zone" of a given altitude and climate. In its most strict form, the Merriam Life-Zone scheme is grossly typological. Its rigid definition is: "A natural biotic unit, definable as an area possessing a distinctive assemblage of species and life forms, both of plants and animals, and usually having objectively determinable climatic and physiographic boundaries" (Jones 1936, p. 16). Users, like Jones, of the Merriam Life-Zone scheme recognized that overlapping of some species can occur at zonal "boundaries." For the Olympic Peninsula of northwestern United States, Jones found the overlap to be small; less than 5 percent of the total species of the area have an anomalous zonal distribution. The Merriam Life-Zone scheme uses, for the most part, an unwieldy and inappropriate nomenclature. Of the six extratropical zones, only the Arctic-Alpine Zone seems cognate in its attributes. Other zonal names, such as Hudsonian, Canadian, Sonoran, and Austral, find little relevance away from their "type localities," except by association to a given altitudinal zone. Since the zones are usually congruent with a climatic climax vegetation, it is to be expected that biogeographers and ecologists would use climax dominants to typify a zone. Franklin and Dyrness (1973), in portraying the vegetation types of Washington and Oregon, gave the name of the climax dominant species to a given zone. Thus their *Tsuga heterophylla* Zone is equivalent to the Merriam and Jones Humid Transition Zone. Regardless of the terminology and the fuzziness at zonal boundaries, the life-zone schemes do reflect a profound effect of montane topography on vegetation (Fig. 4.19).

MOUNTAINS AND CLIMATES

Mountains are powerful geoedaphic surfaces. Their singular topographies affect environments in a host of ways. I have already touched on the interaction between mountains and regional to local climates. Let me elaborate here on several related themes: climates and accordant mountain summits, timberlines, and orientation of mountain systems. In each of these, I will maintain that the mountain system is the primary conditioner of the attendant climatic effects. This is not to deny that some macroclimates that impinge on mountains are only dimly related to landform in origin (e.g., major oceanic weather systems).

My assertion of a dominating influence for mountains on regional climate has been voiced by others (Barry 1992; Harper et al. 1994). Petersen (1994), in the Harper book on the Colorado Plateau, makes the bold statement that "one of the most significant influences on climate in recent earth history has been the thrusting of great mountain chains and plateaus (such as the Tibetan Plateau and the high country of the western United States) into the lower atmosphere. These uplifts resulted in a globe that can be pictured as having two major bulges in opposite sides of the Northern hemisphere."

Accordant summits are a common geomorphic feature on all continents, and

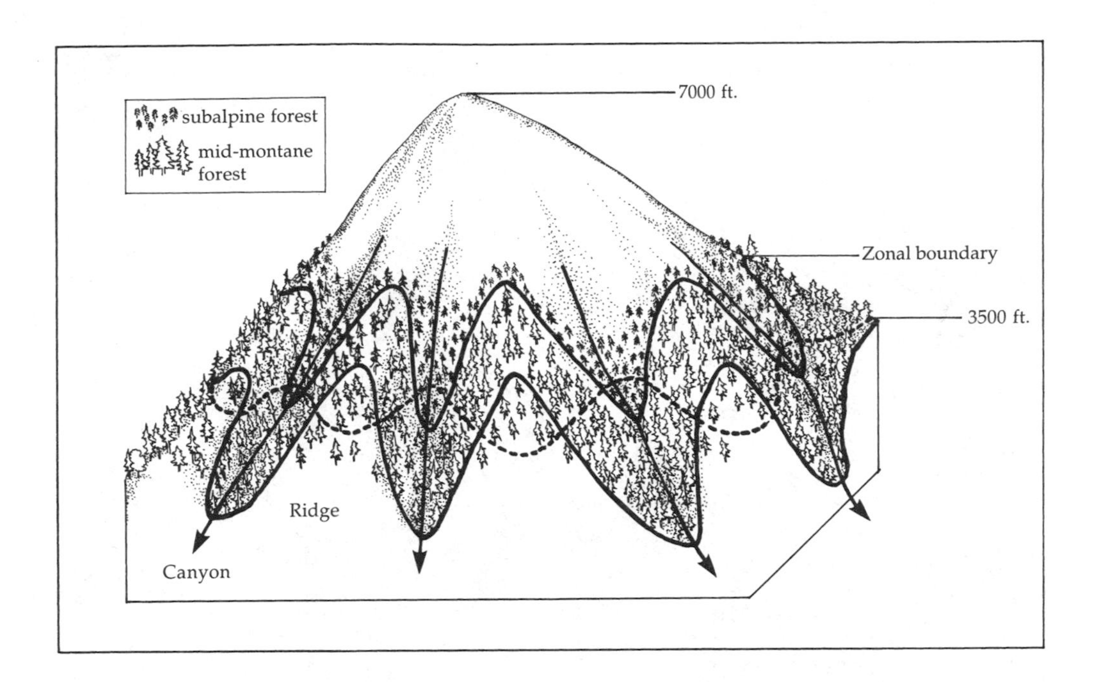

4.19 Effects of montane slope and exposure on temperate forest vegetation. Subalpine forest shifts downward in canyons (vertical arrows), but is confined upwardly on ridges. From Kruckeberg 1991a

are clearly displayed in the midlatitude mountain systems of the Northern Hemisphere. Mountains like the Alps of Europe and the North Cascades of western North America appear to have summits of similar elevations; that is, they are accordant when viewed from afar. One theory attributes the phenomenon of accordant summits to prevailing climates; recommended is the major review of this climamorphic explanation by Thompson (1990). Accordant summits are especially dramatic in high elevation mountains (Hochgebirge) where alp slopes and summits project well above timberline. In them, the peaks extending above timberline have attained similar elevations. The mountain system of the North Cascades of Washington State illustrates both the phenomenon and related features (Fig. 4.20), and also evokes an hypothesis for the cause of these features. While individuality is the telling impression to an alpinist viewing a local cluster of peaks from the ground, an aerial view of the North Cascades reveals the remarkable accordance, with summits averaging 8000 feet in elevation (Thompson 1962, 1964, 1990). Such an accordance evokes a vast jagged "plateau" in aerial views, long recognized by geomorphologists as "summit uplands" or "summit plateaus" (*Gipfelfluren* in European literature). Two other features—timberline and alp slopes above timberline—also show accordance.

Gipfelfluren take on markedly different forms in different climates. Thus mountain systems influenced by coastal maritime climates, such as the North Cascades

4.20 Accordant summits in the Picket Range of the North Cascades Range of Washington State. Accordances (or Gipfelfluren), the altitudinal regularities of summits, can be found in many other mountain ranges; for example, the Rocky Mountains, the White Mountains of New England, the Himalayas, and the European Alps. Photo by W. F. Thompson.

or the Alps of central Europe, have a marked serrated relief. In contrast, the Gipfelfluren in mountains influenced by continental climates have a gentler summit relief, exemplified by the Wind River Range of Wyoming in the Rocky Mountain system, and the Presidential Range of northeastern United States. Thompson (1962) describes this interior type of Gipfelfluren as "a strongly rolling felsenmeer-mantled† garden summit upland." With accordance at timberline and above, alpine summits have a number of climatic and cryogenic attributes that result from high mountain climates. These include glaciation, massive snow accumulations (areas of permanent or seasonal snowfields), mass wasting of rock and soil, freeze-thaw cycles (inducing patterned ground, fracture of bedrock, stone sorting, gelifluction, etc.), and other temperature/moisture effects. It is hypothesized that accordance of summits (Gipfelflur relief) is a product of regional climate extending back into

† Felsenmeer (literally "sea of rocks"): "Any considerable area, usually fairly level or of gentle slope, which is covered with moderate sized or large blocks of rock" (Habeck and Hartley 1968).

the Pleistocene (Thompson 1990). Deviations from accordance do occur: volcanic peaks, recent and ongoing tectonic uplift, and age of the mountain system are notable variables. But where they occur, accordant summits are striking and may well be the products of climate.

However, in my judgment the mountain-climate interface works two ways. It should be expressed as a reversible function: mountains ⟷ climate. Any mountain range is first and foremost a product of tectonic geology; regional climate affects its surfaces and summits only because the landform is there to begin with.

At least three of the five major Mediterranean type landscapes and biomes of the world share topographic similarities as well as the oft-cited similar Mediterranean climate. The likeness in landforms is nicely explained by Thrower and Bradbury (1977):

> Mediterranean areas are defined in terms of climatic similarity. Although their structural origins may be diverse they possess a similarity of landform process and character. Three areas, i.e. most of the Mediterranean type region (except the southern Levant), California, and Chile, are marked by relatively young orogenic systems, exhibiting high sharp, folded and faulted mountains and hills rising close to the coast. These uplands, in the three areas, gained their present gross morphology through violent upthrusting in late Tertiary and early Quaternary periods. Tectonic instability and volcanic activity are associated with these linear mountain chains to the present time. Glaciation, even at maximum extent, was very localized within these areas and probably only modified pre-existing valleys cut before the Pleistocene. . . . [I]n all Mediterranean areas coastal lowlands are more or less discontinuously interrupted by areas of greater relief reaching the shore.

The similarities of the several Mediterranean biomes of the world are often characterized by similarity in climate—a long and hot drought period between early spring and late fall rains. Yet the quotation above points to a consistent geoedaphic cause as well: the similarities of landforms, both their processes and products.

Surface heterogeneity promotes yet another variable for plant distribution. Orientation of landforms in the context of prevailing climate gives rise to what plant ecologists call aspect. Aspect, or slope exposure, is the orientation of landform in relation to compass direction. North slope versus south slope equals contrasting slope exposures. Steepness, as well as compass direction to slope, is another dimension to aspect (Fig. 4.21). Hans Jenny (1980, p. 276) says it well in his treatment of State Factor Topography: "Soil and vegetation react to aspect, whether facing north, south, east, or west."

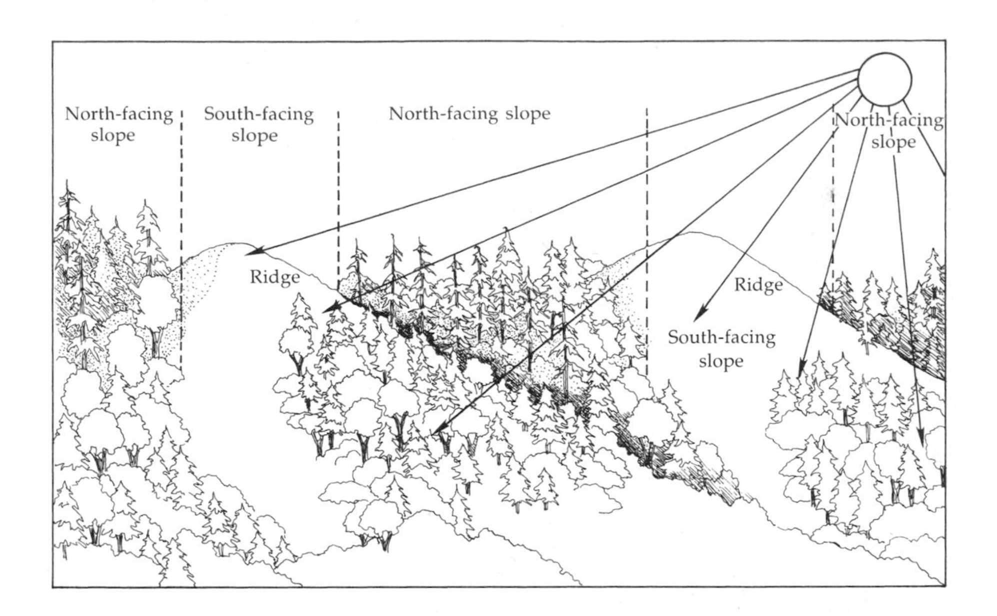

4.21 Aspect (slope, exposure), a function of landscape variation in mountains, yields marked contrasts in vegetation types. Drawing by S. Noel in Kruckeberg 1991a.

Steepness and compass direction of montane slopes conspire with heavy snowfalls to create avalanche conditions (Fig. 4.22). When avalanches of massive snowdrifts occur, their effects on vegetation can be either catastrophic or inconsequential. When forest is in an avalanche track, trees are devastated. But where pliant shrub cover grows in the track, little damage is done. These differences in avalanche impact on vegetation can be linked to the frequency of active avalanche tracks. Ones that occur nearly every year will have the flexible shrub cover, while the rarely occurring avalanche will devastate well-established forest trees. This linkage between periodicity of avalanches and kinds of vegetation is tellingly seen in the Cascade Range of Washington (Cushman 1981). So here again montane landforms influence climate to yield distinctive vegetation patterns. See Figs. 4.23–25 for other slope effects.

Contrasts in aspect also have a scale factor. Every heterogeneous surface, from a clod of soil or a single rock on level ground to the dramatic north-versus-south slopes in mountains, induces the effect of aspect on plant life. Without expanding at length on this self-evident influence, I would like to cite just two or three examples.

An early recognition of the effect of aspect on vegetation was recorded by Göte Turesson, long before his salient contribution to ecological genetics (the ecotype concept). Turesson had his early formal botanical training in Washington State. Observing slope exposure differences in the Spokane area, he noted that Douglas fir (*Pseudotsuga menziesii*) was confined to shady north slopes: "The opposite south-

4.22 Slope and aspect combine with snowpack to trigger avalanches. Such tracks, common in the steep terrain of the North Cascade Range, Washington State, often support distinct vegetation (pliant deciduous shrubs like slide alder, *Alnus incana* and *Spiraea densiflora*). Photo by J. Franklin.

facing bank has a very different flora; in fact, the contrast is so striking as to make it seem almost unnatural. The heavily *Pseudotsuga*-forested north-facing slope faces a vegetation on the opposite bank, which in xerophily, rivals the desert flora!" (Turesson 1914, p. 340). The south aspect supported no trees; rather it was sparsely stocked with xeric shrubs and herbs.

On a smaller scale, in a nonforested ecosystem, Butler et al. (1986) examined a toposequence (landform gradient) along a 75 meter gradient of aspect. Their graphic representation (Fig. 4.26) tells the story better than words.

Mima mound topography exemplifies the aspect phenomenon on yet a smaller scale. The mounds at Mima Prairie (Thurston County, Washington State) are 1 to

4.23 Talus and scree are common features of mountain slopes; such rock debris on slopes, either stabilized or loose, often supports vegetation adapted to these conditions. Pictured is a talus slope of sandstone, upper Beverly Creek, Wenatchee Mountains, Washington State. Photo by author.

4.24 *Athyrium distentifolium*, a common fern inhabiting talus and scree. Photo by author.

4.25 Talus and rock avalanche tracks at sea level. Steep slopes of Lummi Island, Washington State. Photo by Washington State Department of Ecology.

2 meters high and about 13 meters in diameter; an intervening level "trough" separates the evenly spaced mounds. Vegetation differences in species composition and in density appear in different quadrants of the compass. Vegetation of northeastern (lee) slopes differed both quantitatively and qualitatively from the sunny windward southwestern side of the mounds (Giles 1969; del Moral and Deardorff 1976) (Fig. 4.27). Indeed, aspect on a vast range of magnitudes qualifies as yet another facet of the effects of landform heterogeneity on plant distribution.

TIMBERLINE—THE GEOEDAPHIC CONNECTION

"Timberline is a biological boundary which doesn't escape even the most casual observer." So said ecologist Peter Wardle (1973) of New Zealand, where that bound-

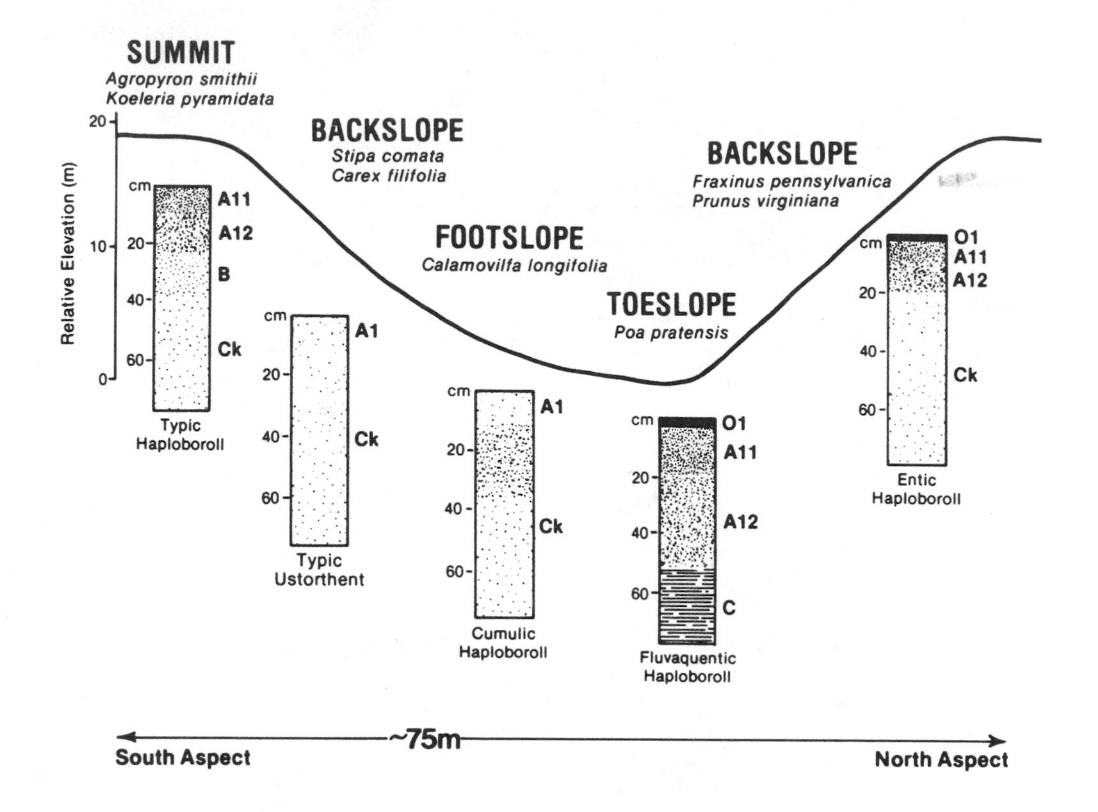

4.26 Topographic variation on a microscale induces variation in vegetation. This illustration shows the topographic sequence of soils and vegetation along a landscape gradient in the North Dakota Badlands. From Butler et al. 1986.

ary is strikingly defined by the upper limit of southern beech (*Nothofagus*). The "land above the trees," aptly named by Ann Zwinger (1972), is marked by a ragged boundary, upper timberline. Timberline itself is a zone where forests upwardly change dramatically, from continuous stands to parkland (tree clumps and subalpine meadows, or "timbered atolls" [Arno 1984]) to alpine scrub and krummholz ("crooked wood"), to isolated dwarfed individual trees (Figs. 4.28–4.30). This telltale vegetation zone, especially on mountains of North Temperate regions, is first and foremost a product of climate. Stressful states of precipitation and temperature limit tree growth. And it is a particular brew of montane climate that creates the timberline zone. Strong winds, heavy precipitation mostly as snow, low annual temperatures, marked diurnal flights of temperature (radiation effects) during the growing season, frost phenomena—all conspire to make timberline (and the alpine zone) a singular climatic world for life. But without the geomorphic grandeur of mountain topography, there would be no timberline (except for polar timberlines) with its influence on climate. Indeed, mountains make climate even as climates shape mountain topography. The prime geoedaphic arbiters in defining the tim-

4.27 Vegetation varies in location (compass direction) on a given Mima mound (ca. 1.5 by 13 m). Douglas fir encroaching on mound topography after first establishing on the lee (north-facing) sector of a mound. Aerial photo (late afternoon) of Mima Prairie, Washington State, by V. Scheffer.

berline and alpine zones are topography and lithology. Diversity of upper montane relief expresses itself variously to create marked contrasts in vegetation (Arno and Hammerly 1984). Slope, north- versus south-facing, in the Northern Hemisphere can yield vastly different vegetational responses. South slopes elevate timberline by several hundred feet because of the higher temperature. A closely related effect is the contrast between ridge tops and similar convex surfaces with concavities like draws and basins. Since the basins are

4.28 At timberline, landform diversity and altitude foster a rich diversity of vegetation types. Pictured here is a krummholz form of *Pinus albicaulis* on the south slope of Mount Adams, Cascade Range, Washington State. Aspect and fellfield terrain favor this conifer. Photo by author.

4.29 Timberline features include solifluction stripes and elfinwood (krummholz) conifer stands. Mount Fremont, Mount Rainier National Park, Washington State. Photo by author.

4.30 Continuous conifer cover gives way here to subalpine parkland (timbered atolls), just below timberline. Landform contrasts include meadowy flats and slopes, swales with late snowpack, and rock outcrops. Sunrise Park, Mount Rainier National Park, Washington State. Photo from Franklin and Dyrness 1973.

colder than the ridge tops, tree limits are higher on ridges. Steepness also influences the display of timberlines. Gentle slopes may reduce the magnitude of the timberline expression, especially at the lower boundary of the zone. On the other hand, mountains with steep to precipitous slopes (nearly vertical surfaces) can suppress the full elevational potential of timberline. Thus, in the Picket Range of the North Cascades, were it not for the near vertical relief near the summits, timberline would extend upward by several hundred feet. A timberline does occur in mountains with sharp relief, but it is defined by topography, not climate, and could be called a "false timberline."

Ecologist Stephen Arno (1984) called attention to another geomorphic control of timberlines. Large mountain masses, in contrast to isolated peaks at similar latitudes, allow the display of timberlines at higher elevations above sea level. This so-called Massenerhebung effect is caused in part by decreasing lapse rates (vertical temperature gradients) that occur on lofty, large mountain plateaus (e.g., in the Rocky Mountains and the Himalayas). The net result of warmer summer temperatures (and less severe winds), especially in continental climates, is to allow an upward extension of timberline. It is the sheer physical mass of the mountain system that brings this about.

The other geoedaphic discriminator in timberline and alpine zones is lithology. The massive surficial presence of rock, either as outcrops of bedrock or as talus, scree, fellfields, and felsenmeers of loose rock, makes its mark on the physiognomy and species makeup of these two vegetation zones. Unencumbered by forest, rock is the ever-present companion of nearly all plant life in the upper montane. Rock outcrops limit plant life to species adapted to fissure and crevice sites. For example, of the nine species in the Dasanthera Section of *Penstemon,* two are crevice plants (*P. rupicola* and *P. davidsonii*), while others like *P. fruticosus* and *P. barrettiae* are mostly found in loose rock of talus and scree. The loose, detached rock of timberline and alpine areas, ranging from angular pebbles to massive angular boulders, also conditions habitats for plants. Smaller rock is subject to cryogenically induced movement (stone stripes and nets); and vegetation, especially herbs and low shrubs, conforms to this patterned ground. Larger detached rocks, in fellfields or felsenmeers, primarily afford microsites for plants: leeward versus windward side of a rock can spell the difference between survival and death. The dominating presence of rock as the primary substrate in subalpine-alpine areas means that soil formation is limited or negligible. Roots are often in direct contact with rock and its immediate but limited weathered product (a lithosol). Besides its ubiquitous presence in all sizes and shapes, rock takes on another attribute that influences timberline and alpine vegetation. Rocks of varying mineral makeup, from acid to ultramafic, as well as igneous, metamorphic, or sedimentary species, take on qualitative differences physically and chemically for upper montane vegetation. Rate and degree of weatherability is an important rock variable. For example, Jenny (1941) reports that limestones are the most resistant of sedimentary rocks while sandstones are the least stable. And for a particular series of igneous and metamorphic rocks, Jenny gives the order of decomposition as follows: Basalt > Gneiss > Granite > Hornblende = Andesite. Color of rock and its immediate breakdown products (gravel and raw soil) can also be selective on species of the subalpine-alpine. One species of *Erigeron* selects light-colored dolomite and another, dark sandstone in the upper White Mountains of California (Mooney 1966).

Chemical composition of rock may also produce differences in floristic composition and vegetation pattern. Arno (1984) suggests that soil and rock chemistry is not as important as high mountain climates in creating vegetation differences. Yet there are instances where dramatic differences in plant composition are caused by differences in rock type. Again, the White Mountains of California provide a telling example. On sandstone, timberline is the upper limit of juniper-pinyon pine woodland. But in places where dolomite rock extends upward beyond sandstones, a second timberline is encountered. A belt of bristlecone pine (*Pinus longaeva*) begins on dolomite where treeless areas above timberline on sandstone border dolomite parent material (Wright and Mooney 1965). I will return in more detail to a consideration of lithology and its influence on montane vegetation in Chapter 5.

ORIENTATION OF MOUNTAIN SYSTEMS: CLIMATE-VEGETATION EFFECTS

A mountain range that lies athwart the path of a prevailing weather system profoundly affects the fate of that weather "engine." Orientation of the topographic barrier can have a cascading effect on a variety of environmental features on either side of the landform. With the dramatic changes in climate, the weathering of rock, soil formation, and genesis of particular soil properties, as well as contrasts in vegetation on either side of the weather barrier—all are consequences of the interposed topography. So striking are the effects of the mountain system's compass orientation with respect to storm tracks that language to express it has inevitably evolved: for example, windward versus leeward, the Hawaiian terms *mauka* and *makai,* the wet side versus the dry side of the mountains, and the rain-shadow effect. Most anywhere along the planet's surface heterogeneity, where mountains interdict prevailing storm tracks, the drama of change in weather and vegetation is played out. Akin (1991, p. 19) describes the rain-shadow effect well: "Where a mountain range lies in the path of prevailing winds, uplift (of air) may be rapid. Rainfall often is heavy on the windward side and, conversely, on the leeward side, where air is subsiding and warming adiabatically, conditions may be dry. A dry area downwind from a mountain range is called a *rain shadow;* if it is very dry it is termed a *rain shadow desert.*" Mountains as topographic barriers to prevailing storm tracks lie mostly in a north-south direction. At least this is the prevailing orientation in the New World. Mountains in North and South America show preponderantly the north-to-south orientation (see Fig. 4.1). And it is in western North America that classic examples of major regional rain-shadow effects can be witnessed. The two major north-south cordilleran systems, the Cascade–Sierra Nevada axis and the Rocky Mountain chain, are textbook examples of grand, regional rain-shadow-producing mountain systems. From Canada to Southern California the lee side of the Cascade–Sierra Nevada system is dramatically drier so as to control vegetational response both in physiognomy and life-form as well as species composition. The east slope of the Sierra Nevada and the terrain eastward beyond the Sierras is rain-shadow country with its telltale vegetation (sagebrush [*Artemisia*]/shadscale [*Atriplex*]) shrub steppe and even true desert flora in Nevada, Utah, and eastern California. And within this vast arid landscape is the Great Basin containing its own rain-shadow contrasts to further embellish the landscapes. Desert mountain ranges, oriented north to south, appear sequentially from eastern California almost to the Rocky Mountains (see Fig. 4.2). Each mountain range not only is a mesic island in a xeric landscape, but can show west slope/east slope rain-shadow effects (Billings 1978; Trimble 1989; Harper et al. 1994). This vast and unique "basin-and-range" country is colloquially known as the Intermountain West.

A similar rain-shadow response by vegetation appears east of the Rocky Mountains, with dry shortgrass prairie and sagebrush. In this same region, the Rocky Mountain weather barrier had its effect in the Pleistocene. Glacial advances terminated farther north in the Dakota area than in the Mississippi lowlands, in part because of the rain-shadow effect of the Rockies (Pitty 1971).

Other notable areas of the world where rain-shadow effects occur include the eastern Mediterranean (Bull 1991, p. 125), the Cape Province of South Africa (Werger 1978), the Himalayas (Mani 1978), the mountains of the South Island of New Zealand (Wardle 1973), the Andes of South America (Moore 1983), and various mountain systems of Europe (Polunin and Walters 1985).

The topographic barrier that can make a rain shadow can be quite local; the windward versus the leeward side of a single ridge can produce contrasts in weather and vegetation. Or a succession of ridges from windward to leeward can effect a progressive decrease in precipitation. The Olympic Mountains of western Washington show this nicely. The lowland western base of the range gets from 80 to 200 inches (2000 to 5000 mm) of precipitation, and the orographic lifting of moisture-laden clouds puts the heaviest precipitation on the western mountain summits, mostly as snow (Mount Olympus at 7965 feet in elevation may have as much as 250 feet [750 m] of snow a year [one inch of snow equals 0.1 inch, or 2.5 mm of rain]). Rainfall decreases dramatically northeastward to reach a low of 17 inches (425 mm) annually in the lowland rain shadow of the Sequim area, less than 50 miles (80 km) from the Pacific Ocean (Kruckeberg 1991a).

What if a mountain system lay parallel to a weather system? Would there only be an altitudinal effect on vegetation, without any manifestation of a rain shadow? Should not a mountain system oriented in an east-west direction in regions of westerly weather direction fit the case? In the first place, few mountain systems have this east-west orientation; then those mainly have rain-shadow effects anyway, reflecting the flow of moisture-laden winds. The Himalayan massif system is the greatest of the east-west orogenies; its rain shadow is in Tibet in the lee of the monsoon storm track. The Atlas Mountains of North Africa have a mesic Mediterranean climate windward and Saharan desert leeward. Similarly the Transverse Ranges of Southern California, roughly at right angles to the north-south Sierra Nevada, have a clearly defined rain shadow—the Mojave and Colorado deserts on their leeward side. The apparent absence of a case where the storm-track weather parallels a mountain system is puzzling. It leads me to conjecture: Major weather systems are products of the mountain systems they encounter. Any topographic barrier or surface heterogeneity, especially Hochgebirge, "captures" a weather system to direct it not parallel but at more or less right angles to the mountain's relief. Thus the rain-shadow effect is inevitable!

TOPOGRAPHIC UPLIFT, MOUNTAIN CLIMATES, AND VEGETATION

The significant interplay of mountains, climate, and vegetation is rooted in variations of atmospheric physics—change in the behavior of air masses as they come in contact with surface heterogeneity. In a word, temperature and precipitation are influenced by mountains. Temperatures decrease with altitude, except for temperature inversions. At timberline and the alpine land above the trees, reduced temperatures become critical for plant life in a variety of ways. Cryogenic perturbation of soil, intense weathering of rock, and subfreezing temperature effects on growth all result from this purely physical change in temperature with altitude. During shorter growing seasons in the upper montane, daily temperature fluxes can be extreme: freezing at night, followed by high daytime temperatures. Solar radiation takes on stressful extremes, especially in subalpine and alpine areas: the amount of both visible and ultraviolet radiation increases at ground level. Further, wind velocities increase with altitude, reaching life-stressing levels in the alpine zone. Billings and Mooney (1968) tabulated these meteorological parameters for three habitats—arctic, alpine, and temperate forest (Table 4.3).

Mountains induce precipitation in singular fashion. Adiabatic cooling** must occur to extract moisture as precipitation from the atmosphere. Air must rise and cool adiabatically to yield precipitation. Mountains promote adiabatic cooling, in that they induce uplift, lying as they do across the path of prevailing winds. This so-called orographic (or topographic) uplift (Akin 1991) may be rapid and can yield moisture with increasing altitude on the windward side. On the leeward (rain-shadow) side of a mountain system where air is subsiding and warm, precipita-

** "This process is called adiabatic since there is no exchange of heat between the rising air mass and its surrounding environment" (Akin 1991, p. 19).

Table 4.3. Comparative environmental characteristics of an arctic, an alpine, and a temperate forest ecosystem (from Billings and Mooney 1968)

Component	*Arctic tundra**	*Alpine tundra*†	*Temperate forest*‡
Solar radiation (1965)	*Source A*	*Source B*	*Source A,* Laramie, Wyoming
Highest daily total (June, July)	760 langleys	780 langleys	704 langleys
Average July daily total	426 langleys	497 langleys	514 langleys
Average July intensity	0.30 cal/cm^{-2} min^{-1}	0.56 cal/cm^{-2} min^{-1}	0.58 cal/cm^{-2} min^{-1}
Quality	Low in short wavelengths, particularly short UV. (2950–3150 Å.)	High in all wavelengths, particularly short UV	High in all wavelengths
Maximum photoperiod	84 days	15 hours	15 hours
Temperature (+l m, °C)	*Source A*	*Source C*	*Source C*
Air			
Annual mean	-12.4	-3.3	8.3
January mean	-26.7	-12.8	-1.7
July mean	3.9	8.3	20.6
Absolute max.	25.6	18.3	37.2
Absolute min.	-48.9	36.6	-33.8
Soil (-15 cm, °C)	*Source D*	*Source C*	*Source C*
Annual mean	-6.2	-1.7	8.3
January mean	-14.5	—	—
July mean	2.5	—	—
Absolute max.	2.5	13.3	31.1
Absolute min.	-15.5	-20.0	-10.0
Precipitation (mm)	*Source A*	*Source C*	*Source C*
Annual mean	107	634	533
Highest monthly (wettest month)	71	203	203
Lowest monthly (driest month)	0	6	0
Wind (km/hr)	*Source E*	*Source C*	*Source C*
Annual mean	19.3	29.6	10.3
Maximum water stress in plants (Source F)	Low (-4 to -5 bars)	Rel. low (-6 to -8 bars)	Rel. high but no data (probably -25 bars or higher)

Table 4.3. *(continued)*

Component	*Arctic tundra**	*Alpine tundra*†	*Temperate forest*‡
Metabolic gases (Source G)			
CO_2 (mg/l)	0.57	0.36	0.44
O_2 (partial pressure in mm)	160	100	122
Soil frost activity	*Source D:* much, active	*Source C:* some, active, on small scale; large, fossil stone nets	*Source C:* none in growing season
Depth of soil thaw	*Source F:* 20–100 cm depending on site; permafrost universally present	*Source H:* 30 cm to bedrock; permafrost rare	*Source H:* no permafrost
Vegetation	*Source D:* Carex-Dupontia-Eriophorum tundra	*Source C:* mosaic of alpine vegetation types with Kobresia meadow tundra as climatic climax	*Source C:* open forest of Pinus ponderosa

SOURCES A-H: Billings and Mooney (1968), table 1.
* Barrow, Alaska (altitude 7 m, latitude 71° 20' N).
† Niwot Ridge, Colorado (altitude, 3749 m; latitude 40° N).
‡ Bummer's Gulch, Colorado (altitude 2195 m, latitude 40° N).

tion drops off markedly. Orographic uplift caused by changes in montane relief obviously can have dramatic effects on vegetation. With increases in moisture with altitude, luxuriance of vegetation can increase—up to a point! Snow, ice, and wind conspire at the highest elevations to impair growth of plants, despite the greater precipitation promoted by orographic uplift. In the highest mountains, summits are enveloped in permanent snow; at these high levels, orographic uplift has produced a form of moisture inimical to most living systems.

Mountains are not simple inclined planes that uniformly exact changes in temperature, precipitation, insolation, and wind with increase in altitude. The impressive surface heterogeneity on any mountain slope (ridges, slopes, canyons, cliff walls, etc.) metes out particular effects on vegetation. For example, high montane vegetation can be displaced to lower elevations in mountain valley bottoms and canyons as a consequence of the cold-air drainage effect. An extreme example can be witnessed in the North Cascades of Washington State. The well-known Ice Caves at the north base of Big Four Mountain are at only 1800 feet (550 m) elevation. Yet because of the locally frigid air off the low-lying permanent snowfield, the surrounding vegetation is subalpine in character—an open shrub-forb meadow with scattered mountain hemlock (*Tsuga mertensiana*). A scant 100 meters north of this locally depressed subalpine habitat one finds the typical midmontane forest of Pacific silver fir (*Abies amabilis*) and associates.

Billings (1954) investigated yet another effect of topography on local climate. Along the altitudinal gradient, from valley bottoms to midslope, in desert mountains of western Nevada, a consistent temperature inversion occurs. Minimum temperatures in valley bottoms were as much as 10 to 15°F colder than those upslope (400–1300 ft higher). The cold valley floors are covered by sagebrush (*Artemisia tridentata*), while the higher thermal belt is coincidental with the pinyon–juniper zone (*Pinus monophylla—Juniperus osteosperma*). Once again, topography makes its mark on climate and in turn influences vegetation. This temperature inversion effect may be widespread in the more arid mountainous regions of western North America (Billings 1954).

MOUNTAINS, GLACIATION, AND PLANT LIFE

For vegetation, the consequences of montane glaciation and to a lesser extent of continental glaciation are of supreme significance for much of our temperate world. Vast areas of the Earth are (and have been) covered with glacial ice; a conservative estimate shows a total of 16.2×10^6 square kilometers as glacial ice (Gerrard 1990, p. 182) at present. While the bulk is in the Antarctic and Greenland ice sheets, a substantial 700,000 square kilometers is mostly mountain glacier ice. Mountain glaciers occur most widely in the Northern Hemisphere (Asia, Europe, and North America); lesser displays are in New Zealand, equatorial Africa, and Australia. There is also substantial glacial ice in the Andes of South America. Further, alpine glaciers were even more extensive in the Pleistocene and other earlier periods of cold-climate accumulations of ice.

Kinds of glaciers, especially their periglacial terrain, affect vegetation in species composition and plant cover. The simplest classification of glaciers is morphological, using their form and relation to surrounding topography (Gerrard 1990). Of the ten types listed by Gerrard (p. 165), we focus on those associated with mountains—the niche (or cliff or apron) glacier, cirque glacier, valley glaciers (both alpine and outlet type), and the mountain ice cap. Mountain or alpine glaciers, especially the valley type, have made in the past (Pleistocene) a fluctuating contact with the continental ice sheets as these lapped up against the flanks of mountains. In mountain valleys, where alpine glaciers were either retreating, advancing, or staying at equilibrium during the advance of continental ice, there came into transient being ice-free areas bordering proglacial lakes. It is unlikely that the

4.31 High altitude and heavy snowfalls perpetuate montane glaciers; advances and retreats of glaciers cause variation in trimlines (see Glossary) of subalpine forest. Nisqually Glacier, Mount Rainier National Park, Washington State. Photo, taken in July 1958 by A. Harrison, from Special Collections, University of Washington Libraries, Seattle.

margins of proglacial lakes were vegetated to any extent.

For our particular geoedaphic bias, it is the periglacial areas of mountain glaciers that reveal dynamic vegetational responses. As any cirque or valley glacier recedes, it leaves a vegetation-free substrate at its perimeter, suitable for eventual plant colonization. Advances and recessions of mountain glaciers have been widely noted in historical times. There is both a general uniformity of change in glacier extent globally and local deviations from more general trends (Harrison 1954). Both advances and retreats have consequences for vegetation. An advance will obliterate a vegetation in its path, whether the "victim" is an early seral stage or a climax forest. With the latter, glacial advance creates a trimline, marking the furthest advance of a glacier into a forested area (Harrison 1954; Habeck and Hartley 1968). Trimlines are readily visible in the upper glaciated valleys on Mount Rainier and Mount Baker of Washington State (Fig. 4.31). Comparisons of old and new forest growth can be used to date the time of the last glacial advance.

In the case of receding mountain glaciers, newly exposed terrain becomes available for colonization. Glacial till and morainal deposits adjacent to the receding glacier begin to support plant life soon after exposure. Initial colonization may be tenuous at first, due to the infertile nature of the substrate (not yet soil) and probably due to cryopedogenic disturbance. Reclamation of bare ground in periglacial terrain depends on recruitment from nearby subalpine-alpine vegetation (Fig. 4.31). From observations (unpublished) I have made at Coleman Glacier on Mount Baker and at Nisqually Glacier on Mount Rainier (both in Washington State), a sequence of vegetation can be discerned. Herbaceous perennials like *Saxifraga tolmiei, Eriogonum pyrolaefolium, Epilobium alpinum, Spraguea umbellata,* and a few grasses, sedges, and rushes, are initial colonizers. Woody species, mostly shrubs (*Alnus sinuata, Salix* spp., and ericaceous shrublets) can appear a bit later. Old trimlines may eventually be reclaimed by younger coniferous forest, of the same species as those in the climax forest just beyond the trimline. Some kind of plant succession can be expected in recently deglaciated areas of other mountains in the Old and New Worlds.

Biological reclamation of deglaciated terrain in the vicinity of montane glaciers can take on several manifestations, each viewed as following one another successionally: establishment of bryophytes, followed by (or concurrent with) the arrival of herbaceous seed plants (grasses and forbs), then the occurrences of lichens, and finally woody plant invasion of the deglaciated terrain. Animals also play a part in this succession, but their role is subordinate to that of the plant life (Stork 1963), and will not be examined further here.

Studies of deglaciated alpine areas (recessional moraines, etc.) are best known in Europe, both for the Alps and for alpine glacier terrain in Scandinavia. A major study by Stork (1963) for montane glaciated terrain in northern Sweden serves to exemplify the successional pattern. Bryophytes are the first plant life to appear, even after only a year following exposure; *Pohlia* and *Bryum* are the pioneer moss genera. Early pioneer seed plants include *Trisetum spicatum, Poa alpina f. vivipara, Cardamine bellidifolia,* and *Saxifraga oppositifolia.* From early pioneer immigration to 150–200 years after deglaciation, the succession is complete; in Stork's words, "the vegetation is closed."

Dating of the sequence from deglaciation to a closed plant community can be quantified critically by the use of lichenometry. Stork (1963) gives a thorough account of the method as refined by R. Beschel in several cited papers. A good example of the lichenometric method comes from Porter (1981) for the dating of substrates at Mount Rainier. *Rhizocarpon geographicum* is the lichen of choice, successfully used elsewhere and one yielding relatively uniform growth rates with the age of thalli. Porter found lichen thalli to increase in size from the tongue of the Nisqually Glacier to the lowest (nineteenth century) limit of maximum glaciation; growth rates were most rapid until the thalli reached 10 millimeters in diameter, then decreased over the next century.

Arrival of woody immigrants to deglaciated terrain may be rapid. Sigafoos and Hendricks (1969) determined the time interval between initial deglaciation and establishment of tree seedlings to be five years on the average. These ages were determined by tree-ring analysis of young trees (mostly conifers) on Mount Rainier. Lichenometric dating is probably more accurate (Porter 1981). Yet both approaches, as well as plant community succession, do tell of the dynamic interplay of high mountain terrain, glaciers, and plants.

The classical succession studies at Glacier Bay National Monument, Alaska, are relevant to the issue of colonization on deglaciated terrain in montane areas. While the revegetated terrain is at sea level, the glacier and its recession are linked to its montane source and thus a facet of montane geoedaphics. A sample of the rich literature on Glacier Bay plant succession can be found in Reiners, Worley, and Lawrence (1971). Certainly the glaciers of southeastern Alaska are exerting a geoedaphic influence on the terrain; the creation of new microtopography and soils from raw glacial detritus, ending in forest, is geomorphological in nature, influencing the course of vegetation change and its composition.

In sum, mountain topography fosters the birth, growth, and decay of alpine glaciers; these glaciers shape alpine landscapes and in turn create habitats for plants. This serves as another instance of the crucial geoedaphic role of topography in affecting plant life.

SAND DUNES

The role of wind in creating landforms is nowhere more striking than in the genesis and perpetuation of sand dunes. They occur on nearly all continents and all are formed by the erosion, transport, and deposition of sand grains (particle size of 2 to 0.063 mm). The instability of these loose aeolian deposits is their most characteristic feature: the forms they take are highly changeable. The loose sand particles, agitated and moved by wind, cause the forms of dunes to change shapes almost incessantly.

Given their instability and their deficiencies in mineral nutrients, dunes would seem to be unlikely habitats for plant life. And in desert dunes of the Sahara, vegetation may be wholly lacking. Yet plants have adapted to dune habitats in many parts of the world, and may often be species narrowly restricted to the dune habitat. With their great variety of shapes, sizes, and horizontal and vertical extent, dunes offer another example of landform heterogeneity. And in turn, vegetation, though often sparse, does accommodate to the variant and changeable dune surfaces. In their book on the physical properties of aeolian sand and sand dunes, Pye and Tsoar (1990, p. 123) define their extent: "Deposits of aeolian sand cover approximately 6% of the global and surface area, of which about 97% occurs in large arid zone sand seas. On average, about 20% of the world's arid zones are covered by aeolian sand, although the proportion varies from as little as 2% in North America to more than 30% in Australia and 45% in Central Asia." Sand seas are extensive aeolian sand deposits, with a minimum area of 125 square kilometers (Pye and Tsoar 1990); smaller areas are defined as dune fields. So vast a landform and its associated plant ecology worldwide is beyond the scope of this book. Hence the approach will be to focus on interior desert dunes of western North America, especially southeastern California and adjacent states. Briefer mention will be made of three other aeolian deposits: loess, coastal dunes, and gypsum "sands."

Loess originates as wind-transported particles of silt (finer than sand). Loess deposits are often derived from wind-transported glacial "flour," and are level or show undulating topography. The Palouse loess of eastern Washington State is of the latter type. Extensive deposits occur in north central United States and in China. Since loess becomes stabilized rapidly by vegetation and forms soils of definable profiles, it usually supports vegetation common to its bioregion.

Coastal dunes occur on countless shorelines of the world. Often they have pronounced surface heterogeneity: foredunes, ridges, lee slopes, and dune slacks. Wind, hydrology, and salt spray join to create a distinctive habitat for plants. The coastal dunes of northern California and Oregon illustrate the geoedaphic impact on vegetation. See Barbour and Johnson (1988) for California coastal dunes and Franklin and Dyrness (1973) for Oregon dunes. On the Oregon dunes, species sort themselves out in time and space to the differences in microtopography. For active dunes the foredune sector overlooking the beach strand is maintained by dune grasses, *Ammophila arenaria* and *Elymus mollis*. Intentionally introduced to bind coastal sands, *A. arenaria* now dominates many coastal dunes and is the chief factor in maintaining foredunes. Newly created dry dunes undergo a well-defined successional sequence from pioneer species like *Glehnia leiophylla, Poa macrantha, Lathyrus littoralis, Carex macrocephala,* and *Convolvulus soldanella*. Proceeding inland from the foredunes, succession leads to dry meadows, a dry shrub type, then to seral pine forest (*Pinus contorta*), and finally to the stable (climax?) forest of western hemlock (*Tsuga heterophylla*). These successional stages ultimately reduce the magnitude of the dune topography (Franklin and Dyrness 1973). Most of the flora of these coastal dunes is recruited from nearby beach strand vegetation.

The nearly pristine coastal dunes region near Eureka, California, the Lanphere-Christensen Dunes, is preserved as a natural area (Faber 1997). Besides common dune species, some are more narrowly restricted: *Erysimum menziesii* subsp. *eurekense* and *Lilium occidentale*.

The coastal dunes along the Pacific coast are not as rich in endemics as the inland desert dunes of California. Yet these coastal dunes portray well the response of vegetation to the changing surface heterogeneity created by active dunes.

Gypsum "sands" often form dunes. Since the "parent material," gypsum, is formed of $CaSO_4$ particles, it is the chemical nature of gypsum habitats that evokes the major response of plants. This subject is addressed in Chapter 5, where gypsum and its distinctive flora are discussed.

We now turn to examples of the substrate-vegetation linkage for desert dunes of the Southwest, especially in eastern California. Most dunes here are composed of quartz sand, so are chemically rather inert. Most dune sand of the California desert is derived from granitic and sandstone rocks. Dunes can form in a variety of ways. Small dunes may start as accumulations of sands around desert shrubs, eventually to overtop the vegetation. Large dunes like the Kelso and Algodones Dunes get their aeolian sand from dry lake beds (Norris 1995). These desert dunes are almost constantly on the move: Small ones move faster than the large ones and some may move more than 30 feet (10 m) a year (Norris 1995). The large Algodones Dunes system of southeast Imperial County is nearly 45 miles long and from 3 to 5 miles wide, with dune peaks 300 feet above the desert floor. Shapes and contours are diverse—dune ridge tops (smooth or rippled) and slopes (smooth or crescentic). Hence the combination of dune topography, dune movement, and the sandy nature of the substrate would seem inimical to plant life. Yet the dunes do support vegetation, often a mix of common with rare, endemic species. Notable among the sand-dwelling species (psammophytes) is the curious root parasite called sand food (*Pholisma sonorae*). Other endemics are Wiggins' croton (*Croton wigginsii*), the Algodones dune sunflower (*Helianthus niveus* subsp. *tephrodes*), and Pierson's milkvetch (*Astragalus magdalenae* var. *piersonii*). All four of these rare dune plants are listed as endangered. The threat to their persistence comes from off-road recreational vehicles. This damaging "sport" has devastated much of the Algodones Dunes. Only 30 percent of the dune has been set aside as the North Algodones Dunes Wilderness (Dice 1997).

CHANGES IN LANDFORMS OVER TIME: EFFECTS ON PLANT LIFE

Thus far in this chapter we have perceived the interplay of landforms and plant life largely as a static phenomenon. Contemporary topography has set the stage for today's biota. For the biogeography and evolution of organisms, the reverse of the uniformitarian dictum holds true: the past is the key to the present. This is true, of course, of changes in landforms and the consequences for the resulting biota. Changes in landforms over time can be traced to at least four sets of causal mechanisms. First, strictly geological, are the tectonically induced changes, nowadays couched in terms of plate tectonics. A second set of causes are the manifold effects—mainly erosional—of climate on landforms; this is the domain of geomorphology (Pitty 1971). Third is a biological causation: some landforms are created by biogenic processes (Westbroek 1991). And a fourth source of landform changes surely must be the interactions among the first three—geology, climate, and organisms.

Hewing to the geoedaphic theme of this book, I limit the scrutiny of causes in landform change to the strictly geologic. After all, without tectonically induced realignments of landmasses, there would be no consequent texturing of the land by particular regimes of climate and by unique suites of organisms.

The significance of plate tectonics to biogeography and evolution has been elegantly stated by Peter Raven and Daniel Axelrod (1974, p. 539): "The isolation of land areas by sea-floor spreading, the uplift of new cordilleras, the emergence of new archipelagos and the disappearance of old ones, and the shifting position of (some) landmasses have both created and destroyed environments to which biota have responded. In this sense, changing physical environments governed by plate tectonics have had a major role in evolutionary history." Indeed, the consequences of plate tectonics for the world's biota are all-pervasive and global. No other geological causation has had such profound physical and biological consequences setting the course of all subsequent evolution of climates, landforms, lithologies, and organisms. Tectonic changes to the Earth's crust take place over stretches of time, from eons to days. While there is a continuum of earth movements from gradual to abrupt in geological terms, it is convenient to view the modal expressions of time-bounded geological events.

Long before the contemporary revolution in geological study, with its new paradigms and vocabulary, early twentieth-century botanists applied Alfred Wegener's theory of continental drift to solving an old riddle (Raven 1972). Floras of the Southern Hemisphere, especially those of Australia and New Zealand and those of South America, share many similar taxa. The affinities, exemplified by *Nothofagus, Weinmannia, Jovellana,* and *Pernettya,* between the two now remote regions made Wegener's ideas attractive to plant geographers perplexed by observed disjunctions in similar austral floras. But without a plausible mechanism to power the drift of continents, Wegener's theory was rejected by his fellow geologists and the plant geographer's riddle remained unresolved. Now, with the advent of plate tectonic theory, biogeographers have a geologic modus operandi to account for past affinities and present disjunctions. We are comfortable with the idea of terrestrial biotas moving not only by migration and dispersal but by having been moved on migrating crustal plates. The whole arena of island biogeography takes on new meaning with the consequences of plate tectonics. Oceanic islands come into being from vulcanism linked to sea-floor spreading. And for mainland biogeographic islands created by isolated mountain ranges, the orogenies linked to plate collisions provide their genesis. We are thus driven to the conclusion shared by Hamilton (1983) that "the biogeographic consequences of plate motions and interactions must be enormous." From the global effects of plate tectonics, we are led to accept the notion that *geology drives biology.*

The realignment of landmasses by plate tectonics is not the only gradual shaper of the Earth's crust. Stemming from the consequences of plate movements are gradual reshapings wrought by climates that are propelled by the displaced landmasses. These are in the realm of geomorphological transformations. Most are erosional effects, like peneplanation, formation of deltas, valleys, and canyons, as well as the accretion of sediments, comings and goings of glaciers, coastline emergences and submergences. Though these phenomena are set in motion by climate, the crustal fabric in which they play out their displays was created by plate tectonics. Again, we must concede the all-powerful and ramifying effects of plate movements and correlated orogenies. Plant geographers were quick to capitalize on the revolution in geology wrought by plate tectonics. Assessment of biogeographic relationships—continuities, disjunctions, endemisms, and so forth—can be found, along with illustrative case histories, in a number of reviews (Raven and Axelrod 1972, 1974; Schuster 1976; Hamilton 1983; McKenna 1983). The biogeographic consequences of plate movements are discussed in Chapter 7.

Catastrophic Genesis of Landforms and the Consequences for Plant Life

Abrupt transformation of the land can also be linked to plate tectonics, directly or indirectly. The most direct connection is with vulcanism at plate margins. Whether by sudden eruptions or by short-duration lava flows from fissures, landscapes are quickly altered, extant biota obliterated, and the newly created terrains become subject to a new suite of erosional processes, to new soil formation, and to colonization by preadapted biota of neighboring floras and faunas.

Besides the local and abrupt consequences of vulcanism, it is crucial to recog-

4.32 Natural catastrophes (floods, earthquakes, volcanic eruptions) have drastic effects on landforms and vegetation. The Kautz Creek mudflow (1947), on the lower west slope of Mount Rainier, killed trees along the stream. Photo by R. and L. Kirk in Kruckeberg 1991a.

nize long-distance effects. Volcanic eruptions can emit vast quantities of ash (tephra) that may blanket landscapes both near and far. Extant vegetation is partly or completely destroyed, and a new substrate affords habitat for recolonization. The 1980 eruption of Mount St. Helens was but the latest in a long series of ash depositions in the Pacific Northwest of North America (Mack 1981; Bilderback 1987; West 1988). Yet another side effect of vulcanism is the significant short-term change in climate. Like the imagined "nuclear winter" after a global nuclear war, a real "volcanic winter" can be witnessed. Airborne particulates from an eruption may persist in the atmosphere to cause significant lower-

4.33 The earthquake of 1965 caused landslides in the Puget Sound area of Washington State. Photo by G. Thorson.

ing of solar radiation and attendant reductions in temperature to cause a drop in primary productivity. The 1982 eruption of El Chichón in Mexico is thought to have caused far-reaching, though short-term, changes in weather patterns well beyond the site of the volcanism (McClelland et al. 1989).

Earthquakes associated with vulcanism and related plate movements may also abruptly alter landforms. Fault scarps, crustal displacements, and landslides are significant byproducts of earthquakes (Figs. 4.32 and 4.33). Sudden geomorphic alterations can also occur with the flooding of vast or local landscapes. Eastern Washington bears witness to just such a catastrophic event, the genesis of scabland and coulee topography from the sudden release of glacial ice and meltwater behind glacial Lake Missoula during the Pleistocene. In a matter of days, the soils of the Columbia Plateau were sluiced away, leaving a vast topography of deep-sided coulees and scoured flood channels (Allen and Burns 1986; Bretz 1959). The new habitats came to be occupied by a biota adapted to life on vertical walls, wet coulee basins, and the lithosolic soils of the scoured flood channels (Franklin and Dyrness 1973; Daubenmire 1970).

A final thought on abrupt, catastrophic events induced by geologic processes. Events beyond our planet can impact the Earth. The most direct and sudden must be meteorite impacts. Large ones may well be causes of major biological extinctions, such as occurred at the Cretaceous-Tertiary boundary (Raup 1986).

4.34 The eruption of Mount St. Helens (1980) devastated vast tracts of mature forest on its slopes; heat, steam blast, and avalanches of mud, rock, and pyroclastic flows were the major destructive forces. Photo by author.

Summary

Landforms of all shapes, sizes, and degree of isolation are ultimately produced by geologic processes. Other physical and biological forces can shape the land, and climate and organisms can produce surface heterogeneity, but without the primary geologic events (orogenies, vulcanism, plate tectonics, etc.) there would be no initial terrain on which the physical and biogenic shaping of the land could take place. Terrestrial vegetation responds in myriad ways to diverse landforms—makeup of floras and their life-forms, ecophysiologically, and evolutionarily.

Mountains take first place among the various surface heterogeneities of the Earth's crust in commanding responses from plant life. Slope, aspect, exposure, elevation, in combination with degree of discontinuity of mountains and mountain systems, are key influences on vegetation. Lithological variety in mountains must be recognized for its selective effect on floras (the biologic effects of types of rock encountered by plants are pursued further in Chapter 5). Montane topography, though initiated by geologic processes and events, is subject to—and creates—climate. Precipitation, temperature, wind, and cloud cover are all influenced by mountain barriers, and in turn these climatic factors shape the contours of maturing mountain systems. Indeed, mountains exemplify two basic dicta: (1) mountain environments are holocoenotic; they form an intricately interconnected fabric; and (2) mountain environments play host to evolutionary, ecologic, biogeographic, and floristic responses by organisms.

Arising from surface heterogeneity, especially induced by orogenies, vulcanism, and continental drift, is the pervasive insularity of landmasses, expressed globally to regionally. Oceanic islands, continental shelf islands, and mainland (habitat) islands (mountains in a "sea" of lowlands) all bear scrutiny as milieus for biologic diversity.

A section of Chapter 7 will examine the role of plate tectonics in the fashioning of biotic distributions. Movements of continents and orogenies at plate boundaries are two prime causes of change in the biogeography of life. And they are, of course, geological in nature.

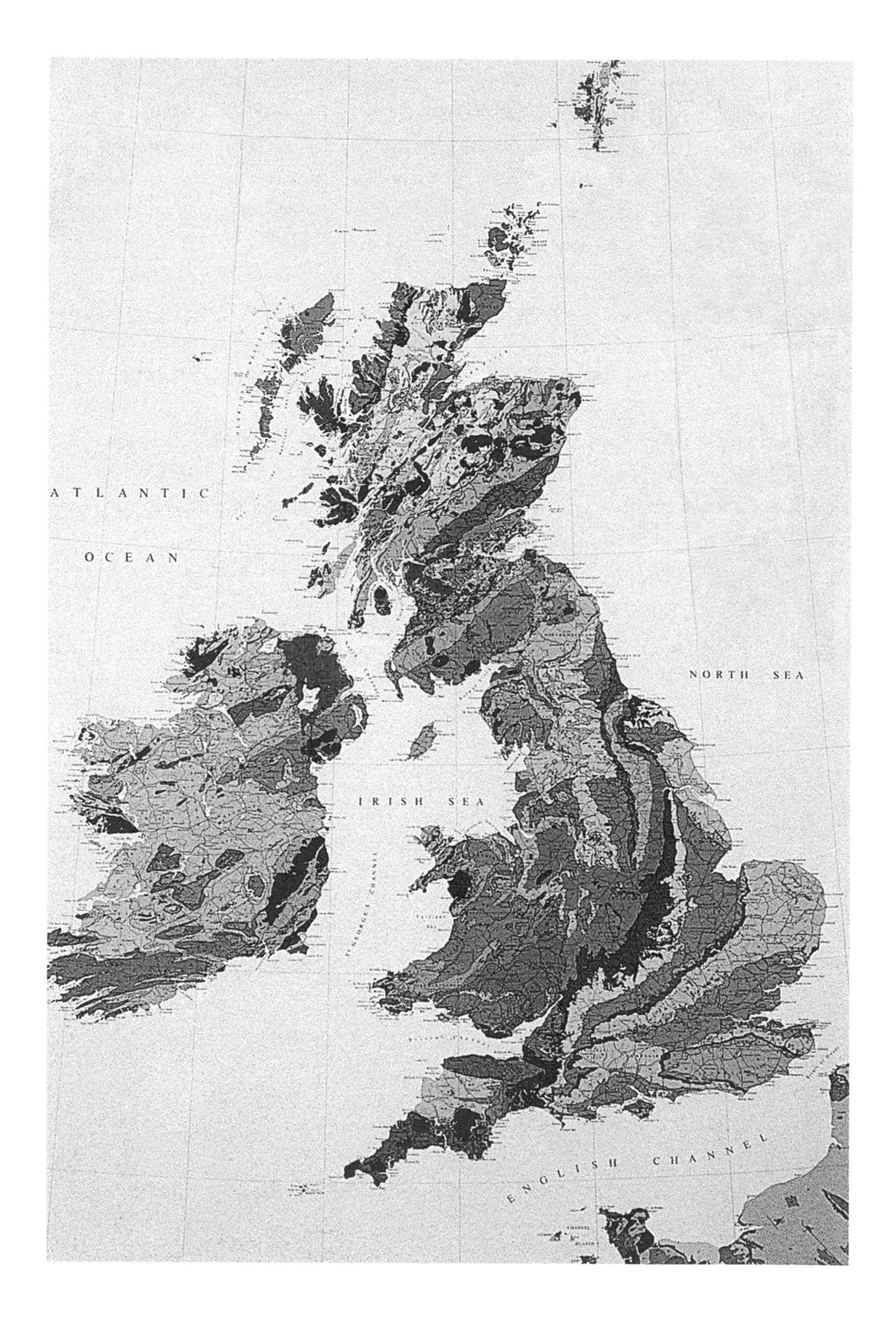
ATLANTIC
OCEAN
NORTH SEA
IRISH SEA
ENGLISH CHANNEL

5 The Influences of Lithology on Plant Life

Of geology's many influences on vegetation, the quality of rock (parent material) and derived soils has made the greatest impression on students of nature. Since early recorded history, the link between substrates and plant cover has caught the eye of observant peoples. Where the underlying parent material (bedrock) produced azonal soils (like those derived from limestones, serpentines, and other ultramafic rocks, shale barrens, etc.), the contrasts between the vegetation found on these exceptional outcrops and what was growing on "normal" ones were recognized as ecological manifestations long before the discipline of ecology was formalized. One such early observation was made by Theophrastos (ca. 380–287 B.C.), an admiring follower and companion of Aristotle (Singer 1950). Theophrastos, a diligent recorder of botanical fable and fact, made the ecological connection between limestone bedrock and the presence of a singular vegetation type, the *phrygana* of calcareous rocky hills of lowland Greece (Dierschke 1975, p. 460). The *phrygana* association (the *garigue* of France) consists of low, xeromorphic, spiny shrubs which contrast sharply with the more mesic oak savanna on adjacent noncalcareous substrates.

Andrea Cesalpino (1519–1603), celebrated Italian naturalist, in 1583 first recorded endemism on serpentine in Italy (*Alyssum bertolonii,* now well known as a nickel accumulator; Brooks and Radford 1978). Cesalpino described the crucifer as growing on "black rock" (*sassi neri*), a Tuscan term for serpentine (Proctor and Woodell 1975). By the nineteenth century, numerous observations on the substrate-

Geologic map of British Isles showing rich lithological diversity. Courtesy British Geological Survey.

vegetation link had been recorded. In Chapter 2, I singled out several geoedaphic pioneers like Thurmann and Unger. It was Franz Unger who saw the dramatic contrasts in vegetation of the Alps where limestone and acid siliceous rocks meet. Unger, a young doctor in the town of Kitzbuhel (he later became a full-fledged botanist of note), gave us the chemical theory (see Chapter 2) of plant restriction, and two terms that succinctly describe the nature of the affinity of species for rock type (Unger 1836). Unger's *bodenstet* ("soil-constant") means edaphic restriction, in contrast to *bodenvag* ("soil-wanderer") for species that are indifferent to soils derived from specific rock types. Of course, Unger did not know that even bodenvag species are often ecotypically differentiated into substrate-tolerant and -intolerant races (Kruckeberg 1951, 1995).

A final historical note on the lithology-plant connection comes from the age of Christopher Columbus (Buck 1949). Needing to replace a mast on a ship of his first fleet, Columbus was counseled to choose a log of pine growing on red soil in nearby Cuba; the red limonitic soils of Cuba are known to produce durable, iron-enriched timbers (Kruckeberg 1959). Here we see the beginnings of biogeochemistry or geobotanical prospecting put to practical use.

Kinds of Parent Materials

The simple taxonomy of rock types—igneous, metamorphic, and sedimentary—hardly reveals the myriad lithological variety that appears at the Earth's surface. Each of these three broad categories embraces many different kinds of rocks with a concomitant range of minerals, chemical elements, and physical properties.

Both plutonic (intrusive) and volcanic (extrusive) in origin, igneous rocks range in composition from acid to highly basic rocks (Jenny 1941); see Figure 5.1. Acid rocks are rich in silica (quartz) and monovalent cations; they are exemplified by granites, granodiorites, and rhyolites. Then comes the intermediate series, rich in the mineral feldspar; these are rocks like the plutonic diorite and gabbro as well as the volcanic andesite and basalt, all with lower levels (ca. 50 percent) of silica. Finally, there are the mafic and ultramafic (magnesium and iron) rocks, often called basic, ultrabasic, or ultramafic; they include olivine-rich basalts and peridotite. A summary of major rock types is given in Tables 5.1 and 5.2.

Plant Responses to Igneous Rocks

Most higher plants do not show preferences for igneous, sedimentary, or metamorphic rocks per se as preferred habitats. It is mostly the weathered products of

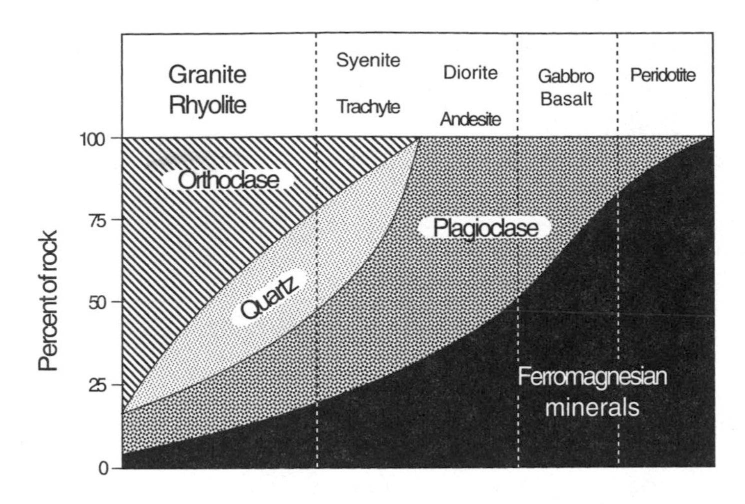

5.1 Proportions of minerals in intrusive and extrusive igneous rocks. Adapted from Skinner and Porter 1992.

rocks—soils—that select certain species from a region's available flora. Yet soils *do* reflect quality of parent material, both in physical properties of soils such as texture, color, and weatherability, and in chemical content. In certain cases, however, especially for plants of rock outcrops, the plant-to-rock linkage is most intimate; saxatile species can live on bare rock or on a very thin soil mantle that closely matches the chemical and physical properties of the parent material.

So what can we say about plants with affinities for igneous rocks? Igneous rocks of the quartz-feldspar mineral type (Tables 5.1, 5.2) weather to become more or less "normal" soils; their physical and chemical properties do not deviate in the direction of some extreme value. Thus soils derived from quartz-feldspar-containing granite or basalt reflect the climate under which they had their genesis; these are zonal (climate-derived) soils. Since granites and basalts are the most widespread igneous rocks worldwide, the zonal soils overlying and derived from these rocks are common almost everywhere (Table 5.3). What variant properties such soils may have are products of the climate under which they have formed. Since zonal soils of different parent materials may meet (intersect) along a lithosequence (Jenny 1980), it is intriguing to inquire: Do adjacent zonal soils of differing igneous origins harbor differences in the plant species or plant communities that occur on them? Such floristic contrasts across a zonal lithosequence, if they exist at all, must be difficult of detection. Little attention has been paid to this possibility—the overt to subtle changes in flora and vegetation going from, for example, granite, a plutonic rock, to its volcanic counterpart, rhyolite (Tables 5.4 and 5.5). I will return to this matter later in this chapter, in the section "Plant Responses to Variations in 'Normal' Lithologies."

Table 5.1. Classification of igneous rocks (adapted from Ollier 1984)

Texture	*Acid*	*Intermediate*	*Basic*	*Ultrabasic*
Coarse-grained	Granite Granodiorite	Syenite Diorite	Gabbro	Peridotite
Medium-grained	Microgranite	Porphyry	Dolerite	—
Fine-grained	Rhyolite Obsidian	Trachyte Andesite	Basalt	Dunite

Table 5.2. Some major igneous rock types: landforms, weatherability, and plant suitability (from Skinner and Porter 1992, Ollier 1984, and Gerrard 1988)

	Landform types (arid regions)*	*Weatherability* (arid regions)*	*Mineral/chemical content*	*Likely plant suitability*
A. Intrusive (plutonic and hypabyssal) rocks formed at depths by cooling of magmas				
Granite and granodiorite	Massive domelike hills; steep slopes; heavily jointed; pinnacles (needles and spines) in vertically jointed rock	Highly resistant (much less so in mesic regions)	Quartz, feldspar, mica	Zonal on gentle landforms, azonal on steep landforms
Syenite and diorite	Ditto granite		Mafic minerals + feldspar (no quartz)	Ditto granite
Gabbro	Irregular-rounded ridges	Readily erodes	Plagioclase feldspar, augite, hornblende	Fairly selective of flora
Ultramafic rocks (dunite, peridotite)	Often jagged, with steep slopes and serrated ridges	Highly variable: often high in mesic regions (olivine readily weathers)	90 percent olivine ($FeMgSiO_3$)	Highly selective; unique (endemic) floras
B. Extrusive (volcanic) rocks (mostly fine-grained), formed at Earth's surface				
Rhyolite (acid igneous and dacite)	Angular with steep walls	Low; yields quartzitic sand	Quartz and feldspar	±Azonal; soils low in nutrients; poor plant cover
Andesite (extrusive form of diorite)	Stepped canyon walls; talus; steep slopes; conical hills, plateaus	Intermediate: between rhyolite and basalt	Mafic minerals (amphibole, pyroxene) and plagioclase feldspar	Zonal; often with good vegetation and rich flora
Basalt (extrusive form of gabbro)	Jagged, well-defined boundaries; level or gently sloping plains, mesas, and plateaus; shield-shaped hills; talus at base of slopes, vertical escarpments; terraced canyon walls	High; along joints, spheroidal; rich clayey soils	Plagioclase, pyroxene, or olivine	Zonal; with good plant cover and high diversity

* Ollier (1984).

Table 5.3. Effect of parent material on chemical composition of Australian soils (averages) (from Jenny 1941)

	CaO, percent		*P_2O_5, percent*		*N, percent*		*K_2O, percent*	
Region	*Granitic*	*Basaltic*	*Granitic*	*Basaltic*	*Granitic*	*Basaltic*	*Granitic*	*Basaltic*
Southern tableland	0.125	0.306	0.100	0.226	0.149	0.125	0.122	0.273
West central tableland	0.135	0.262	0.105	0.170	0.100	0.221	0.113	0.115
Northern tableland	0.211	0.241	0.104	0.192	0.104	0.207	0.159	0.122

Table 5.4. Molecular values of major constituents of selected rocks (oven-dry basis) (from Jenny 1980)

		Molecular values (mmoles)						
Rock type	*Quartz (wt%)*	*Total SiO_2*	*Al_2O_3*	*Fe_2O_3*	*FeO*	*MgO*	*CaO*	*Na_2O plus K_2O*
Acid igneous rocks								
Granite (coarse-grained)	37.3	1240	142	6	3	2	10	98
Rhyolite (fine-grained)	35.1	1267	121	5	13	4	2	112
Granodiorite or quartz diorite (coarse-grained)	27.1	1171	152	6	16	22	57	97
Dacite	25.0	1093	153	13	29	61	65	80
Tonalite	20.0	1060	150	10	59	79	101	69
Basic igneous rocks								
Andesite (fine-grained)	2.2	949	181	23	33	58	76	119
Basalt (fine-grained)	0	815	156	38	64	149	146	87
Gabbro (coarse-grained)	0	786	142	10	192	130	145	63
Ultrabasic rocks								
Peridotite	0	715	16	9	93	1070	25	0.5
Serpentinite	0	693	8	26	29	956	7	6
Sedimentary rocks, averages*								
Sandstones (114 CO_2)	—	1313	47	7	4	29	99	21
Shales (105 CO_2)	—	942	139	26	25	67	109	59
Limestones (945 CO_2)	—	87	8	3	—	197	761	4

* The presence of CO_2 is indicative of carbonates in the rock.

Table 5.5. Chemical composition of basalt and granite, and the average for all igneous rocks (after Hunt 1972)

	Percent values		
	Basalt	*Granite*	*Average of igneous rocks*
SiO_2	50	72	59
Al_2O_3	17.5	13.5	15.5
Fe_2O_3	3	1	3
FeO	7	1	3.75
CaO	8	1	5
MgO	6.5	0.25	3.50
Na_2O	3	3.5	3.75
K_2O	2	5	3
H_2O	1	1	1.25
TiO_2	1.25	0.5	1
Other	0.75	1.25	1.25

SOURCE: Data rounded off from Clarke (1924) and Mason (1958), as cited in Hunt (1972).

Where the effects on flora and vegetation of igneous parent material are most pronounced are with rock of mafic to ultramafic mineral content (for ultramafic terms, see Tables 5.19, 5.20). Mafic rocks, like gabbro and some basalts, can weather into soils that have a selective effect on plants. Whittaker's (1960) classic study of vegetation and lithological diversity in the Klamath-Siskiyou mountain complex along the Oregon-California border included a comparison of the plant communities on diorite with those on olivine gabbro (a mafic rock). Whittaker (pp. 297–299) vividly describes the contrast in vegetation:

> The general description of the diorite vegetation as "mixed evergreen forest" can apply also to the gabbro vegetation pattern at low elevations. Apart from overall physiognomic similarity and the sharing of some species, however, the two vegetation patterns are quite different. The gabbro vegetation is much more open than that on diorite. Average densities of large stems of conifers were less than half as great on gabbro; and density of larger sclerophyll stems is much lower on gabbro in more xeric sites. It is consequently possible to stand on one hillside in the gabbro area and look through the canopy to the soil on another, nearby hillside; in the diorite area one cannot similarly look through the dense evergreen canopies.

5.2 Igneous ultramafic rocks of the massive dunite peaks of Twin Sisters Mountain, North Cascades, Washington State. Photo by M. Randlett.

Furthermore, Whittaker recorded distinct differences in floristic composition, especially toward the upper, more xeric end of the moisture gradient. While the canopy species may be the same on both parent materials, the compositions of the shrub and herb strata are significantly different (Whittaker 1960).

The most striking illustrations of igneous rock supporting a unique flora, vegetation, and physiognomy are the ultramafic rocks, notably peridotite and dunite and their distinctive plant life (Figs. 5.2–5.5; see major reviews in Brooks 1987; Kruckeberg 1985; Proctor and Wooddell 1975; Roberts and Proctor 1992). A cautionary note: It is somewhat arbitrary to deal with the igneous ultramafic rocks apart from their metamorphic counterpart, serpentinite. Since many outcrops are partly serpentinized igneous ultramafic, all degrees of serpentinization may occur. Also, igneous and metamorphic forms are often closely associated both in space and in their geoedaphic properties, and can evoke similar plant responses. Both the igneous and metamorphic version of ultramafic substrates evoke the "serpentine syndrome," Jenny's (1980) term for the set of biological and geochemical attributes that interact. Hence, much of what could be said about plants on igneous ultramafic substrates will appear later in this chapter (the section on metamorphic rocks). Table 5.6 gives the composition of ultramafic rocks.

Ultramafic igneous rocks, and their metamorphic companions, serpentinites, form outcrops worldwide, especially in association with zones of plate subduc-

5.3 The alpine slope of the north Twin Sisters Mountain reveals the exceptional krummholz of *Pinus contorta contorta* on dunite. Photo by author

5.4 Dunite conglomerate on prehistoric landslide, southeast lower slope of Twin Sisters Mountain; stunted shore pine (*Pinus contorta*) and Douglas fir (*Pseudotsuga menziesii*) bordering ultramafic barren. Photo by author.

5.5 Aerial view of Dun Mountain, South Island, New Zealand; the barrenlike scrub vegetation on dunite is surrounded by *Nothofagus* forest on calcareous rocks. Photo courtesy Robert Brooks.

tions. Although instances of purely igneous occurrences of ultramafic rocks are less numerous than those of serpentines, there are some famous examples that produce striking geoedaphic responses in their flora and vegetation. Dun Mountain on the South Island of New Zealand is in its core an igneous ultramafic dunite (a rock of nearly pure olivine mineral and chromite). The mountain presents a characteristic ultramafic landscape: an open scrub vegetation surrounded by dense *Nothofagus* forest on nonultramafic substrates. The mountain not only has a unique vegetation type; it supports several endemics (e.g., *Carex devia, C. traversii, Myosotis monroi, Pimelea suteri*). Brooks (1987) devotes his chapter 21 to the New Zealand ultramafic ecology (Fig. 5.5). On other South Island (New Zealand) ultramafic areas, there are still other edaphic endemics, and also the substrate supports serpentine-tolerant ecotypes from the region's bodenvag species (Lee 1980; Lee and Hewitt 1981; Lee, Mark, and Wilson 1983).

Spectacular displays of flora occur on the igneous ultramafites of New Caledonia (Brooks 1987; Jaffré 1980). Nearly one-third of this large island in the South Pacific had its ultramafic substrates exposed in major massifs or in scattered sills. The rock is mainly a form of peridotite (harzburgite, containing the minerals olivine

Table 5.6. Major elements of ultramafic and other rock types (after Brooks 1987)

							*Source localities**							
Elements (oxides)	*A*	*B*	*C*	*D*	*E*	*F*	*G*	H^a	H^b	*I*	*J*	*K*	*L*	*M*
SiO_2	40.49	70.18	78.66	38.1	39.8	39.0	48.7	42.9	39.8	38.0	41.58	38.7	35.80	39.53
Al_2O_3	1.77	14.47	4.78	2.4	0.7	0.04	15.0	7.5	20.8	0.64	3.48	0.58	0.15	0.93
FeO	4.84	1.78	0.30	—	—	5.0	3.8	—	—	3.39	6.48	7.26	4.80	7.62
Fe_2O_3	6.01	1.57	1.08	3.6	3.1	2.8	1.0	11.2	3.9	3.85	1.81	3.19	2.80	0.65
MgO	37.36	0.88	1.17	41.0	40.3	46.1	10.8	23.5	5.8	39.58	37.50	36.44	44.30	48.83
MnO	0.22	0.12	T	—	—	0.11	—	0.14	0.14	0.11	0.08	—	0.10	0.32
CaO	0.74	1.99	5.52	0.0	0.3	0.00	15.7	7.3	25.7	0.93	3.50	T	0.16	T
Na_2O	0.15	3.48	0.45	0.1	0.1	0.00	1.7	0.34	0.14	0.08	0.47	—	0.05	T
K_2O	0.10	4.11	1.32	0.1	0.1	0.23	0.12	—	—	—	0.16	—	0.03	—
TiO_3	0.65	0.30	0.25	0.1	0.1	0.02	0.29	0.31	0.17	—	0.10	—	0.10	0.013
NiO	±0.3	T	—	—	—	0.35	0.24	(1000)	(80)	0.23	0.17	—	0.38	0.32
Cr_2O_3	±0.5	T	—	—	—	0.44	0.02	(1800)	(75)	0.39	0.19	—	0.27	1.01
CoO	—	—	—	—	—	—	—	(90)	(17)	—	—	—	0.03	—
H_2O	10.18	0.84	1.33	13.9	13.4	5.6	2.5	6.2	2.8	12.25	4.12	—	10.60	0.89
P_2O_5	0.02	0.19	0.08	—	—	0.03	0.02	—	—	0.02	—	—	—	—
Total (%)				99.3	97.9									99.57

Average samples
- A – Average ultramafic rocks (Krause 1958)
- B – Average granites (Krause 1958)
- C – Average sandstones (Krause 1968)

North American ultramafites
- D – Serpentinite, Dubakella soil series, California (Wildman et al. 1968)
- E – Serpentinite, San Benito County, California (Wildman et al. 1958)
- F – Dunite, California (Himmelberg and Coleman 1968)
- G – Gabbro, California (Himmelberg and Coleman l968)
- H – Ha serpentinite; Hb rodingite, Washington State (Coleman 1967)

Europe
- I – Peridotiote/serpentinite, England (Proctor 1991)
- J – Serpentinite, Balkan Peninsula (Tatic and Veljovic 1992)
- K – Serpentinite, Italy (Pichi-Sermolli 1948)

South Pacific Region
- L – Dunite, New Caledonia (Brooks 1987)
- M – Dunite, New Zealand (Coleman 1966)

* Values are in percent, except those in ( ) are ppm.
T = trace.
— = not determined.

and orthopyroxene), with accessory chromium and nickel minerals, the latter of considerable economic value. The New Caledonian Region (Good 1974) is noted for its distinctive plant life; 80 percent of its rich flora is endemic (Brooks 1987), and the ultramafic component looms large and is richly diverse. Tanguy Jaffré (1980) gives a thorough account of this remarkable flora. Of the 1500 species of vascular plants, 60 percent are endemic to the island's peridotite. This surely must be the highest level of geoedaphic endemism in the world. The vegetation on this parent material is of diverse types; an evergreen forest occurs at higher elevations, while a distinctive scrub or maquis type grows at lower elevations (Fig. 5.40). A notable biogeochemical feature of the New Caledonia flora is the high incidence of nickel hyperaccumulating species. The ability to accumulate nickel to levels of 1000 ppm and higher (hyperaccumulation) is one other manifestation of this remarkable geoedaphic adaptation to an azonal, highly selective substrate. Forty-eight species in diverse genera and families are known to be hyperaccumulators of nickel (Brooks 1987); this must be the most concentrated occurrence of nickel accumulators in the world.

A final example of a vegetational response to an igneous ultramafic parent material comes from western North America. The largest deposit of dunite in North America is located in the North Cascades Range of Washington State (Fig. 5.2). Twin Sisters Mountain consists of pure dunite, yet it is surrounded by nonultramafic volcanic and metamorphic substrates (Christiansen 1971). The isolated massif rises from about 1000 feet (308 m) to twin summits of over 6000 feet (1846 m); the north Twin Sister is 6932 feet (2133 m). Nearby lithologies support (or did support before intensive logging) conifer forests typical of the western slope of the Cascades (Franklin and Dyrness 1973). On the Twin Sisters dunite, however, the physiognomy and species composition of the flora changes dramatically. While the conifers growing on the dunite are mostly the same as those found on other nearby substrates, their appearance on dunite is abruptly distinct. Conifer stands here are stunted in stature and more widely spaced. Further, conifer species from adjacent elevational zones may coexist in the same stand on dunite. For instance, at Olivine Bridge, the following conifers from different life-zones coexist: Douglas fir (*Pseudotsuga menziesii*), western white pine (*Pinus monticola*), shore pine (*Pinus contorta contorta*), Pacific silver fir (*Abies amabilis*), Alaska cedar (*Chamaecyparis nootkatensis*), and western red cedar (*Thuja plicata*). This site is an exceptionally sere, open landscape on a massive prehistoric landslide of dunite conglomerate (Kruckeberg 1969b). The openness of this site makes it receptive to chance colonization of species from other life-zones; Fig. 5.4). Elsewhere in the Twin Sisters area another conifer anomaly has been observed: the subalpine zone is marked by the

5.6 Response to dunite (ultramafic) soil by *Prunella vulgaris:* tolerant strain (S) at top and two intolerant strains (NS) at bottom. Photo by author.

unusual presence of *Pinus contorta* as the timberline, krummholz conifer (Kruckeberg 1969b). See Figure 5.3. On nearby Mount Baker, a major volcano, subalpine fir (*Abies lasiocarpa*) and whitebark pine (*Pinus albicaulis*) are the species of the krummholz life-form. Still another conifer anomaly occurs on the Twin Sisters dunite. *Juniperus communis,* usually confined to the subalpine and alpine habitats of the Cascades, is a common understory shrub in the stunted conifer stands at about 1800 feet (557 m), well below its usual elevation.

Yet in this entire massive montane display of dunite there are no endemic species. This is not unexpected, since the Twin Sisters dunite was largely covered with ice in the late Pleistocene, and thus there may have been too little time for endemic species to evolve. Granitic erratics have been found at 5000 feet (1548 m), evidence that the bulk of the Twin Sisters Mountain was open to colonization only since the retreat of continental and alpine ice, about 10–11,000 years B.P. (Kruckeberg 1969b). Vegetation growing under the open conifer canopy is not remarkable. Shrubs typical of the adjacent zonal soils grow on the dunite, such as salal (*Gaultheria shallon*), *Vaccinium parvifolium,* and *Menziesia ferruginea* in mesic sites and *Spiraea menziesii* and *Ledum groenlandicum* in wetter sites. On the most xeric-appearing sites, the herbaceous flora is exceptionally depauperate. Wide-ranging (bodenvag) species like *Cerastium arvense, Arenaria rubella, Fragaria virginiana crinita,* and *Prunella vulgaris* occur as a sparse and patchy ground layer. A notable discovery came from this mix of nonendemic herbaceous species. It was the demonstration (Fig. 5.6) that *Prunella vulgaris,* likely a postlogging, recent introduction, has already evolved tolerance to the preexisting ultramafic substrate (Kruckeberg 1967). High levels of nickel were found in the dunite soils here (1393 and 2643 µg/g, in Kruckeberg et al. 1993). Yet none of the plants appear to be nickel accumulators. The report of hyperaccumulation in *Arenaria rubella* (Kruckeberg et al. 1993) has been shown to be in error (R. D. Reeves, pers. comm.). Like the absence of endemism, the absence of hyperaccumulators supports the notion that too little time for genetic accommodation to nickel in the dunite has been available following Pleistocene glaciation.

Along with the singular physiognomy of open stunted conifer stands is the reliable occurrence on the dunite of three serpentinicolous ferns: *Aspidotis densa, Adiantum pedatum calderi* (= *A. aleuticum,* the serpentine form, *fide* correspondence with C. Paris), and *Polystichum lemmonii.* The latter, a mountain holly fern, is strictly confined to ultramafic substrates almost everywhere they occur in mesic montane sites, from northern California to northern British Columbia. So while the Twin Sisters dunite may not be the floristically rich and unique terrain on igneous ultramafic areas like New Caledonia, it does have a respectable display of attributes displaying the "serpentine syndrome."

Plant Responses to Residual and Sedimentary Rocks

Much of the world's vegetation grows on the more or less consolidated secondary products of erosion, which when lithified become sedimentary rocks. Sediments, the mineral or organic matter transported and deposited by water, wind, and ice, occur in many forms and in all stages of lithification. The regolith of the Earth's crust, the loose, noncemented rock particles, forms the mineral substance of sediments. The regolith encompasses colluvium (scree and talus), soils, and other surficial, lithic detritus (clastic sediments). On consolidation (lithification or cementation) the materials of the regolith are transformed into sedimentary rocks. The change from sediment to rock may involve some chemical change in the mineralization of the materials; often some minerals of the original source rocks dissolve away out of the sediments (some feldspars, etc.) and are removed. Another possibility is for sedimentation and lithification to preserve the original minerals. An additional source of sedimentary rocks is biogenic—remains of organisms or their metabolites that have become lithified. Organisms are the source of the widely occurring carbonate rocks such as limestone and dolomite.

Limestone rocks can form by precipitation and lithification of calcium carbonate (called chemical limestone). Yet the carbonate ion is mostly of biogenic origin—from respiratory carbon dioxide. The other class of carbonate rocks occurs by the lithification of animal (and some plant) remains containing calcium carbonate (called fossiliferous limestone).

Another source is neither regolithic nor organic, but purely chemical. The precipitates from solutes form rock called chemical sedimentary rocks (Skinner and Porter 1992). Deposits of iron, phosphate (as apatite), and various marine evaporites (e.g., sodium chloride and gypsum) are examples of chemical sedimentary rocks.

Many secondary rocks, like those just discussed, are the lithified products of transported sediments following erosion; these are the usual sedimentary rocks. But another part of the regolith may not be transported. Loose, fragmented pieces of older rocks (clastic detritus) that remain in place can become altered, by lithification or by pedogenesis, to form so-called residual (sedentary) deposits (Tyrrell 1929); see Table 5.7 for the relationship of residual deposits to other secondary rocks. Residual deposits can take a variety of forms. The weathering of rock may leave a detritus of angular, unsorted fragments in place. The materials of such deposits range in size from boulders to sand and even fine, flourlike materials. Karst country and its associated terra rossa soils, so well known from the Mediterranean region, are classic cases of residual secondary rock; karst involves the differential dissolution of limestone. Clay-with-flints in the south of England is another residual product of limestone dissolution. Laterite and bauxite, products of intensive oxidizing weathering in the tropics, form vast residual deposits. A final such deposit is soil itself. Since many of these residual deposits are at the Earth's surface, they can be significant and selective substrates for vegetation. I take up particular examples later in this chapter.

The taxonomy of secondary rocks, especially sedimentary rocks, is beset with the difficulties of classifying a highly heterogeneous set of phenomena and materials. Process, product, and time are the major factors in a matrix of classification. Processes are diverse, from tectonic influences, to weathering, transport, and lithification. The products vary in size, shape, and mineral content; and the time frame may extend from the present (ongoing erosion and weathering) to ancient Proterozoic sedimentation. For our purposes, the field classification of sedimentary rocks should suffice. Rock from detrital or nondetrital materials is classified by grain size (in detrital rocks) or nature of origin (chemical or organic in nondetrital rocks). Tables 5.8 and 5.9 list the most common sedimentary rocks and their attributes.

Table 5.7 Types of weathering and their products (from Tyrell 1929)

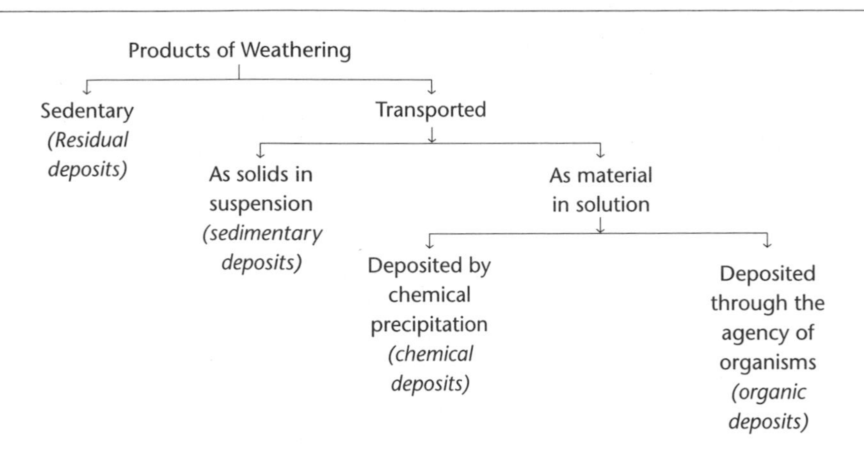

Table 5.8. A classification of sedimentary rocks (from Simpson 1966)

Mainly detrital			*Mainly nondetrital*			
Grain size			*Infinite variation*			
Coarse	*Medium*	*Fine*	*Chemical*		*Organic*	
>2.0 mm	*2.0 mm – 0.05 mm*	*0.05mm – 0.005 mm*	*Precipitates*	*Evaporites*	*Zoogenic*	*Phytogenic*
Conglomerates	Sandstones	Shales	Crystalline	Halite	Shelly coral	Algal
Breccias	Greywackes	Mudstones	limestones	Sylvite	Foraminiferal	limestones
Calcirudites	Quartzites	Siltstones	Chalk	Anhydrite	and crinoidal	Coal
	Calcarenites	Loess	Flint	Gypsum	limestones	Oil
				Dolomite	Diatomite	

A thorough review of accommodation by plants to particular sedimentary parent materials could become encyclopedic, given the worldwide distribution of sedimentary lithology. It should suffice to illustrate the phenomenon of specificity with a few outstanding examples. I have chosen limestone floras and vegetation, shale barrens, and gypsophilous plants for discussion.

LIMESTONE ROCKS: THEIR FLORISTICS AND ECOLOGIES

It should not have been too surprising to find that limestones gave us the earliest recorded observations on the connection between rock type and vegetation. At the beginning of this chapter, I pointed out that it was the pre-Christian era

Table 5.9. Some major sedimentary rock types: landforms, weatherability, and plant suitability
(from Skinner and Porter 1992, Ollier 1984, and Gerrard 1988)

	Landform types (arid regions)	*Weatherability*	*Mineral/chemical content*	*Plant suitability*
A. Clastic sedimentary rocks: reformed and cemented from weathered products of igneous and metamorphic rocks				
Conglomerates (incl. breccias): cemented gravel and rock fragments	High and irregular relief; much talus	Cementing substance dissolves leaving rock fragments intact	Highly variable	Mostly low
Sandstones: cemented sand grains	Layered; bold, massive hills	Readily erodes via frost and solution leaving sand grains	Variable; quartz common	Zonal or azonal depending on soil nutrient levels
Siltstones, mudstones, and shales: cemented silt, mud, and clay	High relief in horizontal bedding; rugged (badlands)	Easily eroded	Clay minerals, mica	Mostly azonal (e.g., shale barrens); local, often endemic floras
Pyroclastics: cemented volcanic ejecta	Irregular, angular micorelief	Indurated; resistant		Mostly azonal; often xeric vegetation
B. Chemical sedimentary rocks				
Chert (solidified amorphous fine-grained silica)	High, jagged relief; resistant to erosion	Slow to weather	Quartz	Low plant cover on shallow, sterile soils
Rock salt	When exposed, rounded, domelike relief	Rapid (by solution)	NaCl	±Devoid of vegetation
Gypsum	Flat to low relief	Rapid (by solution)	$CaSO_4 \cdot H_2O$	Azonal, often with gypsophilous species
C. Biogenic sedimentary rocks				
Limestone and dolostone	High (karst) relief (sinkholes in less arid regions); angular, sharp, narrow ridge crests in tilted rock; resistant to erosion	High in mesic areas (by solution)	Calcite ($CaCO_3$) Dolomite ($CaMgCo_3$)	Azonal, highly selective (calcicoles vs. calcifuges), often rich in endemics
Diatomite			Silica	
Coal oil shale				

botanist Theophrastos who associated the *phrygana* vegetation type with limestones. Other landscapes overlying limestone, like the terra rossa and karst, were known to be related to limestone bedrock from early times. By the beginning of the nineteenth century, botanical observations on limestone vegetation began to proliferate. It was the stark contrast in vegetation between limestone and slate slopes in the Tyrolean Alps that led Franz Unger (1836) to his chemical theory of edaphic restriction (Kruckeberg 1969b).

By the late twentieth century, studies of limestone plant ecology had yielded a wealth of published work, especially in Europe and North America. Kinzel's review (1983) reminds us how complex the reticulate interplay of parent material ⟷ soil ⟷ climate ⟷ plant cover can be. The simple use of pH to distinguish the ecological differences between limestones and siliceous rocks has worked only as an indicator of the edaphic differences. The ecophysiological approach of Kinzel and others has resulted in the recognition that the limestone-plant interface is multifactorial, consisting of such factors as nutrient availability, pH, and solubility of aluminum, iron, phosphate (?), and manganese; single-plant versus plant-competition studies; and arid versus humid climates and limestones. Kinzel's review describes the ecophysiological aspects of carbonate rocks. However, here we examine some examples of landscapes, vegetation, and floristic consequences of limestone and dolomite parent materials.

Limestone Vegetation of North America

The White Mountains, a desert range just east of the Sierra Nevada of California, is the setting of a spectacular vegetation display (Lloyd and Mitchell 1973).Where quartzitic sandstone of the Campito formation abuts Reed dolomite (a calcium-magnesium carbonate rock), an abrupt contrast in vegetation greets the eye. On sandstone, a dense, low stand of sagebrush (*Artemisia tridentata*) borders the celebrated subalpine forests of bristlecone pine (*Pinus longaeva*) at the lithological contact with dolomite (Fig. 5.7). One can make the case for two timberlines in the White Mountains as a result of discontinuities in lithology. The lower pinyon woodland zone (*Pinus monophylla*) gives way upward to the treeless sagebrush community on sandstone; the juncture of the two creates the lower timberline. Then, beyond the "alpine" sagebrush, a second subalpine forest appears; the bristlecone pine stands on dolomite. Both the sagebrush and bristlecone pines occupy elevations from 9500 feet (3170 m) to 11,500 feet (3830 m). Still on dolomite, the bristlecone pines give way upward to a treeless dolomite barren, a sparsely vegetated alpine tundra (11,500 feet to the summit, 14,246 feet or 3540 to 4385 m). The impressive contrasts in vegetation, owing to differences in lithology, make the

5.7 A dramatic contrast in vegetation on different lithologies, with sagebrush (*Artemisia tridentata*) in the foreground on sandstone and bristlecone pine (*Pinus longaeva*) on dolomite. White Mountains, eastern California. Photo by R. S. Mitchell.

White Mountains one of the clearest case histories of geoedaphic influences on plant life, certainly just as striking as the serpentine-volcanic contrasts in the Coast Ranges of California. Besides the striking contrasts in the dominant woody vegetation, the herbaceous flora affords good examples of edaphic restriction. Of the thirteen White Mountains endemics, over half are in the alpine area—on the dolomite barrens (Lloyd and Mitchell 1973). Mooney (1966) recorded a neat edaphic difference for two species of *Erigeron.* On the sharp contacts between the two lithologies, the two fleabanes showed a clear preference. *Erigeron clokeyi* stays on the sandstone while *E. pygmaeus* is restricted to dolomite. The causes of this edaphic partitioning are multiple: substrate color (light on dolomite, dark on sandstone) is linked with temperature and moisture differences; substrate chemistry also plays a part in edaphically separating the two species.

The state of Idaho is richly endowed with limestones. The effects on flora are especially noticeable in montane to alpine areas. D. M. Henderson (pers. comm.

1992) recorded the range of response to limestone for a number of taxa—from calcicole and bodenvag to calcifuge (Table 5.10). Henderson is quick to point out that aspect, slope, stability of rock, and so forth, influence the nature of the limestone addiction/avoidance; thus the landform variables that interface with lithology are given in Table 5.10.

Limestone lithology appears commonly east of the Mississippi, especially in the Southern Appalachian Mountains and in the Ozarks. An elegant case study of a limestone endemic by Erickson (1945) illustrates the close tracking by *Clematis riehlii* to the discontinuous, semi-isolated limestone glades of the Ozarks. Erickson's mapping of this phenomenon (see Fig. 7.7) shows a classic example of edaphically induced insularity; an archipelago of suitable habitats is tellingly portrayed.

Plant communities on rock outcrops in the eastern United States continue to capture the fancy of systematists and ecologists. Endemism and unique plant associations are frequent attributes of these rather barren substrates. Besides floristically distinctive displays on serpentine, sandstone, and shale, it is the limestone "cedar glade" (Fig. 5.8) flora that has stimulated the most serious studies (Baskin and Baskin 1988, 1989; Quarterman, Burbanck and Fralish 1993). The Baskins have made a specialty of the study of cedar glade floristics, mostly in Kentucky and Tennessee; their investigations have focused on the distribution, ecology, and ecophysiology of the cedar glade endemics. All seven taxa of the cruciferous genus *Leavenworthia* are cedar glade endemics. In their 1989 paper, they list 23 taxa as cedar glade endemics for just Tennessee; 41 endemics are known for the unglaciated eastern United States. One environmental factor common to these taxa is an obligate high light requirement (Baskin and Baskin 1988). Though it is assumed that plants of these edaphic sites are lime tolerant, poor competitors, and shade intolerant, there is "nothing unique about the biology of any of them [that] has been identified that explains their restriction to cedar glades" (Baskin and Baskin 1989). Why "cedar glade"? The nearly treeless openings often do contain scattered red cedars (*Juniperus virginiana*); the glades are usually surrounded by hardwood forest.

Limestone Vegetation of Europe

Limestones and kindred carbonate rocks abound in Europe, and are often the dominant lithology of an area. For example, in the British Isles, carbonate rocks are so frequent and manifest in their influence on landscape and flora that they form an integral part of natural and human ecology. Europe's carbonate rocks take many forms, from chalk (the British term for soft, recent limestone), limestone pavement, karst (a geomorphic variant of limestones), and dolomite (with 15 percent magnesium carbonate). The major limestone occurrences extend from the

Table 5.10. Some calcicole-calcifuge plants of Idaho
(D. M Henderson, pers. comm., 1992)

Vertical cliffs of carbonate rock
Kelseya uniflora (restricted to carbonate)
Petrophytum caespitosum (predominantly on carbonate)
(both also occur on similar substrate and angle at lower elevations)

Moist ledges, especially where snow accumulates
Primula parryi (calciphile in this area and strictly alpine)

Talus and scree, unstable
Chaenactis alpina (bodenvag)
Anelsonia eurycarpa (predominantly on limestone)
Collomia debilis (bodenvag but with a strong trend to limestone)

Talus, stable
Polemonium viscosum (bodenvag)
Silene acaulis (strict calciphobe)
Hulsea algida (bodenvag, on quartzite, Challis volcanics, and carbonates)
Hymenoxys grandiflora (bodenvag)

Moist slopes, including solifluction areas
Poa alpina (bodenvag)
Luzula spicata (bodenvag)
Ranunculus eschscholtzii (bodenvag)
Lloydia serotina (strong preference for limestone)
Zigadenus elegans (strong preference for limestone)

Specialized habitats of lower elevations, including substrate and topography
- Moist subalpine meadows where parent rock is other than carbonate
  Agoseris lackschewitzii (regional endemic also in same habitats in adjacent Montana)
- Lower slopes adjacent to Salmon River, vicinity Challis and Clayton
  Challis volcanic deposits (ash, mud-rock flows, etc.)
  Astragalus amblytropis (narrow endemic here)
  Astragalus aquilonius (narrow endemic but disjunct on same deposits in Lemhi Valley)
  Thelypodium repandum (narrow endemic)
  Oxytropis besseyi salmonensis (narrow endemic)
  Eatonella nivea (disjunct Great Basin and Columbia Basin)
  Elymus ambiguus salmonis (narrow endemic *sensu* Hitchcock but subsequently combined with *E. salina*)
  Enceliopsis nudicaulis (Great Basin disjunct)
- Other specialized habitats
  Limestone cliffs and ledges, lower montane
  Astragalus amnis-amissi (narrow endemic)
  Draba oreibata oreibata (narrow endemic)
  Draba hitchcockii (narrow endemic, also grows on talus at base of cliffs and a bit out in the mountain mahogany but always in immediate vicinity of steep limestone!)
- Other edaphic specialists
  Cheilanthes feei (strictly limestone)
  Pellaea breweri (mostly limestone)
  Polystichum kruckebergii (so far [in Idaho] anything *but* limestone)
  Cymopterus douglasii (alpine, stable rocky habitats on limestone—a very narrow endemic!)
  Cymopterus ibapensis (disjunct from northern Great Basin where it grows in shrub steppe but here is known only from high alpine habitats on stable to unstable limestone talus)
  Townsendia condensata (alpine, limestone, at least in a part of its range)

5.8 Limestone openings in forest, southeastern United States, are known as cedar or limestone glades. This glade is at Cedars of Lebanon State Forest, Wilson County, Tennessee. Photo by J. and C. Baskin.

Table 5.10. *(continued)*

Anelsonia eurycarpa (limestone mostly; one population known from quartzite but with limestone nearby)
  Lesquerella carinata (regional endemic on limestone)
  Silene acaulis (alpine—avoids limestone!)
  (all ericads seen—except *Arctostaphylos uva-ursi*—avoid carbonate substrates in Idaho)
  Ribes hendersonii (regional endemic, mostly limestone)
  Phacelia lyallii (regional endemic, not found on nonlimestone—yet)
  Papaver kluanense (high alpine, disjunct from Arctic, known only in Idaho from quartzite "island" atop a mountain of limestone)
  Caltha leptosepala sulfurea (local Idaho endemic—subalpine wet meadows—mostly avoids limestone)
  Saxifraga oppositifolia (alpine—gravelly soil near melting snowbanks and only on limestone)
  Carex rupestris (alpine—always on limestone here)
  Lloydia serotina (alpine—nearly always on limestone; one population known from quartzite but with limestone within 1 m)
  Leucopoa kingii (*Hesperochloa*) (nearly always on limestone)

United Kingdom across central Europe to Greece and nearby Turkey. Especially rich and diverse floras occur throughout this vast display of carbonate rocks. Of particular botanical interest are the limestones of Britain, the Alps, and the Balkans; these three examples will be elaborated upon below.

So intimate is the connection of carbonate rocks to human endeavors and wonderment that a special vocabulary dealing with them and their botanical (including horticultural and agricultural) attributes has evolved. In Britain three or four types of carbonate rocks are recognized (Fig. 5.9). Chalk is the youngest type geologically and is a softer rock than the older British limestone. Clay-with-flints is a reddish clay soil derived from the dissolution of chalk (the flints are angular fragments of gray quartz, impurities in the chalk that resist weathering). Dolomite is a form of limestone so dominant in the Alps as to be recognized as a geographic province, The Dolomites. Other variant forms include limestone pavement (Figs. 5.10–5.13), oolite (rounded grains of calcium carbonate, like fish roe), certain marbles (metamorphosed limestone), travertine, and stalagmite-stalagtite formations. Several calcium-rich soils develop over limestone rocks—rendzinas over pure limestone, marl from limestone with clay, chernozems from calcium-rich loess, terra rossa (and its British equivalent, clay-with-flints), a reddish clayey soil. Terminology arising from the plant's response to limestone ranges from a perjorative and misleading pair, calciphile (lime-loving) and calciphobe (lime-hating), to the more neutral calcicole (lime-inhabiting) and calcifuge (lime-avoiding). Basiphilic and acidiphilic for base- and acid-tolerant plants are terms often used synonymously with calcicole and calcifuge, respectively. But Kinzel (1983, p. 204) points out that from the standpoint of mineral nutrition the two sets of terms are not truly congruent: "In several plants of the Caryophyllaceae, for example, ion metabolism seems to be impaired by an abundance of calcium. For these species, then, the term calcifuge is more suitable than acidiphilic. . . . In several members of the Ericaceae, on the other hand, root resistance to acid and ability to utilize ammonium in an acidic media are probably the decisive factors determining distribution. Thus, they would be more appropriately termed acidophilic, than calcifuge."

In the British Isles, limestones range all along the geologic column, from Paleozoic (Devonian limestone of south Devon) and Mesozoic to the calcareous drift of the Quaternary (Lousley 1950); see map of Figure 5.9 above. Lousley's engaging book on the flora of chalk and limestone singles out Box Hill in Surrey as a fine example of limestone flora and vegetation. The low chalk hills of Box Hill are a mosaic of woody vegetation and downs (upland grassland on chalk). Several trees and shrubs are good indicators of limestone: box (*Buxus sempervirens*), whitebeam (*Sorbus aria*), beech (*Fagus sylvatica*), and yew (*Taxus baccata*) are among the sev-

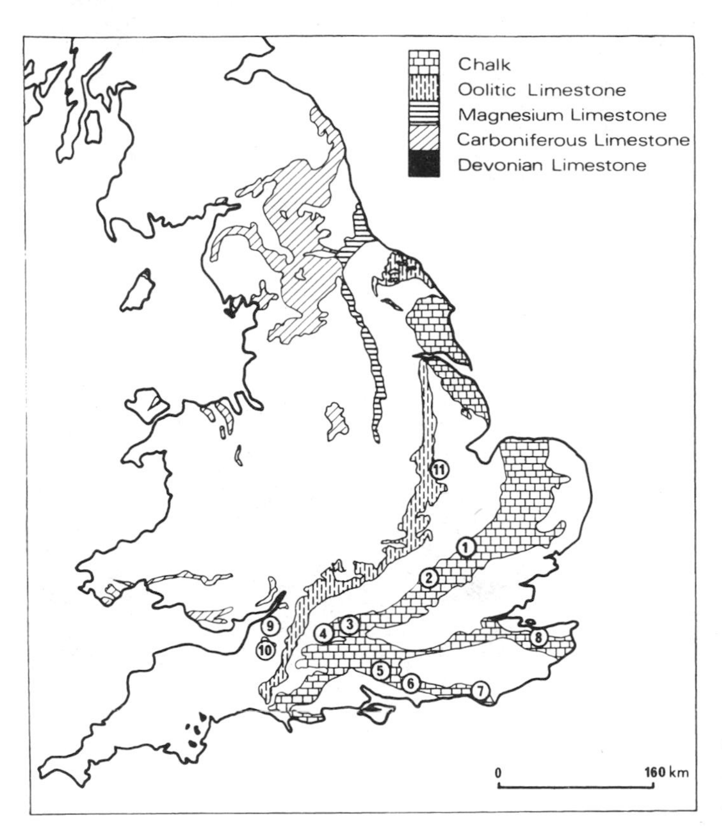

5.9 Distribution of chalk and limestone in England. The common occurrence of carbonate rocks in the British Isles has given the region a rich natural and cultural history. Permission of The Nature Conservancy

eral woody plants of the local chalk. It is the terrestrial orchids of the chalk downs for which Box Hill and other limestone localities are famous. Lousley lists six species for the Box Hill area. Of these, the man orchid, *Aceras anthropomorphum,* is a faithful calcicole here at Box Hill and throughout Europe (Tutin et al. 1980).

The many studies of British limestone vegetation provide a well-rounded exposure to the nature of the plant response. Grimes and Hodgson (1969) caution against a simple calcicole-calcifuge taxonomy of reaction to limestone. While they identify some species as strongly calcicole (e.g., *Scabiosa columbaria*) and others as strongly calcifuge (*Deschampsia flexuosa*), there is actually a continuum of response by other downland species. Some taxa, like *Trifolium repens,* have been shown to be ecotypically differentiated into races tolerant or intolerant to limestone (Snaydon 1962). Further, mineral nutrition studies have focused on lime chlorosis of calcifuges and aluminum toxicity for calcicoles (Grimes and Hodgson 1969), on the availability and/or solubility of cations (manganese, iron, aluminum) in limestone versus siliceous soils, and on differences in nitrification on the contrasting soils (Kinzel 1983). The role of competition in determining the success of species

5.10 A rich surface heterogeneity on level terrain can support a unique limestone flora. This side view of limestone pavement with scattered trees is from Yorkshire Dales National Park, near Whernside, England. Photo by A. L. Kruckeberg.

5.11 Crevices in limestone pavement (Yorkshire Dales) often harbor carbonate-tolerant plants. Photo by Karin Loeffler.

5.12 Solution-widening of joints, on limestone pavement at The Burren, western Ireland. Photo by W. B. White.

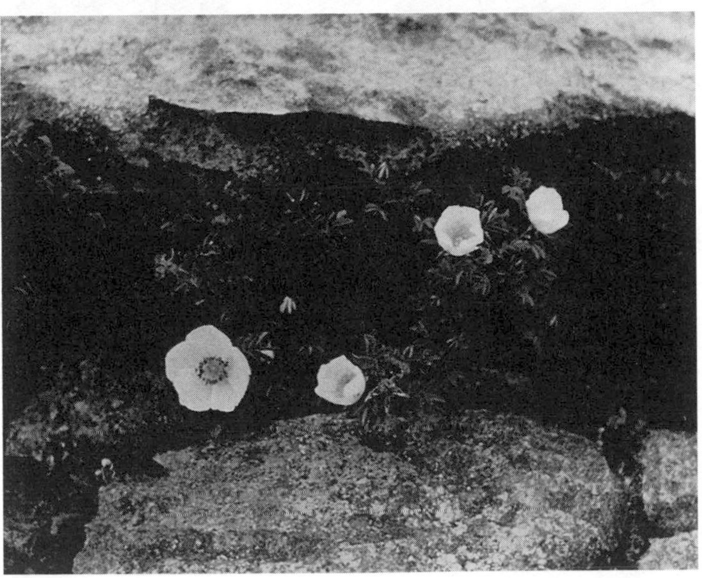

5.13 *Dryas octopetala*, typically an arctic-alpine species, here at sea level at The Burren, western Ireland. Photo by D. Drew.

on lime or acid soils has been underscored by several British workers (e.g., Harper 1977). A much cited classic study by Tansley (1917) demonstrated that species interactions greatly altered the effect of soil type on plant growth. The two bedstraws, *Galium saxatile* and *G. sylvestre,* were grown in pure and mixed culture on acid peat and calcareous soils. "Both species can establish and maintain themselves—at least for some years—on the other soil," but "the calcicole species is handicapped as a result of growing on acid peat and therefore is reduced to subordinate position in competition with the calcifuge rival which is less handicapped," and "the calcifuge species (*G. saxatile*) is heavily handicapped especially in the seedling stage, as a direct effect of growing on calcareous soil and is thus unable to compete effectively with its calcicole congener, *G. sylvestre.*" Tansley's study thus emphasizes the significance of competition in modifying edaphic responses in culture; his results are a warning against reading too much ecological significance into results obtained from plants studied in artificial isolation.

Limestones of the European Alps. It was the limestones of the Alps, and their dramatic floristic-lithological contrasts with nearby rocks, that set in motion the translation of observations into geoedaphic theory. And the theorizing began in the eighteenth century, first with the observations of H. F. Link in 1789, followed by those of G. Wahlenberg in 1814 and then Franz Unger in 1836, who saw the lithological contrasts in terms of chemical differences. Those early papers were cited in Braun-Blanquet (1932) and Gigon (1971). The Alps are noted not only for great stretches of limestone and dolomite, but for many other lithologies (e.g., gneisses, slates, and the serpentines of the "Toten Alpen"). Lithological diversity with sharp discontinuities between rock types is a hallmark of Alpine landscapes, geology, and vegetation. The German phrase "Vielfältigder Gesteine" (Gigon 1971) says it well: "rock diversity."

It was the limestone-silicate rock contrasts that captured Unger's attention and resulted in his classic 1836 paper, *Über den Einfluss des Bodens auf die Verteilung der Gewächse, nachgewiesen in der Vegetation des nordostlichen Tirols.* Ever since then, the "Kalkfrage" (the limestone question) has kept generations of plant ecologists, physiologists, and taxonomists busy seeking answers to the question: How (and why) do plants differ in their accommodation to limestone? The plant response to carbonate rocks of the Alps has been exhaustively studied for over two centuries, and can still evoke questions in ecology, floristics, physiology, and evolutionary genetics.

The detailed phytosociological study by Gigon (1971) exemplifies the ecological contrasts between carbonate and siliceous substrates. In his study area, the Davos

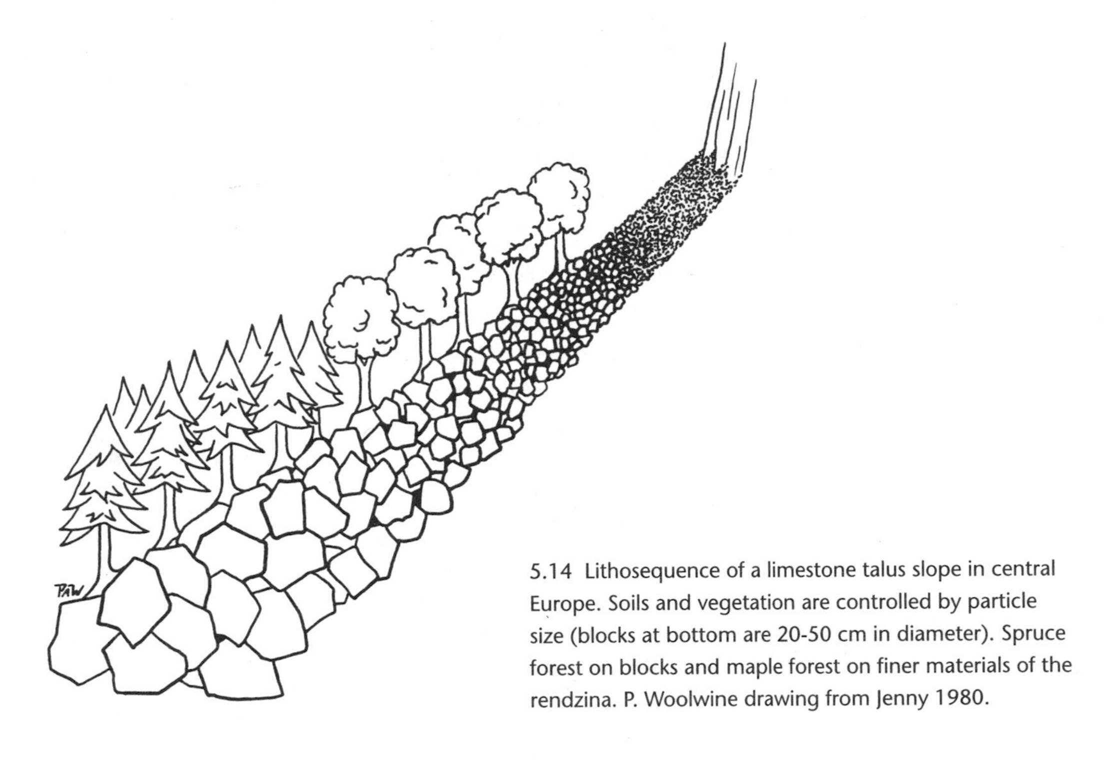

5.14 Lithosequence of a limestone talus slope in central Europe. Soils and vegetation are controlled by particle size (blocks at bottom are 20-50 cm in diameter). Spruce forest on blocks and maple forest on finer materials of the rendzina. P. Woolwine drawing from Jenny 1980.

region of Switzerland, the acid glacial till and the dolomite limestone were considered the main variables, with climate, topography, and time—since glaciation—held constant (*sensu* Jenny's soil forming factors [1941]). Two contrasting vegetations show a well-defined substrate specificity. On limestone, the Selserion community supports seven calcicole species. Similarly, the Nardetum community on siliceous rocks, dominated by *Nardus strictus,* harbors twenty-three calciphobe species, plants absent from the adjoining limestones.

A well-defined talus lithosequence on pure limestone has been described by Bach (1950), for the Jura Mountains of Switzerland (Fig. 5.14). The largest, angular stones are at the base of the talus and support a spruce forest (*Picea abies*) with *Hylocomium* moss on the forest floor. The upper slope of progressively finer limestone fragments is a humus carbonate soil (rendzina) and supports a forest of mountain maple (*Acer pseudoplatanus*), with a shrub layer of mountain ash (*Sorbus* sp.) and hazelnut (*Corylus* sp.), and a variety of herbs. The talus is fed from above by cliffs of limestone rock. So in this instance, as with many other lithosequences, the graded nature of the rock debris adds another dimension to the plant response to the rock itself.

Botanists for years have been tabulating the floristic differences between limestone and siliceous rocks. For the Swiss Alps, the lists by Landolt (1960) reveal both the faithful occurrence and the avoidance of carbonate rocks. First Landolt listed vicariant species pairs; the nineteen pairs of calcicole-calcifuge species include the two Alpenrose species, *Rhododendron hirsutum* (calcicole) and *R. ferrugineum* (calcifuge), early recorded as geoedaphic specialists by Unger in 1836. In the same handbook, Landolt also provided two long lists—one for limestone indicator species and one for lime-avoiding species. Certainly these three rosters reveal the major influence of limestone substrates on the Swiss Alpine flora.

Limestones of the Balkan Peninsula. As in the Alps, limestones abound in the Balkans. A most singular form of limestone topography, karst, has its type locality in the Balkans. Karst, a germanized version of the Slavic *kars,* is typified in the limestone plateau country north of Trieste. Karst topography (see Chapter 4) is created by the differential weathering of tilted limestone to create sinks, caves, ravines, and underground streams. Since karst is such a remarkable phenomenon worldwide, we need to come back to it in greater detail in a separate section following this account of limestone floras.

W. B. Turrill's classic floristic study *The Plant Life of the Balkan Peninsula,* though published in 1929, is still the most comprehensive account for that locality. Turrill's association with the region started during World War I, when he was attached to the British Salonika Forces in Greek Macedonia. Three later visits of long duration gave him a firsthand familiarity with other Balkan countries. He was fully aware of the significance of lithological diversity for the region's rich flora and varied vegetation types: "[T]he soil conditions of the Balkan Peninsula show a larger range than is to be found in any other area of smaller or equal extent in Europe, except perhaps in the British Isles. This is partly because of the many kinds of rocks which outcrop, but even more because of the very varied climatic conditions in different parts and the broken physiography." Of this lithological variety, limestones play a major part: "Calcareous rocks outcrop over an enormous area in the Balkan Peninsula, and their weathering and the resulting products vary according to local conditions" (p. 25). Since so much of the Balkan limestones show karst features (dolines, sinks, ravines, underground streams, etc.), their floristics and ecology will be discussed in the section on karst.

Vegetation accommodates to Balkan limestones in a variety of forms, depending on elevation and climate. Both deciduous and coniferous forests can occur on limestones. But the most characteristic vegetation types are what Turrill (1929) calls "brushwoods communities" (p. 144), reflecting the region's Mediterranean climate. Classical names for these shrub communities, like maquis (*macchie*), garigue, and *phrygana,* attest to their recognition since ancient times. Contemporary ecologists view these as Mediterranean forms of sclerophyll vegetation; several variants are named (e.g., twelve garigue types have been described in Polunin and Walters 1985). Several occur on limestone; examples include Kermes oak garigue in the western Mediterranean and rosemary garigue and sage garigue of the eastern Adriatic coast. The most ancient recognition of a garigue type, by Theophrastus, gave rise to the Greek name *phrygana* (also called eastern thorny garigue). The *phrygana* is characteristic of the coastal areas and low hills of Greece and Crete (Polunin and Walters 1985), where limestones predominate. This low spinescent vegetation type has an abundance of aromatic labiate, rosaceous, and leguminous shrubs: for example, *Thymus capitatus* and *Ballota acetabulosa* (Labiateae); *Sarcopoterium spinosum,* thorny burnet (Rosaceae); and such dwarf legume shrubs as *Anthyllis hermanniae, Astragalus massiliensis, Calicotome villosa,* and *Genista acanthoclada.* So, under the dual influences of calcareous substrate and Mediterranean climate, a microcosm of unique plant life has evolved. It must be said that some brushland communities are the outcome of human activity—timber removal and overgrazing, acted out over centuries of human occupancy. Such shrub communities are thus quite evidently successional but maintained as such by continued human intrusion. Hence, to attribute several of the garigue types on limestone just to the lithology is too simplistic.

All degrees of restriction to limestones can occur—from no restriction (indifferent, ubiquist, or bodenvag species) to narrow endemism. This is particularly true of Balkan limestones, where the Mediterranean climate and absence of glaciated areas offer ample opportunity for the parent material to directly affect the flora. Every Balkan country can boast long lists of endemic species, locally or regionally restricted to carbonate rocks. Good accounts of limestone flora can be found in Turrill (1929), Polunin and Walters (1985), and the fine general review of European limestone floras by Shimizu (1962–1963); see Table 5.11 for lists of some Balkan limestone endemics.

A remarkably rich flora on limestones has been lavishly portrayed by A. Strid (1980) in his *Wild Flowers of Mount Olympus.* His opening geological statement sets the lithological scene: "Almost the whole of Olympus is made up of limestone and marble, representing various formations." He goes on: "As the mountain consists chiefly of well-jointed limestone through which water readily percolates the area is dry, and the grey colours of the rocks and screes in the alpine zone add to the impression of desolation. Water is rarely to be found above 1100 m except for snow-melt water. Although the dry rocks and scree fields may not appear very inviting to a botanist, these are in fact the habitats where many of the rare and endemic

Table 5.11 Some limestone plants of the Balkan Peninsula (from Polunin and Walters 1985, Turrill 1929)

Taxon	*Distribution*	*Remarks*
FERNS		
Asplenium viride	Widespread calcicole	
CONIFERS		
Pinus heldreichii	Central and western Balkan Peninsula; central Italy; to timberline on Mount Olympus, Greece	Polunin and Walters (p. 18)
P. nigra subspecies	Some subspecies on limestone	
Abies borisii-regis	Mount Olympus, Greece	
A. cephalonica	Southern Greece	Polunin and Walters (p. 120)
MONOCOTYLEDONES		
Sesleria tenuifolia	Istria	Turrill (p. 170)
Gagea pusilla	Istria	Turrill (p. 170)
Veratrum nigrum	Istria	Turrill (p. 170)
Scilla bifolia	Mount Olympus, Greece	
DICOTYLEDONES		
Quercus species (e.g., *Q. cerris, Q. frainetto*)	Karst woods of Dalmatian coast	Polunin and Walters (p. 110)
Haberlea and *Ramonda* species	Balkan Peninsula	Relictual? *fide* Turrill (p. 170)
Silene otites	Istria	Turrill (p. 170)
Delphinium fissum	Istria	Turrill (p. 170)
Phillyrea latifolia	Istria	Turrill (p. 170)
Myrtus communis	Istria	Turrill (p. 170)
Digitalis laevigata	Istria	Turrill (p. 170)

Table 5.12. Some limestone endemics of Mount Olympus, Greece (from Strid 1980)

Taxon	*Distribution and habitat*	*Remarks*
MONTANE ZONES		
Genista sakelliardis	Conifer woodland	
Silene oligantha	Conifer woodland	
Centaurea grbavacenis	Conifer woodland	Also known from few sites in southern Yugoslavia
Janakea heldreichii	Mixed forest	This gesneriad is the best-known endemic of Mount Olympus
Allium heldreichii	Mixed forest	Very occasional elsewhere
Achillea grandifolia	Ravines of montane/subalpine	It and several other species are Balkan endemics (Strid, p. xxiv)
Carum adamovicii	Ravines of montane/subalpine	Olympus endemic
Asperula muscosa	Panzer pine woodland	Besides these two Olympus endemics in the *P. heldreichii* woodlands, there are several Balkan endemics here
Silene dionysii	Panzer pine woodland	
ALPINE ZONE		
Cerastium theophrastii	Alpine screes	On dryish fellfields, with other Olympus endemics
Potentilla deorum	Stabilized screes and cracks of frost-riven limestone	Widespread at summit area
Campanula oreadum	Rock crevices at summit area sunny site	Strid: "Although the highest peaks may appear to be exceedingly dry and barren, no less than some 55 species have been recorded above 2800 m" (p. xxviii)
Viola striis-notata	Mobile alpine screes	Scattered and never abundant

5.15 Limestone is the major rock type of Mount Olympus, Greece, and supports many endemics. View of the mountain from the north; limestone towers of summit area. Photo by A. Strid.

5.16 *Viola striis-notata*, a limestone endemic. Photo by A. Strid.

species are to be found." One encounters endemics along the ascent of Olympus beginning with the Coniferous Woodland and Forest (500–1500 m). But it is in the Alpine Habitat (Figs. 5.15 and 5.16) that endemism both regionally and locally is best displayed. Strid reckons that there are about 150 species in the alpine (above 2400 m), over half of which are Balkan Peninsula endemics. A dozen or so of these are endemic to Mount Olympus. Since the summit area of Olympus is all limestone, these local species are limestone endemics (see Table 5.12). The question arises: Is it the limestone parent material per se that elicits the local endemism, or is it the isolated position—the alpine "island"—of Olympus? I see no easy way to answer the question. If there were nearby a comparable isolated mountain, but of siliceous (acid igneous) bedrock, its proportion of endemics could be contrasted with those of Olympus.

Extensive limestone outcrops are to be found in Turkey (Figs. 5.17 and 5.18), ranging from sea level to the high peaks of the Taurus Mountains (e.g., the Bolkar and Aladag ranges). These Anatolian substrates foster a rich flora with many limestone endemics. During a botanical excursion to Turkey in 1998, colleagues and I observed that limestones and serpentines are intimately associated with outcrops side by side; they display marked contrasts in flora (Kruckeberg, Adiguzel, and Reeves 1999).

5.17 Limestone slopes near Seydisehir (Konya Province), Turkey. The pyramidal shape is unique to the *Juniperus* species here. Photo by A. Guner.

5.18 *Muscari muscarimi* endemic to limestone of southwestern Turkey. Photo by A. Guner.

Asian Limestones, Especially in Japan and Taiwan

Temperate Asia is another haven for limestone floras. For Japan and Taiwan, we have a thorough record of carbonate-inhabiting plant life. Shimizu's studies (1962, 1963) of Japan and Taiwan limestones focus mainly on the floristics and biogeography of the plants. But he also provides a comprehensive review of the limestone plant literature, not only for Japan and Taiwan but for other temperate regions. Limestones in Japan range right down the archipelago from Hokkaido (Oshima Prefecture) to Kyushu. The frequency and areal extent of limestones in Japan are highest from central Honshu to the southern islands, Shikoku and Kyushu. The crystalline limestones of these outcrops are old, belonging to the Permo-Carboniferous system, called the Chichibu Paleozoic group. In Taiwan, limestones outcrop massively and extensively between the Pacific Ocean side and the central mountains of Taiwan; these limestones appear as precipitous cliffs running north to south, east of the island's mountainous spine. While some limestones outcrop at low elevation, Shimizu's (1963) survey concentrates on montane to alpine areas. He groups the limestone flora by degrees of fidelity: Group I, "exclusives," to Group V, "strangers" (accidentals). The first three groups have an abundance of taxa of varying fidelity to limestone. Thus in Group I (exclusives, or strict limestone endemics), 43 taxa are Japanese and 35 are from Taiwan. Only four in Group I are common to more than one or a few limestone outcrops. Members of Group II, though widely distributed, are rare and discontinuous on limestone cliffs. There are about 40 taxa (species and varieties) in this group. Group III contains those taxa that are common to limestone but can occur on other substrates. These come close to matching the indifferent or bodenvag grouping; over 125 taxa belong to Group III. Shimizu names the specialists for each of these groups. Table 5.13 extracts from the Shimizu lists some of the more noteworthy calcicoles.

It is not surprising that Shimizu sees both similarities and affinities between plants of limestones and plants of serpentine. Both substrates foster unique floras and vegetation; both elicit all degrees of fidelity, from narrow endemism and regional endemics to bodenvag taxa. While no narrow endemics of limestone are found on serpentine (Shimizu's Group I species), closely related taxa occur on either limestone or serpentine. Thus var. *iwatense* of *Thalictrum foetidum* is a limestone endemic while var. *apoiense* is restricted to serpentine. Two Japanese edelweiss taxa show the same affinity: *Leontopodium hyachinense* is on serpentine and *L. hyachinense* var. *miyabeanum* is on limestone. Certain plants of Group II, taxa faithful to limestones on a regional scope, are also relictual on serpentine. Shimizu gives five examples: *Gymnocarpium jessoense, Juniperus sargentii, Potentilla fruticosa, Spiraea nervosa,* and *S. nipponica.* Several Group II taxa also show this dual proclivity for both

Table 5.13. Notable limestone plants of Japan and Taiwan (from Shimizu 1963)

Taxon	*Habitat and distribution*	*Remarks*
FERNS		
Asplenium ruta-muraria	Limestone crevices of Japan (Hokkaido to Kyushu) and Taiwan	On limestones and other base-rich rocks in Europe and North America
Gymnocarpium jessoense	Conifer woods, Hokkaido, Honshu, and Shikoku	Also on serpentine (= *G. robertianum* var. *longulum* in Ohwi 1984)
Polystichum deltodon	Shaded limestone crevices, Honshu, Shikoku, and Kyushu	Also in western China (on limestone?)
GYMNOSPERMS		
Juniperus formosana	Small "alpine" tree of central mountains of Taiwan	
J. sargentii	Procumbent in "alpine" from Hokkaido to Taiwan	Also on serpentine
DICOT. ANGIOSPERMS		
Betula chichibuensis	Local on limestone cliffs and rubble, central Honshu (Saitama Prefecture)	
Carpinus turczaninovii	Shrub, on sunny limestone ridges, western Honshu	
Cerastium calcicola (and *C. kaoi*)	Limestones of Taiwan	
Clematis williamsii	Honshu, Shikoku, and Kyushu	Three other species are also calcicoles
Thalictrum foetidum var. *iwatense*	Local in crevies of sunny limestone bluffs, Iwate Prefecture, Honshu	*T. foetidum* var. *apoiense* restricted to serpentine; *T. urbanii* var. *majus* on limestones of Japan
Berberis chingshuiensis	Local at summit of Mount Chingshui, Taiwan	
Corylopsis matsudai	Limestones of Mount Tencho, Taiwan	*C. spicata* restricted to serpentine on Shikoku
Eriobotrya japonica	Limestone hills, Honshu, Shikoku, and Kyushu	On siliceous rocks only on two islands; also in China and Burma (substrate?)
Raphiolepis impressivena	Another endemic of Mount Chingshui, Taiwan; sunny limestone ridges	
Rhodotypos scandens	Limestone bluffs, Okayama Prefecture, Honshu	
Buxus microphylla var. *insularia*	Mountains of Honshu and Kyushu	Mount Kosho (Fukuoka Prefecture), famous for its boxwood trees
Rhamnus yoshinoi	Sunny cliffs or ridges of limestone, Honshu, Shikoku, Kyushu	Rare in Japan
Viburnum propinquum	Sunny limestone ridges, Taiwan	Also in China (substrate?)
Leontopodium hyachinense var. *miyabeanum*	Cliffs and rubble of limestone, Hokkaido	var. *hyachinense* endemic to serpentine; var. *miyabeanum* known only from the type locality. *L. japonicum* var. *perniveum* also on limestone, Honshu
MONOCOT. ANGIOSPERMS		
Lilium bukosanese	Limestone cliffs, Mount Buko, Honshu	It and *L. maculatum* are limestone species
Tofieldia coccinea	3 vars., all on limestone, Honshu and Shikoku	
Tricyrtis macrantha	Limestone crevices, Shikoku	N.B.: Shimizu opines that liliaceous plants are rare on limestones; the same may be true for grasses and sedges

limestone and serpentine; for example, *Berberis amurense* vars. and *Buxus microphylla* var. *japonica* can be found on both substrates, from Hokkaido to southern Honshu.

The spatial proximity of limestones to ultramafic areas is more than coincidental. Both rock types are near one another in ophiolite suites (see Glossary). Lowermost is the ultramafic (mantle) rock, with the oceanic sediments, often limestone, appearing near the top of the sequence (Coleman and Jove 1992; Malpas 1992). Associations of limestones with ultramafic rocks are known from many places; for example, Cuba, southeastern Europe, Turkey, Japan, and southeastern United States.

This shared affinity for the two base-rich substrates is likely to be reflected in similar ecological effects. At least two attributes held in common may contribute to the causes of the similarities. First is the highly selective nature of the two substrates, one rich in calcium, the other rich in magnesium, while deficient in varying degrees in other mineral nutrients. Serpentines are probably more demanding as an agent of natural selection; ultramafic substrates require an adaptive syndrome that can cope with high magnesium, high levels of nickel, and an overall paucity of other essential elements. A second commonality is the discontinuous occurrence of the two substrates; limestones and serpentines are usually insular in distribution—mainland edaphic islands in a "sea" of normal substrates. Geoedaphic insularity as an isolating factor for speciation has been reviewed recently (Kruckeberg 1991b) and will be the subject of discussion in Chapter 7.

KARST TERRAINS

Currently the term *karst* embraces a whole series of geomorphic and weathering phenomena, all associated with carbonate rocks, especially limestone. Because karst landforms are so widespread, both in temperate and tropical climates, and because the remarkable karst lithology may support particular vegetations and flora, it merits special attention. Karst as landforms has been dealt with in Chapter 4; here I examine lithology and landform as they jointly influence plant life.

Viles (1988) defines karst as "terrain produced mainly by solutional erosion, which usually produces an important underground drainage system." Other definitions specify limestone as the major substrate: "Karst [develops in areas of] weathered limestone, distinguished by ridges, clefts and caves, and formed as a result of erosion by water" (Polunin and Walters, 1985, pp. 220). The definition in Skinner and Porter (1992) emphasizes the topographic attributes of karst: "Karst topography [is] an assemblage of topographic forms resulting from dissolution of carbonate bedrock and consisting primarily of closely spaced sinks" (p. 551). All these definitions barely hint at the striking, often spectacular, landforms that greet the eye whether in temperate or tropical lands.

Table 5.14. A classification of karst forms according to size and situation (from Viles 1988)

Situation	*Maximum dimension*	*Examples*
Subaerial surfaces	<10 m 10–1,000 m	Karren, limestone pavements, tufas
Soil-covered surfaces	<10 m 10–1,000 m	Subsoil karren, closed depressions, residual hills, cones, and towers
Subsurface	<10 m	Scallops, small speleothems
	>10 m	Cave passages and galleries

Viles (1988) points out that karst phenomena and the word's conceptual roots have outgrown the narrower origins as in the Dinaric karst of northwestern and coastal Yugoslavia. Karst is a germanized Slavic name originally applied to "a large area of limestone with distinctive surficial and underground geomorphology, found in Yugoslavia" (p. 319). The term now encompasses a multiplicity of phenomena, from landforms of a great variety and nomenclatural richness to processes of karst formation, both biotic ("phytokarst") and abiotic.

Geomorphologists have had a special attraction to karst. They not only have been adept at naming many different karst types but also have attempted to explain its causes. Some of the vast geomorphologic literature on karst is provided at the end of Viles's review (1988); her classification of karst is reproduced here (Table 5.14), where the attributes of bare rock, soil-covered surface, and underground area form the basis of her classification. Process is more complex and contentious. Most geomorphologists have invoked the physical causation, a hydrologic weathering of carbonate rock to form karst. But Viles reminds us of the other potent formative processes, all biogenic. What she calls biokarst (called phytokarst by earlier authors) forms under the biologically induced weathering by such organisms as algae (especially blue-greens) and lichens. It is not clear from Viles's review if higher organisms, especially vascular plants, are also agents of biokarst production.

It is appropriate to ask here: Is karst the result of geologic, climatic, or organic processes? Or is it the result of an interplay of all three? Then, if its origin is more than geological, is it beyond the scope of this book? Indeed karst is a product of all three natural forces—geologic, climatic, and organic. Thus it aptly exemplifies the interconnected processes that make landforms, taking us back to the holocoenotic view of landscape-making. An excursion into the interconnectedness of karst formation reveals: (1) limestones from which karst is formed are largely organic in origin, the deposits of plants and animals that produce molecules of calcium (or calcium-magnesium) carbonate; (2) once deposited, the organic origins give way to inorganic change—the geological processes of lithification, metamorphism, and tectonic deformation of limestone beds; and (3) the next phase of the geomorphic cycle is the creation out of limestone beds of myriad karst forms; the prevailing climate and organisms then assume a dominant creative role. Karst is, indeed, a holocoenotic phenomenon!

Geography and Formation of Karst

Karst topography is likely to occur anywhere that highly soluble rocks are exposed to weathering. Although limestone rocks are those most commonly converted to karst topography, the same landforms can develop on deposits of dolomite, gypsum, and salt (Skinner and Porter 1992). Since karst terrain can develop in both temperate and tropical environments, it is not surprising that karst is found on every continent, almost from pole to pole. Table 5.15 gives a global summary of major karst occurrences—the most well-known are the Dinaric karsts of Yugoslavia, the sinkhole karsts of Kentucky and Tennessee, and the spectacular tower karsts of southern China.

Students of karst landforms and their causes (e.g., Jennings 1985) deal with a paradox in geomorphology. A single process—solution—acts mainly on one kind of lithology, carbonate rock. Yet many landforms result. The vast, often local nomenclature of karst types is a symptom of its varied manifestations. Given the single process, the simple explanation of karst's diversity should rest with climate. High temperatures and precipitation of the tropics should produce a karst of greater topographic contrast (e.g., tower karst with its steep limestone hills arising above adjacent flat plains). Hence temperate or even high arctic karst under regimes of lower temperatures and rainfall should yield a more subdued topography (e.g., "doline karst where depressions of various kinds are sunk below the general level of the limestone surface"; Ollier 1984, p. 206). But the simple variable of climate does not explain the diversity. Jennings (1985) critically evaluates the simplistic explanation and finds it wanting hard data, especially on solution (erosion) rates. The long-standing controversy over whether erosion is faster in the tropics or in temperate regions is not yet resolved, according to Jennings. Other factors besides solution rates complicate the picture—geotectonic movements, glaciation, rapid versus slow runoff, and so forth. But the fact remains that tropical karsts form the grandest, most spectacular landforms of all—tower karst, cockpit terrain, and others. Since here we are emphasizing the botanical consequences of karst, we must leave unresolved the conflicting scenarios for its formation.

Table 5.15. Notable occurrences of karst, worldwide

Region	*Type of karst*	*Vegetation/Flora*	*Remarks and references*
Yugoslavia, Adriatic, and adjacent Balkan areas	Dinaric karst Dolines	Temperate (Mediterranean climate) flora; some limestone endemics	Herak and Stringfield (1972); "kars" (-karst) originated here
Italy: northwest of Trieste; Apulia on the Adriatic	Dolines, lapiés (karren), debris plains, poljes, blind valleys	Temperate (Mediterranean climate) flora; some limestone endemics	Belloni et al. (1972); Poldini (1989)
France: Paris Basin, Massif Central; Massif Armoricain	Dolines, caves, gorges, and steep valleys	Temperate (Mediterranean climate) flora; some limestone endemics	Avias (1972)
Germany, Austria, Hungary, Czechoslovakia, Poland, and Russia with major karsts	Gorges; solution corridors, and sculpturing (karren)	Temperate floras; some endemics and indicators	Herak and Stringfield (1972); White (1988)
UK: Ireland, England	Pavement karst; gorges, dry valleys, dolines	Temperate floras; some endemics and indicators	Etched limestone at The Burren (Ireland); Jennings (1985)
North America: Appalachians, Kentucky Highlands, Ozarks, Texas	Sinkholes, dolines, cutters, caves	Temperate floras; some endemics and indicators	Davies and LeGrand (1972); White (1988); Jennings (1985)
Mesoamerica: Yucatan, Mayan Guatemala	Cenotes (vertical karst), dolines, caves	Subtropical or tropical flora	White (1988)
Caribbean (Antilles): Jamaica, Cuba, Puerto Rico	Mogotes (Cuba), cockpits (Jamaica); dogtooth karst, cone karst	Tropical; many endemics in Cuba and Jamaica	Cuba (Borhidi 1991); Jamaica (Proctor 1986)
South America	Limited karst (e.g., Venezuela)	—	Jennings (1985)
Asia: China, Japan (?)	Guilin tower karst (Kwangsi); many other karst types	Subtropical to temperate floras	China has largest continuous karst lands in world (Daoxian 1993)
Australia: Nullarbor Plain, Cooleman Plain, Kimberley	Gorges, karst valleys, caves, etc.	Arid to mesic vegetation	Jennings (1985)
New Zealand	Doline fields; gorges, caves	Warm temperate	Jennings (1985)
Africa: North Africa	Dayas (small depressions) in Algeria; caves and ponds; limestone fynbos (South Africa)	Mostly arid (xeric) Fynbos endemics	Jennings (1985) Heydenrych et al. (1994)

But before we embark on an exposition of the plant life on karst, a special kind of karst formation yielding biokarst or phytokarst should be mentioned. Given the ease with which limestone weathers to make karst by purely hydrologic (climatic) means, it is not surprising to find that substantial karst formation can be caused by organisms. Viles (1988), in reviewing the evidence for biokarst (or phytokarst) genesis, emphasizes the role of microorganisms (bacteria, blue-green algae, and fungi) as the major creators of biokarst. But higher plants can also form what she calls "root karst"; examples are known mainly from the tropics (Yucatan Peninsula and West Malaysia). Even animals have been identified as agents of limestone abrasion. Both marine and terrestrial limestones can be significantly abraded or etched by animals, especially invertebrates (snails, crabs, echinoderms) and even tortoises (Viles 1988; Butler 1995).

Plant Life on Karst

If, as we have just seen, plants can make karst, then would not karst serve as a substrate for plant life? At the very least, we know that karst formations exposed at the Earth's surface support vegetation (Turrill 1929; Crowther 1987; Borhidi 1991; Howard and Briggs 1953; Howard and Proctor 1957). But does this karst vegetation have any unique attributes? Are there species endemic or at least largely restricted to karst? And at the community level, are there karstic plant associations? Unlike nonkarstic limestone substrates for which a wealth of geoedaphic information exists, there appears to be little on the existence of special floristic and vegetational attributes of plant life on karst. It seems that karst had been the province of the geomorphologist whose primary concerns are with the physical and historical attributes of karst. Little as it is, we now review our knowledge of karst floras and vegetation, first for temperate and then for tropical regions.

The classic (type) locality of temperate karst in the Istrian to Montenegran portions of the Balkan Peninsula has prompted only a modest record of its plant life. Turrill (1929, p. 165) gives a short yet intriguing account of the vegetation and flora on Istrian karst: "Where forest and brushwood have been destroyed the open type of stony or rocky ground vegetation predominates. It is often impossible to draw any sharp line between the vegetation of stony ground and that growing in clefts and joints of the exposed limestone rock. The herbaceous vegetation is most luxuriant in spring and early summer, but as a whole even in summer is much less xeromorphic than the 'Felsentriften' (rock pastures) of Dalmatia." Turrill lists a number of characteristic genera for the Istrian karst—for example, many Labiatae (*Thymus, Marrubium, Teucrium, Satureia*), many bulbous and other geophytes

(*Asphodelus, Asphodeline, Anthericum,* and *Allium*), as well as many other herbaceous perennial dicots. Turrill goes on to comment:

> In Istria and Montenegro . . . exposed rock [of true karst] occurs on hill and mountain slopes and escarpments, forming the sides of some dolines, and as flat limestone pavement. . . . [On the rocky peak of Mount Maggiore, 1396 m], the beech forest reaches nearly to the top, the trees dwarfing considerably. The peak is of nearly bare limestone, but the following plants may be found in the clefts: *Cerastium arvense* var., *Silene saxifraga, Arabis alpina, A. scopoliana, Kernera saxatilis, Saxifraga aizoon, Globularia bellidifolia, Micromeria thymifolia, Senecio nebrodensis, S. abrotanifolius, Galium purpureum, Scrophularia laciniata, Rosa pendulina* and *Sesleria tenuifolia.* The dolines so characteristic of the karst are of many sizes, shapes, and degree of steepness. They vary consequently as to the nature of the sides and vegetation clothing them. When the sides are steep and rocky, trees and shrubs of the karst wood and brushwood associations may occur in the clefts. Thus *Quercus lanuginosa, Juniperus communis, Corylus avellana, Carpinus orientalis, Cornus sanguinea* and *Crataegus monogyna* are some of the trees . . . on the sides of rocky dolines. Numerous plants of the karst heath occur where the woody vegetation is absent. Expanses of limestone pavement, often with widened and deepened joints, are very frequent in the karst country. Low bushes of *Juniperus communis, Cotinus coggygria,* and *Rhamnus* spp., with *Sedum* spp., *Tunica saxifraga, Cyclamen europaeum, Andropogon ischaemum, Anthericum liliago* and *Satureia* sp. occupy the cracks where earth has accumulated.

Turrill does not identify endemics to the Istrian karst, but does remark that many of the species occur in other stony habitats and that the number of endemics is less than in similar habitats farther south. Poldini's (1989) detailed study of karst vegetation in the Trieste region mentions several species locally or regionally restricted to the karst. Most occur as karst endemics both in the Trieste area and southward. "Only *Centaurea kartschiana* is authentically endemic to the karst of Trieste," he further notes. "All other [karst endemics] are 'subendemic' because they also grow in former Yugoslavia. Among these I would also include *Onosma javorkae,* which grows exclusively on the Dalmatian coast but also reaches as far as Trieste, as do *Senecio lanatus* and *Euphorbia wulfenii*" (Poldini, pers. comm., 1994).

Karstification of limestones in the tropics progresses from limestone pavement to dogtooth and cockpit karst, thence to the spectacular cone and tower karsts vividly pictured in accounts of Southeast Asian and Antillean limestones. A thorough study of Cuban cone and tower karst (locally known as "mogotes") reveals a rich and highly endemic flora (Borhidi 1991); see Figures 5.19 to 5.21. The mogotes of the Sierra de Nipe in eastern Cuba (the Nipe-Baracoa subprovince) support about 15 endemic species (Borhidi, p. 366), such as *Thrinax compacta* (a palm), *Tabebuia mogotensis,* and *Gesneria lopezii* in a steep-sloped karstic woodland. On Monte Libano, karstic (mogote) limestone interfaces with serpentine to serve as a meeting point between the floras of the two substrates. At this lithologic contact, some floristic overlap occurs; thus two serpentinicolous species (*Pinus cubensis* and *Agave shaferi*) can occur on the mogote limestone (Borhidi 1991, p. 368). Given the singular topography and carbonate substrate of tower and cone karst elsewhere in the tropics (Malaysia, China, Puerto Rico, Mesoamerica, etc.), one would anticipate characteristic vegetations and exceptional, even endemic, floras. Yet there seems to be a paucity of floristic studies of these fascinating landforms. Borhidi's *Phytogeography and Vegetation Ecology of Cuba* (1991) is the best work I have encountered that deals with plants on tropical cone and tower karst.

Another type of limestone landform occurs in the tropics. Less extreme in geomorphic expression than cone or tower karst is the sharply serrated limestone surfaces (dogtooth karst) or the sinkhole-doline topography locally called cockpit terrain in the Antilles. Dogtooth (*dientes de perro*) limestone supports a xeric thorn woodland in Cuba (Howard and Briggs 1953) and on other Antillean islands (Howard 1955). Woody species predominate in these well-drained, dryish habitats—a tree-shrub scrub, usually of low and dense stature, and often stocked with species bearing thorns or spines. Cockpit terrain on limestone is well known in Jamaica (Figs. 5.22 and 23); its unique vegetation is described by Asprey and Robbins (1953): "The Cockpit country is so-called because the whole area is made up of circular depressions (dolinas) up to 500 ft. deep filled by bauxitic soils with accumulated humus from the surrounding rim of limestone rock. The area is typical karst country where underground drainage, subterranean rivers, sinkholes, and caves are common. Many of the depressions are the result of sinkholes from underground streams" (p. 348). Bare limestone rock on the slopes of the cockpits (dolines) supports xeric to mesic forest depending on exposure and aspect; the floors of the cockpits are of a different forest composition, growing on well-developed bauxitic soils (Asprey and Robbins 1953; Howard and Proctor 1957).

The "cockpit" karst flora of Jamaica is rich in endemics. Proctor lists 105 endemic species. "Of these, one is a fern (*Thelypteris trelawniensis* Proctor), 14 are monocotyledons and 90 are dicotyledons. Five species are not endemic, but in Jamaica occur only in the cockpit country. . . . Of the dicotyledons, 12 are herbs, 4 are herbaceous vines, 1 is a subwoody vine, 4 are woody scramblers, 39 are shrubs, 15 are shrubs or small trees and 15 are trees. Of the trees, 2 are large and valuable for

5.19 Elephant-shaped mount ("mogote"); tropical karst, Sierra de los Organos, Cuba. Photo by A. Borhidi.

5.20 Limestone gorge of the Abra Mariana, Cuba. *Agave albescens* is dominant here. Photo by A. Borhidi.

5.21 Water-accumulating bottle-shaped trunk of *Bombacopsis cubensis* on limestone slopes of "mogotes." Photo by A. Borhidi.

5.22 Aerial view of cockpit karst in Jamaica. Fault lines have opened up some of the karst depressions; these openings are often cultivated. Photo by J. Tyndall-Biscoe.

5.23 Forest on cockpit karst in Jamaica, with numerous endemic woody and herbaceous plants, listed by Proctor 1986. Photo by J. Tyndall-Biscoe.

their timber. These two trees are *Manilkara excisa* and *Terminalia arbuscula*" (Proctor 1986, App. II, p. 143). This listing, in an appendix to the symposium volume *Forests of Jamaica* (Thompson, Bretting, and Humphreys et al. 1986), has the species by family; well represented are species in the Bromeliaceae (4), Orchidaceae (7), Euphorbiaceae (6), Myrtaceae (6), Rubiaceae (12), and Compositae (9). Proctor's paper is the only one I have found that tabulates in detail karst endemics for tropical regions.

Another Antillean habitat with karstified limestone merits mention. The flora of the Cayman Islands has been exhaustively surveyed by Proctor (1984). All of the islands between Cuba and Jamaica are limestone, from recent coral to a massive core of Tertiary limestone. Some of the upland terrain of Grand Cayman Island is karstic: "deeply dissected Tertiary limestone" (Proctor 1984). It supports a dry evergreen woodland, often degraded by wood gathering and agriculture to a dry evergreen thicket. Twenty-one species and varieties are endemic to the Cayman Islands, all of which must be calcicoles. But it is not certain which ones are restricted to karstic topography. Furthermore, some of the islands' endemism may be due to insular isolation (*sensu* "island biogeography") rather than to close tracking of edaphic preferences. The natural history and biogeography of the Cayman Islands are treated in a recent work edited by Brunt and Davies (1994); see especially the chapters by Proctor and Brunt for accounts of the Cayman floristics and vegetation.

OTHER SEDIMENTARY ROCKS AND THEIR PLANT LIFE

Hydraulic reworking of the regolith and the subsequent lithification of clastic sediments make a diverse array of sedimentary rocks. All are variants of the four types based on clastic (particle) size: conglomerates from boulders, cobble, and pebbles; sandstones from sand; siltstones from silt; and shale from clay (Skinner and Porter 1992).

Shales, as substrates for plant life, offer some intriguing insights into the interplay of rocks and plants. Shales, or their metamorphic equivalents (phyllites, slates, or schists) are composed of the finest grade of clastics, nearly pure clay, compacted and indurated by pressure to form a sedimentary rock. Shale's chemical composition, roughly 47 percent silica (SiO_3), 40 percent aluminum (Al_2O_3), and 13 percent water (Hunt 1972), offers little in the way of essential nutrients for plants. In the process of becoming sediment, the original clay may lose most of its adsorbed exchangeable cations, which can be replaced by hydrogen ions. In a word, shales are often sterile. Further, their weatherability to make soil can vary, depending on the dip of the strata. In horizontal shales, water penetrates with difficulty; but in

5.24 The shale barrens of Virginia support a unique scrub vegetation and some rare species, in the midst of deciduous forest. Shale barren slope above Cowpasture River, Alleghany County, Virginia. Photo by J. and C. Baskin.

steeply inclined shale beds, water penetrates with ease. Shale erodes more readily than other rocks, mostly because its clay matrix is soft and readily weathered. Also, as horizontally bedded shales are impermeable, runoff is greater; thus its weathered products are readily removed. Another distinctive feature of shales is their fissility—the tendency for the rock to split into thin flat shards. Shales are worldwide in distribution from temperate lands to the tropics. The shale barrens of eastern United States are prime examples of the shale–plant life connection, telling a fascinating geoedaphic story.

Platt's classic study (1951) of the mid-Appalachian shale barrens links rock and vegetation in a telling ecological context. Shales, outcropping in a southwesterly direction from southern Pennsylvania to western Virginia, are ancient. Devonian in age, they have been repeatedly peneplaned to expose the long and narrow outcrops of Barallier shale, interspersed among ridges of hard sandstone. High resistance to weathering and uniformity of composition characterize the striking and unique surface displays of Barallier shale (Fig. 5.24).

The fissile, highly siliceous Barallier shales are associated with three vegetation types: north slopes with a good vegetation cover of trees, shrubs, and herbs; talus slopes at the base of the barrens with a richly wooded cover; and the upper slopes and crests dominated by the shale barrens. Platt's soil analyses revealed significant differences between the vegetation of north or lower slopes and the barren. The former have higher base exchange capacities and higher cation values; the levels for the shale barrens are lower. The pH values were roughly the same for all three habitat types. But it is the physical attributes of the barrens, not their nutrient status, that impart character to the habitat. In contrast to vegetated slopes, barrens have a thin mantle of shale rock fragments at the A soil horizon. The layer of rock fragments is precariously stable: undisturbed it stays in place, but it is easily mobile if disturbed by man or beast. The steeper the slope the more barren its appearance. Moisture and temperature are also key actors in the shale barren drama. High insolation and consequent rapid drying of the surface layer of the substrate were judged by Platt as prime conditioners of the barrens habitat. In terms of life-form diversity, *sensu* Raunkiaer (1934), geophytes and hemicryptophytes (all herbaceous perennials) dominate, with 73 species. Only 5 species of annuals (therophytes) grow on the more favorable sites of the barrens; and woody plants (phanerophytes: trees and shrubs) numbered 21 species, but were markedly stunted on the barrens. A 300-year-old juniper (*Juniperus virginiana*) was only 15 feet high, and a 108-year-old oak only 6 feet tall.

The herbaceous perennial flora of the barrens is of the greatest taxonomic and phytogeographic interest: 19 are indicator species, and of these, 8 are strict endemics. Platt's view of the nature of this narrow endemism merits quoting: "The unique environment is the result not only of combinations of factors, but also of rhythmic sequences of the combinations which vary from point to point in intensity. Factors of substratum, exposure and slope are responsible for an edaphic situation in which the C horizon, adequate for continuous plant growth, is overlain by a layer of rock fragments which becomes drastically less favorable as the growing season progresses. These factors, coordinated with growth processes, are responsible for the sparse vegetational cover with its accompanied reduction in competition" (Platt 1951, p. 295). This stressful environment leads to an evolutionary response—endemism: "The real causes of endemism are resolved in the interaction between environmental factors and genetically determined physiological processes of the individual and the population. The endemics, in their evolution have become obligate to high light intensity, to a soil adequate for their extensive root systems and a low level of competition. They are restricted to the shale barrens because no other habitat in the general region possesses this unique combination of soil and light conditions which permits them, but does not permit stifling competitors to grow" (Platt, p. 299).

Not only have the shale barrens of the mid-Appalachian region been a cradle for evolutionary diversification and edaphic adaptation, but Wood (1971) remarks on the vicariant and distinct nature of some of the shale barren endemics, linking them with their nearest relatives in western United States. Thus the endemics *Eriogonum allenii* and *Senecio antennariifolius* have close congeners in the West.

Platt comes down firmly on the side of the physical environment of shale barrens as the prime cause of plant restriction. He found no strong indications that exceptional soil nutrient levels promoted physiological adaptation. But Cannon (1971), in her review of plants as indicators of mineral deposits, suggested that several of the genera Platt listed as shale barren indicators are also know accumulators of sulfides (e.g., *Allium, Arabis, Oenothera, Eriogonum,* and *Senecio*). Cannon proposed that "possibly high sulfur in the shales is another restrictive factor." The sulfur content of the Barallier shale, or its derived soil, needs to be determined. See Keener (1983) for the most recent review of the endemic flora on the Appalachian shale barrens. It is very likely that similar "azonal" floristics and vegetation occur on other shale barren exposures elsewhere in North America and beyond.

Floras of sandstone areas, unlike shale barrens, are not particularly unique; their soils are usually zonal, reflecting regional climate. Only when the sandstone is sculpted into unique landforms (Fig. 5.25) does the flora take on special, often endemic character. It is then that the factor of isolation (e.g., the Inselberge phenomenon—islands in mainland terrain), not lithology per se, becomes the critical biologic arbiter. Notable are the floras of sandstone Inselbergs, like the Grand Canyon of western North America and the tepuis of Venezuela (see Chapter 4). In the latter case, two lithologies meet in the tepui terrain; the massive sandstone is intruded by the igneous rock, diabase. Huber (1992) notes that the two substrates differ in the nutritional status of their derived soils: The sandstones with much quartz are low in inorganic nutrients; their soils are essentially organic. The soils over the diabase have higher levels of inorganic nutrients. I could find no mention of any floristic differences between the two soil types.

GYPSOPHILY: PLANTS RESTRICTED TO GYPSUM DEPOSITS

In contrast to sedimentary rocks composed of clastic material, some sedimentary rocks are formed by lithification of chemical precipitates. Chert (a hard, fine-grained silica rock), iron and phosphate deposits, and a variety of evaporites are sedimentary rocks of this nature. One of the evaporite rocks, gypsum, is widely known in geobotany for its distinctive indicator flora. Gypsum ($CaSO_4 \cdot 2H_2O$) precipitates from saline waters often to form vast beds of nearly pure gypsum. Gypsum occurs widely in areas where marine or other saline waters evaporated to create the deposits. While gypsum has been found in Great Britain and the European

5.25 Sandstone barren in glade, Davis Rock, Saint Clair County, Missouri. Photo by J. and C. Baskin.

continent, the deposits in xeric areas of the American Southwest and adjacent Mexico are especially noted for their unique floras (Cannon 1971; Johnston 1941; Parsons 1976; Turner 1973; Turner and Powell 1979; Waterfall 1946). Like so many other azonal substrates, gypsum has a generous share of indicator plant species and several narrow endemics (Figs. 5.26 and 5.27). Waterfall (1946) and Johnston (1941) list many obligate gypsophiles, some widespread in southwestern Texas, New Mexico, and Mexico, while others have more restricted ranges. Species of the genus *Coldenia* (Boraginaceae) display an intriguing fidelity to gypseous substrates. *Coldenia hispidissima* is confined to gypsum, while *C. canescens* is on limestone. The substrates and their respective species are often intimately associated, but fidelity to limestone or to gypsum is high (Waterfall 1946). Another species pair in the ocotillo genus, *Fouquieria,* shows a similar fidelity; *F. shrevei* is restricted to gypsum while *F. splendens* is widespread on nongypseous soils (Daubenmire 1947). Some of these authors speak of a gypseous soil, but none of them elaborate on its particular properties *as a soil*. One wonders if gypsum rock actually weathers to produce even a skeletal (azonal) soil. Or do the plants take root in unaltered gypsum? B. L. Turner (pers. comm.) supports the latter view.

5.26 The plant response to gypsum (hydrated $CaSO_4$), called gypsophily, manifests itself as unique plant communities and gypsophilous endemics. Pictured here are gypsum dunes at Cuatro Cienagas, Coahuila, Mexico. This site is rich in plant and animal endemics. Photo by B. L. Turner.

5.27 Gypsum deposits may form karstic landforms. Pictured here is a shallow doline on gypsum near Vaughn, New Mexico. Photo by W. B. White.

Metamorphic Rocks and Plant Life

The birth of new rocks from old ones spells special fascination for geologists and geobotanists. Transformation of parent rock into metamorphic progeny rocks involves textural and mineral changes wholly within the solid state of the rock. Particular combinations of pressure, temperature, and chemical environments transform a granite into gneiss, a basalt into greenschist, a limestone into marble, or a shale into phyllite, slate, or schist (Tables 5.16 to 5.18). The transformation of one rock into another, *without meltdown,* is often associated with plate tectonics. As plates move, crustal fragments collide, and rocks get squeezed, so that the physical and often the mineral makeup of the rock is changed (see chapter 5 of Skinner and Porter 1992, or Tyrrell 1929, for good discussions of metamorphic rocks).

Changes in textural and mineralogical properties progress according to intensities of temperature and pressure (low to high grade metamorphism). The most prevalent textural change is orientation of the minerals perpendicular to the direction of stress. The parallel orientation of minerals, called foliation, results in rocks that split into leaflike flakes. This new textural state has consequences for plant life, since weathering of metamorphic rocks to form soils may be more rapid and complete than for the original igneous rock. Thus gneiss weathers more readily than its igneous parent, granite. Types of textural changes induced by foliation include slatey cleavage (in fine-grained rock), schistocity (in coarse-grained rocks), and gneiss (high grade metamorphism to yield a coarse-grained rock with well-defined foliation). Minerals are also transformed during metamorphism. And along the continuum from low to high grade metamorphism, there can be a sequence of mineral species. Some of these new minerals are rarely, if at all, found in igneous or sedimentary rocks. Examples of the new minerals are kyanite, sillimanite, and andalusite. While the elemental, chemical composition of the minerals may not change during metamorphism, the new minerals may have different properties (rates of weathering, clay-forming properties, etc.).

In contrast to the persistence of chemical species during metamorphism, chemical composition of minerals can change under conditions producing metasomatism. In metamorphism, small water-to-rock ratios prevail (e.g., 1:10); thus fluid volume is not sufficient to dissolve the rock or to change its composition. But if large water-to-rock ratios prevail (10:1 or even 100:1), the fluid can alter the mineral composition of the rock by a gain or loss of ions. Thus a limestone may not change into marble if the rock is bathed with fluid; rather, it becomes an assemblage of garnet, diopside (green pyroxene), and calcite. When the fluids causing

Table 5.16. A classification of metamorphic rocks (adapted from Simpson 1966)

Type of metamorphism	*Dominant agent*	*Geological situation in which most likely to occur*	*Nature of process*	*Typical minerals formed*	*Characteristic Rocks (by texture)*
Thermal, including contact	High or moderate temperature +Igneous emanations (liquids and gases)	Close to igneous intrusions and extrusions	Recrystallization Replacement from invading liquids and gases	Andalusite Al_2SiO_5 Grossularite (Garnet) $Ca_3Al_2Si_3O_{12}$ Wollastanite Ca SiO_3	Nonfoliates Hornfels Quartzite Marble
Cataclastic	Directed pressure at low temperature	Belts of folding crush zones Thrust planes	Crushing Pulverizing Brecciation	Micas: KAl+2+ $(AlSi_3O_{10}OH)_2$	Breccia Mylonite
Regional	Strong directed pressure at high temperature	Regions of great earth movement (tectonic belts)	Recrystallization under different values of temperature and pressure Progressive recrystallization	Micas: $KAl_2(AlSi_3O_{10})$ $(OH_2$+Na K.li Mg.) Kyanite Al_2SiO_5 Staurolite $FeAl_4$ $Si_2O_{10}(OH)_2$	Foliates Phyllites Schists Gneiss
	Strong hydrostatic pressure High temperature	In deeper parts of tectonic belts	Recrystallization	Augite Feldspar Garnet	Gneiss
				Serpentine minerals and rocks: Lizardite $Mg_3Si_2O_5(OH)_4$ Chrysotile $Mg_3Si_2O_5(OH)_4$ Antigorite $Mg_{2.82}Si_2O_5(OH)_{3.65}$	Serpentinites
Injection	Magma, gases, and liquids at high temperature	Regions invaded by batholithic intrusions	Recrystallization Replacement of minerals in invaded rocks by material from invading substances	Feldspars Augites Hornblendes Micas Garnets	Banded gneiss Migmatites

Table 5.17. Progressive metamorphism of shale and basalt: mineral assemblage and foliation change as a result of increasing temperature and differential stress (from Skinner and Porter 1992)

Shale	*Not metamorphosed*	*Low grade*	*Intermediate grade*	*High grade*
Rock name	Shale	Slate	Phyllite	Schist Gneiss
Foliation	None	Subtle, slaty cleavage	Distinct; schistosity apparent	Conspicuous; schistosity and compositional layering
Size of mica grains	Microscopic	Microscopic	Just visible with hand lens	Large and obvious
Typical mineral assemblage	Quartz clays, calcite	Quartz, chlorite, muscovite, plagioclase	Quartz, biotite, garnet, kyanite, plagioclase	Quartz, biotite, garnet, sillimanite, plagioclase
Basalt	*Not metamorphosed*	*Low grade*	*Intermediate grade*	*High grade*
Rock name	Basalt+H_2O	Greenschist	Amphibolite	Granulite
Foliation	None	Distinct schistosity	Indistinct; when present due to parallel grains of amphibole	Indistinct because of absence of micas
Size of mica grains	Visible with hand lens	Visible with hand lens	Obvious	Large and obvious
Typical mineral assemblage	Olivine, pyroxene, plagioclase	Chlorite, epidote, plagioclase, calcite	Amphibole, plagioclase, epidote, quartz	Pyroxene, plagioclase, garnet

metasomatism are hot (>250°C), hydrothermal alteration of the parent rock results. The serpentine family of minerals is one such group of hydrothermally altered products.

SELECTED METAMORPHIC ROCKS AND THEIR PLANT LIFE

The properties of metamorphic rocks influence the makeup of their derived soils. And in turn those soils exert influences on the plant cover. Those dogmatic statements need qualification. The magnitude of effects—from weathering of rock to make soil, to the soil as substrate for plants—inevitably varies in manifestation from the subtle to the obvious. While there may be little detectable differences in soil and vegetation, say between a granite and its metamorphic counterpart gneiss, hydrothermally altered andesite differs in soils and vegetation dramatically from the unaltered andesite (e.g., Billings 1950). See Figure 5.63. This contrast in magnitude of effect could be due to the differences between metamorphism of "nor-

Table 5.18. Some major metamorphic rock types: Landforms, weatherability, and plant suitability (from Skinner and Porter 1992, Ollier 1984, and Gerrard 1988).

	Landform types (arid regions)	*Weatherability*	*Mineral/chemical content*	*Likely plant suitability*
1. Metamorphics from shales and mudstones				
Slate (low grade metamorphism)	Rounded, steep-sided hills and sharp-crested ridges; highly eroded	Highly erodible	Quartz, clays, muscovite, chlorite	Azonal: low fertility; poor plant cover; some endemism
Phyllites (medium grade)	—	—	—	—
Schist and gneiss (high grade metamorphism)	High relief, rugged, no talus (gneiss); moderate relief (schist)	Schist: easily weathers (foliate) gneiss; fairly resistant	Quartz, feldspar, amphiboles, pyroxene	Mostly zonal; fairly rich flora
2. Metamorphics from basalt				
Greenschist (low grade)	Moderate to low relief	Easily weathered (foliate, schistose)	Chlorite, epidote, plagioclase, and calcite	Zonal; good plant cover
Amphibolite and granulite (intermediate to high grade metamorphism)	Moderate relief	Deeply weathering	Hornblende	Fertile soils and good plant cover
3. Metamorphics from other sedimentary rocks				
Quartzite: from sandstone	High relief	Highly resistant	Increase in silica and recrystallization	Low; chemically inert
Marble: from limestone and dolostone	Massive rounded; smooth ridge crests	Readily weathers by solution	Calcite or dolomite	Azonal; like limestone floras
4. Other metamorphics				
Serpentinite	Winding ridges between elongate, smooth to cone-shaped hills	Easily eroded	Ferromagnesian-silicates low Ca, high Mg, Ni (Cr, Co)	Low, but with unique floras: ∞ endemics; exclusion of ∞ nearby species

mal" (nutritionally balanced) rocks (e.g., granite to gneiss) versus metasomatism, the hydrothermal alteration of an extrusive igneous rock—no chemical change in the former, but significant change in the latter.

My bias in seeking out striking geoedaphic case histories with their vivid contrasts must be admitted; such examples are often visually spectacular and readily explained on grounds of geoedaphic differences. Yet it is worthy of contemplation that geoedaphic responses range along a continuum of lithological differences. I intend to pursue the more subtle, less manifest rock-plant interfaces later. For now, we look at some examples of metamorphic rock where the plant response is clear-cut.

SERPENTINITE ROCK AND SERPENTINE SOILS: THE PLANT RESPONSE

No other rock-to-soil-to-plant syndrome is more self-evident for its singular attributes than the "serpentine syndrome," so named by soil scientist Hans Jenny (1980). Nearly everywhere that the metamorphic version of ultramafic rock, serpentinite, outcrops one can behold remarkable landscapes—stunted forests or sere shrublands, and in the extreme manifestations of the "syndrome," barrens with scarcely any plant cover (Fig. 5.28). And the plant cover on serpentine often consists of faithful indicators and narrow edaphic endemics. It is no wonder that this paradigm of geoedaphic uniqueness has provoked two centuries of recorded observations. Fortunately, major reviews of the late twentieth century obviate the need to fully treat the serpentine story here: see Proctor and Woodell (1975), Kruckeberg (1985), Brooks (1987), Shaw (1989), Roberts and Proctor (1992), Baker et al. (1992), and Jaffré et al. (1997). Moreover, many attributes of the igneous forms of ultramafic rocks, discussed earlier in the chapter, fit the serpentine case. Hence I will pursue the serpentine case history in limited fashion, focusing on serpentinite as a metamorphic rock, its global geological context, and some recent works on its flora and vegetation.

The word serpentine continues to be used broadly—and loosely—to describe the rock, the minerals, the soil, and the vegetation associated with the serpentinized ultramafic rocks (Fig. 5.29; Tables 5.19 and 5.20). Geologists confine their use of the word to the three constituent minerals of the serpentine mineral family—lizardite, antigorite, and chrysotile. The rock composed of the serpentine minerals then is called serpentinite. The unaltered (unmetamorphosed) ultramafic protoliths (precursor igneous rocks), such as peridotite, dunite, and harzburgite, are often constituents of ophiolite suites in fold-mountain belts of the world, are of mantle origin, and are usually associated with plate tectonics (Coleman and Jove 1992; Malpas 1992). The geoedaphics of igneous ultramafic rocks were discussed early in this chapter. The common shared chemical makeup of both the igneous and metamorphic forms is the presence of magnesium and iron as the ferromagnesian silicate elements; other common elements in these rocks include nickel, chromium, and cobalt. The making of serpentinite rock from the igneous

5.28 The extreme version of the serpentine syndrome is the barren, nearly devoid of plant cover, here shown in the Wenatchee Mountains, Washington State. Serpentine barren, fringed by *Abies lasiocarpa* (also on serpentine); upper DeRoux Creek, Teanaway River basin. Photo by author.

protoliths is essentially the process of metasomatism—hydrothermal transformation, at low temperatures and pressures suitable for the genesis of the serpentine minerals (Malpas 1992).

Texture and color of serpentine rock vary widely. California geologist A. Knopf said it elegantly: "These rocks show an infinite variety of forms. They are like Cleopatra—never stale" (in Dietrich and Skinner 1979). The rock can show strong foliation or not, can be highly fissile or solid, and often

Table 5.19. Nomenclature of serpentine rock types (based on Brooks 1987)

Term	*Minerals or rocks described*
Serpentine	Antigorite and/or chrysotile minerals: $Mg_3Si_2O_5(OH)_4$
Serpentinite	Rock composed of serpentine minerals
Ultramafic rock	Rock containing high concentrations of magnesium (Mg) and iron (Fe), hence the term ultramafic; contains more than 70% mafic minerals
Ultrabasic rock	Rock with less than 45% silica
Ophiolite	Originally serpentinite but later modified (see text) to include ultramafic rock assemblages some of whose constituents are not necessarily ultramafic
Ultramafites	Ultramafic rocks, igneous and metamorphic (serpentinitic)

Table 5.20. Terminology for serpentine and other ultramafic rocks, soils, and flora/vegetation

ROCKS

Ultrabasic rocks: Rocks with less than 45% silica; igneous or metamorphic, with ferromagnesian minerals

Ultramafic rocks (ultramafites): Igneous or metamorphic rocks containing more than 70% mafic (ferromagnesian) minerals

Igneous (mostly intrusive) ultramafic rocks:

Dunite: Contains more than 90% olivine mineral

Harzburgite (and lherzolite, websterite): In peridotite group of ultramafites, containing variable mixtures of olivine and pyroxene minerals

Gabbro: Medium- to coarse-grained, with variable amounts of plagioclase, pyroxene, and olivine

Peridotite: Group name for ultramafic rocks with variable amounts of olivine and pyroxene; includes dunite, harzburgite, and others

Olivine basalt: Extrusive form of gabbro

Metamorphic ultramafic rocks:

Serpentinite: Ultramafic rocks containing serpentine minerals

Rodingite: Altered gabbroic rock, containing abundant calcium silicates; often at contact zones between serpentinite and nonultramafic rocks

is lustrous (glistening) and slippery (soapy) to the feel. Color ranges from nearly black to off-white, with mottled gray-green most common; it can be blotchy or mottled like a snake's skin, whence the names serpentine and ophiolite (Faust and Fahey 1962).

Table 5.20. *(continued)*

MINERALS (Serpentine family of ferromagnesian minerals; geologist's strict use of the term "serpentine")

Minerals in igneous ultramafites:

Olivine: Mg_2SiO_4; also $(Mg, Fe)_2SiO_4$

Pyroxene: Various silicate minerals ($XYSi_2O_6$); e.g., diopside, $CaMgSi_2O_6$

Minerals in metamorphic ultramafites:

Chrysotile: $Mg_3Si_2O_5(OH)_4$ fibrous form (asbestos)

Lizardite: Same formula, but platy in texture

Antigorite: Similar to chrysotile, occurring as corrugated plates or fibers

Talc: $Mg_3Si_4O_{10}(OH)_2$

SOILS

Serpentine soils: Derived from ultramafic parent materials

Nickeliferous soils: High nickel content (>1000 mg/g Ni) of soils mostly from ultramafic parent materials

Siderophile: Describing a chemical element having an affinity for iron (Fe)

Smectite: Common clay mineral in serpentine soils; hydrous aluminum silicate $Al_4Si_8O_20(OH)_4$ and containing Mg and Fe

FLORA AND VEGETATION

Serpentine: Plants associated with serpentine and other ultramafic substrates; also landscapes of the same plant cover and origins

Serpentine syndrome: The manifestation of combined physical, chemical, and biological factors associated with serpentine soils (Jenny 1980)

Hyperaccumulators: Plants (species, infraspecific variants, or races) that accumulate more than 1000 ppm of a heavy metal; mostly nickel on serpentine soils

OTHER

Ophiolite: Originally meant serpentinite. Currently the term is applied to a suite of rocks, including ultramafites, associated with plate tectonics; ophiolite belts are found worldwide at oceanic-continental plate margins

Serpentine: Besides its use as the term for ferromagnesian minerals, it is used broadly (and loosely) to describe any physical or biological phenomena associated with ultramafites (e.g., serpentine soils, serpentine vegetation, serpentine flora, etc.)

SOURCES: Brooks (1987); Coleman and Jove (1992); Jenny (1980).

Rates of weathering of ultramafic rocks to make soil depend on the kind of parent rock and other physical and biological factors (Coleman and Jove 1992). Though serpentinite rock fractures more readily (greater fissility) than does peridotite (the igneous form), the latter protolith weathers more readily than serpentinite. Robert

5.29 Serpentine "still life": samples of serpentine rock and soil from the North Coast Range, California. Author's studio photo.

Coleman, specialist in ultramafic geology, points out (pers. comm.): "It is generally assumed that peridotites weather at a more rapid rate because olivine, the main mineral, is quite unstable under normal atmospheric conditions. . . . On the other hand, serpentine minerals are stable at the earth's surface and will persist within soils for a much longer time than olivine. However if the waters associated with weathering have a low pH, both olivine and serpentine minerals will weather at the same rate. Most of the nickel-laterites are derived from peridotite under tropical weathering conditions whereas serpentine minerals are much more stable in this environment and rarely give rise to nickel-laterites."

Soils derived from serpentinite rock usually retain the chemical character of the minerals in the parent material, though concentrations of elements may change. Where serpentinization abuts country rock of nonultramafic composition, localized enrichment by $Ca\ SiO_4$ can occur; the rock of these reaction zones is called rodingite (Coleman 1967). Soils in these contact areas then may have higher calcium levels than in the adjacent serpentines. Since serpentine soils form under a wide range of climates (from subarctic to tropic), the weathering products, soils, will differ in pH, clay content, and cation concentration. However, Brooks (1987) finds four attributes that are shared in common by serpentine soils: (1) They have high levels of the heavy metals iron, nickel, chromium, and cobalt. (2) They are low to deficient in plant nutrients, especially nitrogen, phosphorus, and potassium. (3) They exhibit low calcium/magnesium quotients; calcium is usually deficient (Walker 1954). (4) They tend to have lower levels of clay colloids, and the clay minerals have a low exchange capacity. In brief, then, serpentine soils are infertile for crop plants and may be toxic as well (high levels of nickel, chromium, and cobalt). Though toxic, chromium is highly insoluble in the soil and thus may not affect plant growth. The clay minerals of serpentine soils may be newly derived species of colloids (e.g., chlorites or smectites) or else the minerals of the parent rock may persist in the soil as the cationic exchange colloids (Kruckeberg 1985, 1992).

Since serpentinites and other ultramafic rocks occur widely in both temperate and tropical regions, their derived soils will differ markedly in response to the prevailing climatic regimes. For the tropics, the reviews by Brooks (1987) and Proctor (1992), and for Cuban serpentines Borhidi (1991), give detailed accounts of these soils. Often laterites develop over ultramafic rocks, mostly if the parent material is the serpentinized form. Extensive laterites and other oxisols are known from Cuba, New Caledonia, Indonesia, and Brazil. Ultramafic laterites usually contain significant amounts of nickel (low-grade ores like garnierite) and are major industrial sources of the mineral. Laterization results in leaching of silica and magnesium and in concentrating iron. Ultramafic laterites are acid and have low cation exchange capacities, and the net result is low levels of exchangeable cations and other essential inorganic nutrients; in contrast, they have high levels of iron and nickel (Birrell and Wright 1945).

In temperate regions, laterization of serpentinites may have occurred; one such laterite is reported for Oregon (Hotz 1964). However, this is an ancient laterite formed under tropical conditions in the Tertiary (R. Coleman, pers. comm.). But most serpentinitic soils are entisols (also called lithosols or azonal soils), with little profile development. In residual (upland) serpentines, the parent materials are present in all fractions from angular rock fragments (spalls) and gravel to silt and clay. The pH values for temperate serpentine soils hover at about neutrality (7.0) or are in the alkaline range, especially in low rainfall areas. Cation exchange capacity (CEC) can be low to high, depending on the clay fraction in the soil. Calcium/magnesium quotients are invariably low. Nutrient deficiencies are common, and toxic levels of nickel and cobalt often occur. In western North America, soil surveys recognize as soil series those soils of both residual and of alluvial origins (Kruckeberg 1985); both types retain the calcium/magnesium imbalance and are infertile. Tables 5.21 and 5.22 give selected analyses of serpentine soils from tropic and temperate areas.

Besides their singular mineral nutrient properties, serpentine soils have physical and biological properties that are key symptoms of the serpentine (geoedaphic) syndrome (Fig. 1.1). Texturally, residual or colluvial serpentine soils are stony, with a major rock and gravel component, and often possess only a minor silt and clay fraction; most have only A and C horizons. Their scanty plant cover makes them

Table 5.21. Chemical analyses of serpentine and contrasting nonserpentine soils

Serpentine sites	*CEC*[1]	*Ca*[2]	*Mg*[2]	*Ca + Mg*[2]	*Mg/Ca*	*pH*
A	20.1	6.5	13.9	20.4	2.1	6.6
B	27.8	4.9	16.0	20.9	3.3	6.4
C	16.1	1.5	16.5	18.0	11.0	7.1
D	25.0	2.3	15.2	17.5	6.6	6.6
E	15.0	2.4	11.4	13.8	4.8	7.2
F	16.0	2.8	11.8	14.6	4.2	7.0
G	19.5	2.9	13.9	16.8	4.8	—
H	—	7.5	18.9	26.4	2.52	6.7
I	—	0.85	10.4	11.25	12.2	6.9
J	27.8	3.35	29.50	32.85	8.8	6.9
K	33.6	1.5	32.7	34.2	22.5	6.9
L	10.4	0.5	0.8	1.3	1.8	5.2
M	—	2.6	185.0	187.6	71.0	6.8
Averages	—	3.05	28.9		12.0	6.7
Non-serpentine sites	*CEC*[1]	*Ca*[2]	*Mg*[2]	*Ca + Mg*[2]	*Mg/Ca*	*pH*
N	34.0	52.5	2.7	55.2	0.05	7.5
O	14.2	7.4	0.1	7.5	0.01	6.1
P	80.8	34.4	6.6	41.0	0.19	6.2
Q	26.1	13.1	3.4	16.5	0.25	5.6
Averages		26.9	3.2		0.13	6.3

A – Tulameen Dist., British Columbia (Kruckeberg 1992b)
B – Kittitas Co., Washington (Kruckeberg 1992b)
C – Skagit County, Washington (Kruckeberg 1992b)
D – Grant County, Oregon (Kruckeberg 1992b)
E – San Benito County, California (Kruckeberg 1992b)
F – Lake County, California (Kruckeberg 1992b)
G – West Newfoundland (Roberts 1992)
H – England (Proctor 1992a)
I – Scotland (Proctor 1992a)
J – Portugal (De Sequeira and Da Silva, 1992)
K, L – New Caledonia (Proctor 1992b)
M – New Zealand (Lee 1992)
N – San Luis Obispo County, California (Kruckeberg 1985)
O – Kittitas County, Washington (Kruckeberg 1985)
P – Hawaii (Kilmer 1982)
Q – Iowa (Kilmer 1982)

[1] Cation Exchange Capacity

[2] Expressed as milli-equivalents $100g^{-1}$, 2mm fraction, oven-dry soil

Table 5.22. Siderophile elements* in ultramafic rocks and soils

	Ultramafic soils ($\mu g/g^{-1}$ or %)				Ultramafic rocks (%)			
Source	*Ni*	*Cr*	*Fe*	*Co*	*Ni*	*Cr*	*Fe*	*Co*
New Zealand	1,650	1,750		190				
New Caledonia	4,700	12,700		570				
New Caledonia	10,400	25,500		800				
Italy	2,500	3,900	(6.40)	312				
Balkans					(0.17)	(0.19)	(6.48)	—
Portugal[†]	110	0.3	230	—				
UK	2,170	60	—	110	(0.18)	(0.37)	(9.52)	
UK[†]	0.7	—	—	—				
Newfoundland	2,278	1,447	—	271				
California	2,780	1,640[‡]	35,000	—				
Ave. ultramafic					(0.30)	(0.50)	(6.01)	(0.02)
Zimbabwe	4,600	480	—	270				
Japan	2,600	5,200						
Ave. sandstone					(0.001)	(0.001)	(0.167)	(0.002)

* Elements of the iron family (Fe, Co, Ni, Cr, V).
[†] Available amounts.
[‡] Cr = ppm.
SOURCES: Various articles in Roberts and Proctor (1992).

highly erodable, especially on sloping terrain. Red is their predominant color, owing to the iron oxides throughout the A and C horizons. High insolation and low plant cover conspire to make the serpentine soil environment hot and dry, at least on the soil surface. It has been observed (Hardham 1962) that despite the xeric look of serpentine landscapes, subsurface moisture is often present. Serpentine slopes are both catchments for runoff and sources of groundwaters that rise to the surface. In western North America it is not uncommon to find springs and bogs or fens dotting an otherwise xeric serpentine landscape. And drainages in serpentine areas may run with water longer in summer months than do their neighboring drainages on nonserpentine substrates.

The interconnections of chemical, physical, and biological vectors acting within the serpentine syndrome are inevitably multidirectional; the system is a dynamic web of matter and energy flows, even though the magnitudes may be low. The biologic component manifests both cause and effect. Although the deficient nutrient status is derived primarily from the serpentine parent material and its weathered product (the soil), the resultant unique vegetation cover, often sparse to nearly barren, has a strong biologic feedback component. The stressful chemical and physical environment fosters a low productivity. In turn, the limited recycling of the low levels of biomass keeps the soils at low nutrient levels. Besides the meager biomass conversion back to the soil, undoubtedly the microbial biota of the serpentine soil will be depauperate, paralleling the scanty vegetation that it must process. Admittedly, this is only an educated guess, for next to nothing is known of the microbial world of serpentine soils. The role played by soil fungi, including those making mycorrhizae, as well as by other soil organisms is scarcely known (Kruckeberg 1993).

Jenny's (1980) presumed purpose in creating the term "serpentine syndrome" was to emphasize the dynamic interplay of factors affecting plants on serpentine sites. Embellishing the syndrome idea, I conceive of it functioning as a feedback loop among such key factors as soil properties (physical and chemical), water and temperature variations, seasonal growth and reproduction, community structure, recycling of biomass, herbivory, and disturbance. Such feedback loops are undoubtedly at play in other azonal (as well as zonal) habitats.

The linkage of rock to soil to flora is a clearcut and simple trajectory anywhere that serpentine outcrops. The truth of this epigrammatic utterance is well substantiated by the rich literature on serpentine plant life. The recent exhaustive reviews mentioned at the beginning of this section amply illustrate the rock-soil-plant connection in its most singular manifestation. So what is left for the present discussion that will acknowledge this remarkable natural syndrome? What seems profitable in this context of global geoedaphics is to offer a sampling of the rich serpentine literature. First I review the salient features of serpentine flora and vegetation. Then we examine a remarkable case history of serpentine plant life in Cuba (Borhidi 1991).

So distinctive is the plant cover over serpentine outcrops that it has taken on the role of "pathfinder" for geologists, soil scientists, and land stewards. Indeed, it is *the* milieu par excellence on which to practice the skills of geobotanical prospecting: using vegetation to locate mineral deposits (e.g., Brooks 1972, 1987). So what are the hallmarks of a serpentine landscape? The most eye-catching displays are where nonserpentine substrates and serpentines meet. Here the contrasts are so sharp as to impress the most casual observer. First to register is the dramatic shift in the physiognomy—the "look" of the contrasting vegetation. Forest gives way to serpentine chaparral or maquis, mesic grassland to thinly populated herb cover, or in extreme instances a dense nonserpentine vegetation yields at the contact to a serpentine barren nearly devoid of plant life. Such commanding contrasts can be found worldwide, in the tropics as well as in temperate areas. Vivid contrasts in the tropics are exemplified by proximate serpentine-nonserpentine displays in New Caledonia, Cuba, Zimbabwe, and Brazil. For temperate areas, starkly contrasting contact zones appear in North America (the Pacific Coast states, Newfoundland, and the serpentine barrens of Pennsylvania and Maryland), in northern Sweden, the Alps, Italy, the Balkan Peninsula, New Zealand (South Island), and Japan. References to writings on these spectacular geoedaphic displays are found at the beginning of this section. A collage of photographs can tell the story better than words (Figs. 5.30–5.36).

On closer scrutiny, the vegetational contrasts between serpentine and nonserpentine landscapes (Fig. 5.37) can be resolved as the consequences of several attributes: differences in life-form and community structure, in morphological terms (shifts from mesophylly to xerophylly), in stocking density, and in species composition. Robert Whittaker (1975, p. 277) captures tellingly the physiognomic differences: "some contrasts of serpentine versus non-serpentine communities in the same climate are: tundra versus taiga in Quebec, pine woodland versus Douglas fir forest in Oregon, chaparral versus oak woodland in California, savanna and scrub versus tropical forests in Cuba and New Caledonia, tussock grassland versus southern beech forest in New Zealand." (Such contrasts are illustrated in Figs. 5.38 to 5.44). Further, the serpentine side of a lithological contact harbors woody plants adapted to the serpentine syndrome—sclerophyllous shrubs with small leathery leaves and of compact, low stature. The European literature formalizes the various morphological modifications as serpentinomorphoses (Ritter-Studnicka 1968; Krause 1958; Pichi-Sermolli 1948): "1. xeromorphic foliage (sclerophylly, glaucousness, size reduction, reduced or increased pubescence, anthocyanous coloration, etc.); 2. reduction in stature (shrubbiness of arborescent species, dwarfing and plagiotropism of herbaceous species); and 3. increase in root system" (Kruckeberg 1985, p. 24).

Not only do individual plants of a serpentine community take on a characteristic morphology, their sparse spacing (stocking density) adds to the telltale sere look of a serpentine community. The extreme in reduced density occurs on serpentine barrens where there may be no plants at all, or where they are diffusely scattered over a predominantly barren surface (Figs. 5.30, 5.34–5.36). The low densities in serpentine communities have been quantified in a few cases: in Portugal (Menezes de Sequiera and Pinto da Silva in Roberts and Proctor 1992) and for southwestern Oregon (Whittaker 1960).

5.30 Aerial view of serpentine barrens, Cle Elum River drainage, Wenatchee Mountains, Washington State, either with no forest whatsoever or sparse cover of Douglas fir and subalpine fir. In contrast, the west-facing slope (upper left) of Mount Hawkins with dense forest on greenstone (metadiabase). Mount Stuart, a granodiorite batholith, is on the far right horizon. Photo by Len Miller.

5.31 Mesic (foreground) to xeric (upslope) gradient on serpentine above Eldorado Creek, North Fork Teanaway River, Wenatchee Mountains, Washington State. The rare *Salix brachycarpa* grows with *Ledum glandulosum* and *Polystichum lemmonii* in the mesic site. Photo by author.

5.32 Subalpine ultramafic landscape: fellfield and talus of serpentinized peridotite support a sparse cover of *Larix lyallii* (light-colored trees) and a stand of subalpine fir (dark). Upper Jack Creek near Mount Stuart, Wenatchee Mountains, Washington State. Photo by author.

5.33 Nearly barren serpentine exposure on north slope of Baldy Mountain (summit, 7,373 ft), Grant County, central Oregon. Scattered Douglas fir and western white pine (*Pinus monticola*), with indicator ferns *Aspidotis densa* and *Polystichum lemmonii*, on serpentine talus. Photo by author.

5.34 Two views of the same stark serpentine landscape show little change over forty years. Upper Clear Creek, New Idria area, San Benito County, California. Upper photo by A. Wieslander in 1932; lower photo by J. Griffin about 1975. Scattered individuals of *Pinus sabiniana* (grey pine) and *P. jeffreyi* (Jeffrey pine). From Kruckeberg 1985.

Opposite page:

5.35 Much-cited serpentine locality, Hill 1030 (USGS benchmark), four miles northeast of Middletown, Lake County, California. Three species of *Streptanthus*, *Phacelia egena*, and *Aspidotis densa* on the barren; serpentine scrub oak, *Quercus durata*, and *Pinus sabiniana* (grey pine) in the background. From Kruckeberg 1985.

5.36 Serpentine chaparral with the serpentine endemics *Quercus durata* and *Ceanothus jepsonii* as codominant shrubs, Tehama County, California. From Kruckeberg 1985.

5.37 Contacts between ultramafic and other substrates often show up in sharp contrast because of differences in plant cover. This serpentine barren is flanked by subalpine forest on metadiabase in the Cle Elum River drainage, Wenatchee Mountains, Washington State. Photo by author.

5.38 Serpentine exposures occur in scattered localities in eastern North America. Here treeless grass-forb serpentine sward contacts a hardwood forest on normal substrate. Nottingham, Chester County, Pennsylvania. Photo by J. and C. Baskin.

Opposite page:

5.39 Dramatic contrast of ultramafic barren (right) and subalpine forest on schist (left), Simonin Pass, Red Mountain, South Island, New Zealand. Photo by L. Malloy.

5.40 Tropical forest is replaced by a species-rich maquis vegetation in New Caledonia. Maquis scrub at Chute de Madeleine. Photo by author.

5.41 *Pinus caribaea* on serpentine, in Cuba. Photo by A. Borhidi.

5.42 Secondary savanna with endemic palm, *Gastrococos crispa*, in Cuba. Photo by A. Borhidi.

5.43 *Melocactus matanzanus* on serpentine, Cuba. Photo by A. Borhidi.

5.44 *Euphorbia heleniae* subsp. *grandifolia*, serpentine endemic, Cuba. Photo by R. Reeves.

Biodiversity and Floristics of Serpentines

Besides the characteristic physiognomies taken on by serpentine vegetation, there is often a distinctive species composition—indeed, a serpentine flora. This floristic makeup may have a diversity greater or lesser than that of neighboring nonserpentine floras; I particularize this observation a bit later. Serpentine floras can consist of any (or all) of the following taxonomic attributes (Kruckeberg 1985): (1) narrowly endemic species wholly restricted to a local or regional serpentine outcrop(s); (2) ranges of nonserpentine taxa extending as disjuncts or outliers on serpentine; (3) indifferent taxa (also called bodenvag or ubiquist taxa), which range widely on and off serpentine; and (4) excluded taxa—those species of adjacent nonserpentine substrates that faithfully avoid serpentine. A closer look at these four categories follows.

Serpentine Endemics. Varieties or subspecies and species are the most common class of endemics, though in some floras (e.g., New Caledonia, Cuba, and the Balkan Peninsula) even some genera are narrowly endemic to serpentine. The endemics are usually distinctive variants of a region's flora; that is, serpentine taxa appear to have evolved from related species in the generic flora of a region. The taxonomic status of the endemics varies: some are relictual (or paleoendemic) taxa, while others appear to be neoendemics—newly evolved. The relative ages and taxonomic affinity of serpentine endemics are expanded upon in Brooks (1987) and Kruckeberg (1985). Figures 5.41 to 5.47 and 5.49 to 5.53 illustrate the taxonomic diversity of serpentine endemics. In western North America, species of conifers, ferns, monocots, and dicots may be restricted to serpentines. Similar diversity among endemics holds for other serpentine areas of the world.

Some serpentine floras have a remarkably high number of endemics. The tropical antipodean island group of New Caledonia surely tops the world list. One-third of the main island is made up of ultramafic rocks, from the large and continuous displays in the Massif du Sud to the northwest-trending, more discrete and isolated outcrops (Fig. 5.40). There 1176 species (39 percent of the indigenous flora) are strictly limited to ultramafic areas, and 98 percent of these are endemic to New Caledonia (Jaffré et al. 1987). The large New World tropical island of Cuba also has been a cradle of evolutionary diversification. The total for Cuban endemics comes to 3178 taxa, nearly 50 percent of the native flora. Though only 7 percent of the land area of Cuba is serpentinitic (Figs. 5.41–5.44), it boasts a rich serpentine flora. Borhidi (1991), in his encyclopedic treatment of the ecology and phytogeography of the Cuban vegetation, says: "[O]ne third of the endemic flora of Cuba (920 species, or 31.2 percent) has developed on serpentine areas . . . [and] 14.6 percent of the total flora is endemic, exclusively, to serpentine. Of the 72 endemic phanerogamous genera, 24, i.e., 33.3 percent, live in serpentine areas." We must come back to this remarkable Cuban serpentinitic flora later.

The California Floristic Province comprises cismontane California to southwestern Oregon (Fig. 5.48). While only one percent of this botanically rich, Mediterranean type floristic region displays serpentine outcrops, that area of 2860 square kilometers has spawned a significant number of endemics: 215 species, subspecies, and varieties of native flora are restricted wholly or in large part to serpentine soils (Figs. 5.49–5.52). They include conifers, woody and herbaceous dicots, monocots, and ferns (Kruckeberg 1985). I calculated that 9 percent of the total endemic taxa for the California Floristic Province (198 taxa out of 2133 endemics) are restricted to ultramafic areas (Kruckeberg 1992).

While no figures for total serpentine endemism in the Mediterranean region can be given, the many serpentine localities from Portugal to Turkey would yield a significant cumulative total for serpentine endemics. Notable centers are Italy with 13 taxa (called serpentinophytes by Vernano-Gambi 1992), and 45 serpentinophytes for the Balkan Peninsula (from Tutin et al., cited in Tatic and Veljovic 1992). According to the latter authors, the Tutin (1980) datum may be inflated, since not all are confirmed as obligate serpentinophytes. Professor Arne Strid, who has studied the flora of Greece for many years, provided me with unpublished data from his Flora Hellenica data base: 4000 taxa have been reported to occur on Hellenic serpentines, the vast majority of which are certainly not obligate serpentine endemics. Strid (pers. comm.) estimates that 50 to 80 species are confined to serpentines of the mountainous areas of northwestern Greece and southern Albania. I am led to the conjecture that there are over 100 serpentine endemics in the Mediterranean basin, ranging from the Iberian Peninsula in the west eastward to Anatolia.

In South Africa, witness the Barberton serpentines (Fig. 5.46) and the Great Dyke, Zimbabwe. On the latter, one of the world's largest ultramafic outcrops, the flora is considered to be relatively species-poor, but does include about 20 endemics (Proctor and Cole 1992). In Japan, ultramafic areas appear along nearly the full length of the Japanese archipelago, from Hokkaido south to Shikoku and Kyushu; outcrops occur from sea level to alpine level (Fig. 5.47). Mizuno and Nosaka (1992) estimate that there are about 50 species of serpentine-character plants in Japan, of which 37 are recognized as typical serpentinophytes and the remainder are relics.

Other notable centers of serpentine endemism and serpentinitic vegetation have been pointed out by Brooks (1987) and Roberts and Proctor (1992). In the eastern United States, the serpentine barrens of Pennsylvania and Maryland, as well as isolated outcrops from Vermont to Georgia, are often cleanly delimited from nonserpentine landscapes, but their floras are not especially distinctive. Kevin Dann (1988) has given us a charming and richly rewarding natural history of serpentines in eastern North America; his *Traces on the Appalachians* discusses the geological, biological, and cultural tracings on these ferromagnesian landscapes. In eastern Canada, two areas have well-developed serpentine vegetations. The serpentines of the Gaspé Peninsula (Quebec) are best known for their many disjunct taxa and shifts in life-form spectra to more herbaceous community types (Scoggan 1950). The plant communities of Newfoundland serpentines have distinct physiognomic character as well as interesting species disjunctions; some are known to accumulate nickel (hyperaccumulators like *Arenaria humifusa* and *A.*

5.45 Serpentine endemic cycad, *Macrozamia miquelii*, at Glen Geddes, Queensland, Australia. Photo by R. Reeves.

5.46 In South Africa, serpentines are mostly in the Transvaal, shown here above Barberton, the rich mineral district. This open serpentine slope of *Themeda* grassland contains scattered *Protea* species and the cabbage tree (*Cussonia paniculata*). Photo by author.

5.47 Serpentines occur intermittently the full length of the Japanese archipelago. Here on Hokkaido, Mount Yubari has a rich serpentine flora. Pictured is *Viola yubariensis*, a local serpentine endemic. Photo by author.

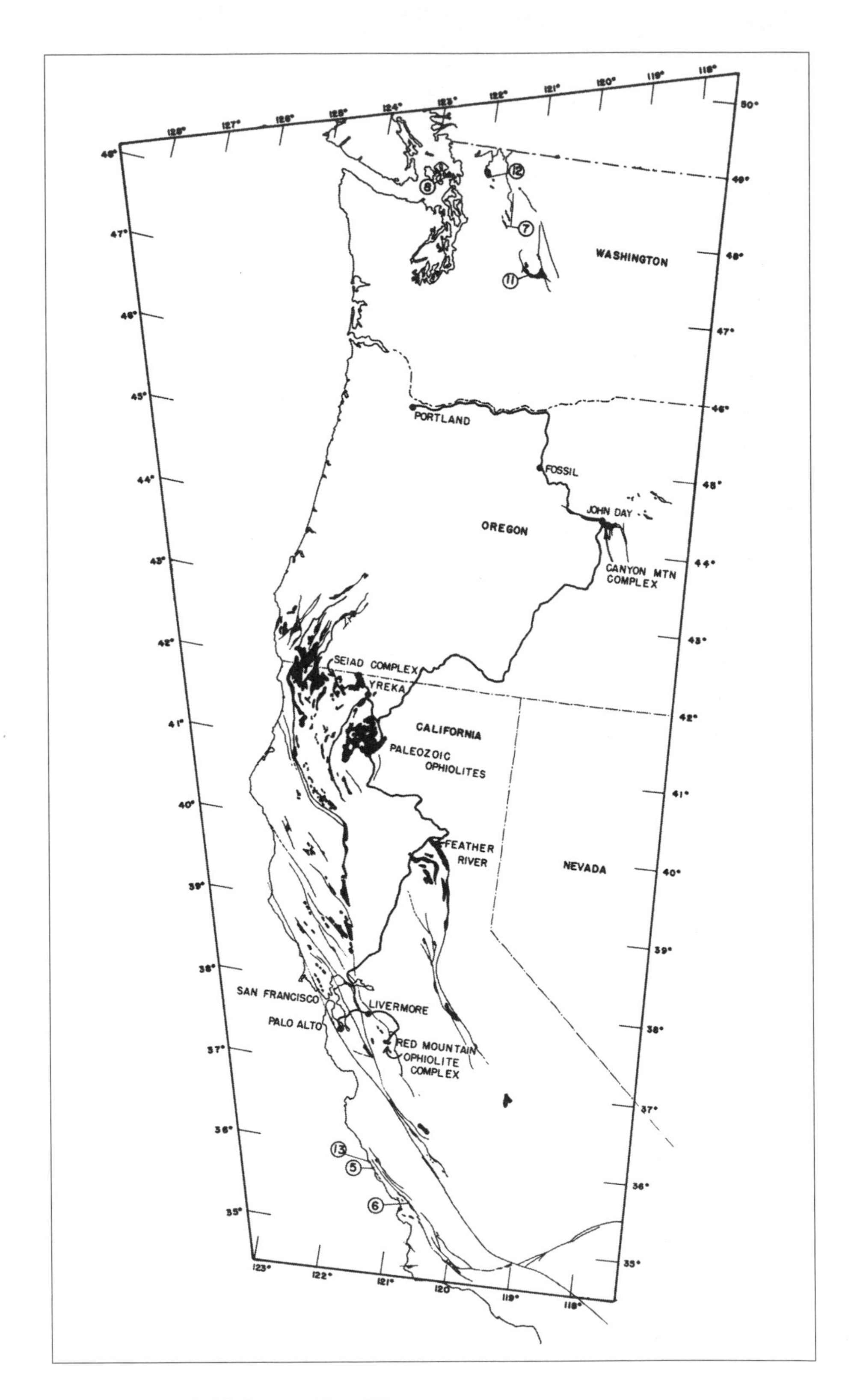
WASHINGTON
PORTLAND
FOSSIL
JOHN DAY
OREGON
CANYON MTN COMPLEX
SEIAD COMPLEX
YREKA
CALIFORNIA
PALEOZOIC OPHIOLITES
FEATHER RIVER
NEVADA
SAN FRANCISCO
LIVERMORE
PALO ALTO
RED MOUNTAIN OPHIOLITE COMPLEX

marcescens, Roberts and Proctor 1992). Yet, despite the strong serpentinitic characters of these eastern North American displays, no serpentine endemic species are known.

I have already given an account of the igneous ultramafic rock (dunite) on the South Island of New Zealand. On the same island is the spectacular alpine serpentine barren of Red Mountain, with its striking vegetation contrasts with normal lithology (Fig. 5.39). Serpentine vegetation also is found at the northern extremity of New Zealand's North Island. Substantial serpentine areas occur in Australia; see Figure 5.45 for the remarkable endemic cycad. A recent summary of serpentine ecology for these Australasian lands is in Baker, Proctor, and Reeves (1992).

Besides the rich serpentine flora of Cuba (discussed later), serpentines and associated vegetation are known in Puerto Rico, as well as smaller displays in the Caribbean and on the mainland of South America—in Brazil. In fact the serpentines of Brazil are about the only significant ultramafic occurrences in South America (Brooks 1987). Geochemical and floristic reconnaissance in the Goiás Province of Brazil has turned up a host of new finds (Brooks et al. 1990). In addition to finding serpentinophytes in several genera, Brooks and his colleagues identified the occurrence of nickel hyperaccumulators in the ultramafic flora.

Taxonomy of Serpentine Endemics. Serpentine endemism holds a particular fascination for all manner of botanists, from taxonomists to evolutionary biologists. The rarity and narrow fidelity of the endemics to the substrate not only capture the fancy of the field botanist, they pose questions of origin and relationship which I address at length later (Chapter 6). But here we offer a cautionary observation. Narrow endemics, in general, are subject to taxonomic "inflation." Their position in the taxonomic hierarchy has been called into question repeatedly (Kruckeberg and Rabinowitz 1985). Just because a population of plants is found on an unusual substrate or ecological island, does it merit being awarded taxonomic recognition? A recent example is in the genus *Hastingsia.* Becking (1986) recognized as a dis-

Opposite page:

5.48 Ultramafic areas in western United States are primarily in the Pacific Coast states. In Washington, there are four ultramafic regions, the first three in the Cascade Range: Stillaguamish River drainage (7), the largest in the Wenatchee Mountains (11) bordering the Stuart Range, and Sumas Mountain (12); the fourth site is in the San Juan Islands (8). In Oregon, the east-central serpentine region is near Canyon City in the John Day River country; in southwestern Oregon there are extensive serpentines in the Klamath-Siskiyou Mountains. In California, serpentines are in two regions: the North and South Coast Ranges west of the Great Valley and in the Sierra Nevada western foothills; the greatest concentration is in northwestern California. From R. Coleman and in Kruckeberg 1985.

5.49 Serpentines of the Pacific Coast states harbor many endemic species. The number is highest in California, followed by Oregon and then Washington. *Darlingtonia californica* frequently occurs in serpentine seeps inland in California and Oregon. Photo by M. Denton.

5.50 One of several serpentine endemic species of *Streptanthus, S. polygaloides* is found along the western Sierra Nevada foothills and is the only hyperaccumulator of nickel in the genus. Photo by author.

5.51 *Eriogonum alpinum* is a local serpentine endemic near the summit of Mount Eddy, Siskiyou County, California. Photo by M. Denton.

5.52 *Cupressus sargentii*, often in dense stands, is endemic to California serpentines, ranging from Santa Barbara County to northern California; here shown at Cuesta Summit, San Luis Obispo County. Photo by author.

5.53 A plant characteristic of Wenatchee Mountain serpentine, *Douglasia nivalis,* a common indicator of serpentine, is also found on nonserpentine sites along the east slope of the Cascade Range, Washington State. Photo by author.

tinct serpentine species *H. atropurpurea,* while Lang and Zika (1997) argue that Becking's taxon falls well within the infraspecific variation in *H. bracteosa,* also on serpentine in southwestern Oregon.

To be accorded the status of species or as an infraspecific variant, the entity should have other attributes besides local habitat restriction. A set of morphological characters that distinguish the endemic from its nearest relative is the surest claim to taxonomic recognition. Still, the taxonomic level at which the entity is to be recognized can and does fluctuate. Examples from serpentine floras nicely illustrate the question of taxonomic recognition and rank of endemics. In the California Floristic Province, one entire section (Euclisia) of the cruciferous genus *Streptanthus* is serpentinicolous and several of its taxa are local endemics (Kruckeberg and Morrison 1983). One of them, *S. polygaloides* (Fig. 5.50), clearly stands apart from all the others, both in the exceptional morphology of its flower and in its singular geochemical attribute: it is the only nickel hyperaccumulator* in the genus (Reeves, Brooks, and MacFarlane 1981; Kruckeberg and Reeves 1995). Its uniqueness caused E. L. Greene (1904) to place it in its own genus, the monotypic *Microsemia*. Given its exceptional attributes—morphological *and* geochemical—a case can be made for singling it out as the monotypic *Microsemia polygaloides,* as was suggested by Reeves, Brooks, and MacFarlane (1981). Yet by usage or benign neglect it remains in *Streptanthus* (Munz and Keck 1959; Hickman 1993). Another example in *Streptanthus* tells a different story. The two species in the section Biennes, *S. brachiatus* and *S. morrisonii,* are absolutely faithful to serpentine barrens in central California and are strikingly distinct from their nearest congeners. There is no question about their status as species. But should infraspecific variants be recognized, as was done by Dolan and La Pré (1987)? Nearly every local population of the two species has some mark of distinction that might merit an infraspecific name. It so happens that the several subspecies proposed by Dolan and La Pré are accorded protection as rare and endangered taxa. If they had not been given names, they might not have been listed as endangered. Only one of the several infraspecific variants proposed by Dolan and La Pré are recognized in *The Jepson Manual* (Hickman 1993).

Another California example illustrates inflation at the genus level. One group of *Linum* species is largely endemic to serpentine; it was monographed as a distinct genus, *Hesperolinon,* by Sharsmith (1961). Some serpentine endemics, including some monotypic genera, have been kept at the genus level: *Japonolirion* (*J. osense*), a rare Japanese endemic in the lily family, and *Halacsya* (*H. sendtneri*), of Balkan serpentines, have not yielded to deflation. Where else but in New Caledonia would one expect to find endemic genera with several to many species, all largely restricted to ultramafic areas? *Pancheria* (Cunoniaceae) has 27 species and *Nephrodesmus* (Leguminosae) has 6 species; many more ultramafic genera in New Caledonia (38 in number!) are mostly monotypic (Jaffré et al. 1987). I am not aware that any of these are suffering from taxonomic inflation, although *Boronella* has been shifted into the larger genus *Boronia* by Mabberley (1987). Students of species relation-

* Until recently, the role of nickel hyperaccumulation has been a mystery. Now it has been clearly established that it serves as an herbivore deterrent against insects. This adaptive function is best known for *Streptanthus polygaloides*, a serpentine endemic crucifer of California (see Boyd et al. in Brooks 1998). R. S. Boyd and associates have also tested the hypothesis that high tissue nickel can inhibit pathogens and that soils under hyperaccumulators may be allelopathic.

Localization of nickel in epidermal subsidiary cells of leaves has been determined by Heath et al. (1997).

Serpentine endemism is a paradigm par excellence for probing the general question: What is the origin of (and impetus for) species diversity—both in the edaphic milieu and beyond? There appears to be a correlation between the amount of diversity on serpentines (as well as richness in endemics) and the diversity in the floras of nonserpentine habitats nearby. Rich nonserpentine floras occur where there is a rich serpentine biota. Examples come easily to mind: New Caledonia and Cuba are prime tropical instances, while the Mediterranean climate floras of the Balkan Peninsula and the California Floristic Province fit the model. Conversely, there are cases where the reverse holds. High North Temperate floras like the Gaspé, northern Sweden, and the Pacific Northwest of North America have low species richness in both nonserpentine and adjoining serpentine habitats; some do not have any endemics at all (Sweden and the Gaspé). A reasonable explanation of these differences in species diversity can be derived from the taxonomic affinities shared between serpentine species and those on nonserpentine sites. Such affinities occur mostly at the level of genera but also can be discerned at the family level. Examples from the California Floristic Province are plentiful: *Clarkia* and *Epilobium* (Onagraceae), *Streptanthus* (Cruciferae), *Linum* (Linaceae), and *Calochortus* and *Fritillaria* (Liliaceae) have all attained substantial diversity on and off serpentine (Kruckeberg 1985). In the Balkan Peninsula, one finds several genera with the same pattern (Strid 1989; Tatic and Veljovic 1992): *Alyssum, Bornmuelleria,* and *Cardamine* (Cruciferae), and several genera of the Caryophyllaceae (e.g., *Arenaria, Gypsophila, Saponaria, Silene*). In fact, most common Mediterranean genera have their serpentine endemics too. Even the monotypic genus *Halacsya,* with its striking serpentine endemic *H. sendtneri,* may have an affinity with such Balkan borages as *Echium* or *Anchusa* (Tutin 1980). (Others have argued that *H. sendtneri* has no close relatives and is thus considered to be a relic species; see Brooks 1987.) The botanical El Dorado of New Caledonia is replete with cases of serpentine endemics having species of the same genus on other substrates. Jaffré et al. (1987) give a long list of such genera (their table 3); outstanding among them, especially for the large numbers of serpentine and nonserpentine species, are *Phyllanthus* (Euphorbiaceae), *Psychotria* (Rubiaceae), *Eugenia* (Myrtaceae), and *Alyxia* (Ochnaceae). The large number of endemic genera on New Caledonia serpentines poses a special problem. Do the serpentine endemic genera have their closest relatives in nearby nonserpentine habitats, or are their closest kin outside New Caledonia? As far as I am aware, no genetic or phylogenetic evidence for kinship has been clearly demonstrated. Yet Jaffré et al. (1987) do suggest within-island recruitment from nonserpentine sites for the origins of serpentine endemics.

Other Floristic Elements on Serpentines

Rarely does one find serpentine outcrops populated only with narrow endemics. The usual serpentine flora is stocked with both obligate serpentinophytes (endemics) and particular elements of the region's flora that span a variety of substrates. Those species that range across edaphic boundaries have been called ubiquist (or indifferent) species. Franz Unger first recognized the category, calling those species of the Austrian Alps that range across the silicate-limestone boundaries *bodenvag* species ("soil wanderers"). Indeed, on some serpentine areas the bodenvag species dominate in cover and in biomass. This is evident on the serpentines of the Wenatchee Mountains of Washington State, where serpentine landscapes with a sparse, low-site conifer cover are stocked with species from surrounding nonserpentine areas (del Moral 1972, 1974; Kruckeberg 1969b). I have witnessed the same physiognomy and composition of the vegetation elsewhere: (1) In northern Japan, on the ultramafic areas of Mount Hyachine and Mount Yubari (Hokkaido), regional conifers (e.g., *Pinus pumila, P. parviflora,* and *Abies mariesii*) as well as hardwoods (*Betula ermanii* and *Quercus mongolica* var. *grosseserrata*) are plainly bodenvag species. (2) At Dun Mountain, New Zealand, a subalpine scrub made up of such bodenvags as *Leptospermum scoparium, Dracophyllum* (three species), *Metrosideros umbellata,* and *Cyathodes* (two species), displaces a *Nothofagus* forest; the scrub vegetation gives the landscape a serpentinitic look, even though these shrubs are not endemic. (3) The mineral belt near Barberton, South Africa, has a predominance of bodenvag shrub species but nevertheless differs in physiognomy from the nearby nonserpentine vegetation.

There are at least three different components of this bodenvag floristic element on serpentine: (1) The bodenvag species are recruited from adjacent nonserpentine sites. (2) They are not in the local flora but have their nearest occurrences often miles from the serpentine. Foothill pine (*Pinus sabiniana*) and chamise (*Adenostoma fasciculatum*) occur on many serpentine sites in the Coast Ranges of California, but are often not present at all in the adjacent nonserpentine, mixed conifer forests (Kruckeberg 1985; Hanes 1977). (3) Bodenvag species, though not even varietally distinguishable from their nonserpentine populations, are often ecophysiologically and genetically distinct. In fact, bodenvag serpentine races were early recognized examples of ecotypic differentiation in response to an edaphic factor (Kruckeberg 1951, 1967). Serpentine tolerance of bodenvag species has been determined by progeny and transplant tests. Some indifferent species faithfully display tolerance or intolerance, depending on their native substrate; examples include *Achillea lanulosa, Pinus contorta, Gilia capitata, Fragaria virginiana, Senecio pauperculus, Prunella vulgaris,* and several other western North American herbaceous and woody bodenvag species, tested by the author (Kruckeberg 1951, 1954, 1967; Figs. 5.6, 5.54–5.56). So they are really not "indifferent" to substrate. They simply do not reveal their genotypic response to the serpentine habitat by those attributes conventionally used by taxonomists. Their inherited infraspecific variation is chiefly expressed physiologically as tolerance to the serpentine habitat. This is not to say that there are no morphological differences between the races. Cymerman (1988) was able to demonstrate that there are quantitative differences in morphology between races for two herbaceous perennials (*Achillea millefolium* and *Senecio pauperculus*) on Wenatchee Mountain ultramafic sites. These morphologic variants were not recognized taxonomically by Cymerman. No doubt the same could be demonstrated for other bodenvag species; their serpentine races have a certain serpentinomorphic "look" about them that is part of the "serpentine syndrome." See also Cooke (1994) for a similar study.

Other examples of racial differentiation have been

5.54 Response of serpentine (S) and nonserpentine (NS) strains, demonstrating tolerance to serpentine by S races and intolerance by NS races, is shown in California strains of *Achillea millefolium* subsp. *lanulosa* grown on serpentine soil (above) and nonserpentine soil (below). The S strains are 142, 164, 184, and 135. See Kruckeberg 1951 for details. Photo by author.

5.55 Pot tests for tolerance of serpentine (S) and nonserpentine (NS) strains of Pacific Northwest species. Seedlings from seed germinated on serpentine soil (Twin Sisters dunite). *Achillea millefolium* subsp. *lanulosa*: NS strain (*right*) dying at cotyledon stage and S strain with vigorous rosettes (*left*). From Kruckeberg 1967.

5.56 *Pinus contorta* subsp. *latifolia*: three NS and two S strains, with S strains showing greater vigor. From Kruckeberg 1967.

reviewed in a recent symposium (Kruckeberg, Walker, and Leviton 1995); several European herbs are cited as well as the racial response to serpentine for the widespread western North American ponderosa pine (*Pinus ponderosa*).

Not all bodenvag species that have been subjected to progeny or clonal testing on serpentine and nonserpentine soils have proved to have tolerant and intolerant races. The first such case was demonstrated for *Pinus sabiniana* (foothill pine) by Griffin (1965). Progeny of this California conifer from serpentine and nonserpentine sites grew equally well on serpentine. A similar genotypic indifference was reported by McNaughton et al. (1974) for *Typha latifolia;* various strains grew equally well on mine spoils contaminated with heavy metals. Why some bodenvag species have genetically determined ecotypic variants and others do not is still a mystery. A working hypothesis presents itself in the form of Baker's "general purpose genotype" (Baker 1965). As he has argued for certain weedy species, he also argues (Baker 1995)—and I agree—that some bodenvag taxa can have broad edaphic tolerance endowed by a single genotype. This may account for the absence of edaphic races not only in *Pinus sabiniana* but also in ruderal adventives that find their way onto serpentine (Kruckeberg 1993).

Serpentine Indicators

The notion that species are indicators of particular environments is a time-honored one in plant ecology (Clements 1928; Whittaker 1954a; Ellenberg 1988). Species with narrow ecological amplitudes can show a high fidelity to a given environment. Whittaker (1954a) reminds us, though, that "no species is simply an indicator for a factor." Thus the serpentine endemics like *Halacsya sendtneri, Asplenium adulterinum,* or *Quercus durata* show the narrowest of fidelity to serpentine, but not just to tolerance to high magnesium or low calcium. Their fidelity is to the "serpentine syndrome"—the environmental complex that constitutes the serpentine environment. The indicator value of a species goes beyond its utility in the identification and classification of floristically based plant community types. Indicators also have practical value in identifying land-use options. And edaphic indicators, especially those that track particular lithologies (serpentine, limestone, gypsum, etc.), are the pathfinder plants for geobotanical prospecting—using plants as indicators of mineral deposits (Brooks 1972; Cannon 1960, 1971).

Species as indicators of serpentine can be either narrow endemics or those with high but not absolute fidelity to serpentine (Figs. 5.57–5.59). When not narrowly restricted to serpentine, species of high fidelity often act as outliers—appearing on serpentine extending beyond the central range of the species as peripheral populations. Such distributions may be discontinuous; the gaps between the

5.57 Serpentine indicator species, though not endemic to serpentine can be highly faithful to ultramafic soils. *Aspidotis densa,*a telltale indicator fern on serpentines from the Gaspé Peninsula (eastern Canada) to the Pacific Coast. Although occasionally found on nonserpentine soils, it is on nearly every ultramafic outcrop in the West. Photo by author.

core populations and the outliers may even be substantial in ecological and spatial distances. Some examples will illustrate these modes of plant indication of serpentine.

Endemic Indicators. When serpentine endemics occupy a prominent place in the landscape, they provide telling character to the habitat. Two such endemics in California, *Quercus durata* and *Cupressus sargentii,* define the physiognomy of the community. Leather oak is a codominant shrub with the endemic *Ceanothus jepsonii* in serpentine chaparral, and the cypress defines the chaparral woodland (Kruckeberg 1985). In the Balkans, the endemic herbaceous perennial *Halacsya sendtneri* is the character species of several serpentine communities (Brooks 1987). So rich is the serpentine scrub of New Caledonia that no one endemic can serve

5.58 The serpentine form of *Adiantum aleuticum*, with imbricate pinnae, is abundant on wetter serpentine sites in the Pacific Northwest. Photo by author.

as the indicator species. Rather, endemics in certain genera characterize a particular type of maqui; for example, *Grevillea* and *Phyllanthus* define a low-elevation scrub type, while species of *Knightia* and *Hibbertia* characterize another scrub formation (Jaffré in Brooks 1987).

Bodenvag Indicators. Two California conifers, *Calocedrus decurrens* and *Pinus jeffreyi,* nicely illustrate the type of indicator that can show high fidelity to serpentine yet is widely distributed elsewhere in the same bioregion. Both species are common in the Sierra Nevada, mostly on granitic substrates. But in northwestern California, both serve faithfully as indicators of serpentine (Kruckeberg 1985). The North American fern *Aspidotis densa* (Fig. 5.57), while bodenvag in its propensity

5.59 The wild buckwheat, *Eriogonum pyrolaefolium*, while frequent on sterile nonserpentine soils (pumice and slate scree), is common on serpentines in the Wenatchee Mountains, Washington. Photo by author.

to occupy a variety of rock outcrops, comes into its own on serpentine: from the Gaspé of eastern Canada to the Pacific Coast, it faithfully tracks serpentine from sea level to the subalpine (Kruckeberg 1964, 1985). This fern occurs on fairly sterile sandstones and other sedimentary and igneous rocks; but it is far more gregarious on, and a good indicator for, xeric serpentines. A great many community types have been described for the Balkan Peninsula's serpentines (Tatic and Veljovic 1992). Most often the character species are bodenvag, but when they are in particular species combinations, they define a serpentine community. Three examples of such communities (phytocoenoses of the aforementioned authors) include the Calluno-Quercetum serpentinicum, the *Cotinus coggygria–Satureja thymifolia,* and the *Erica verticillata–Pinus halepensis* communities.

Serpentine Ferns as Indicators. Wide-ranging serpentine endemics present us with a special conundrum. Since the substrate is often discontinuously arrayed—serpentine islands or archipelagos intercalated among nonserpentine substrates—the question arises: How have spatially separated serpentine habitats become colonized with the same endemic or bodenvag indicators? For seed plants the question is vexing (and unresolved) since dissemination often requires large seeds; seed of endemic *Cupressus, Ceanothus, Umbellularia,* acorns of endemic oaks, and other large seeds are most likely dispersed locally (precinctively). The problem of colonization by serpentine ferns takes on a different character. Dissemination by airborne spores gets the ferns widely dispersed. But not so easily explained is the ability of spores of serpentine ferns to "find" a serpentine site. We may assume that these spores are part of the universally distributed load of airborne spores—a heterogeneous mix of spores of mosses, ferns, and other cryptogams with a wide range of substrate tolerances. Samples of airborne spores settle out everywhere on the planet, but only the serpentine fern species get established on serpentine—a "sieving" phenomenon of a highly selective kind. There is no proof that such a scenario is actually played out in nature. Only the circumstantial evidence of ferns occupying particular substrates across widely disjunct terrains justifies the hypothesis. Examples are readily at hand. In the New World there are several taxa that fit the discontinuous serpentine pattern: *Polystichum lemmonii* (Fig. 5.60) is discontinuous and yet obligate on serpentines from northern California to central British Columbia; *Aspidotis densa* (Fig. 5.57) is mainly on serpentine from Quebec to the Pacific Coast states; and the serpentine form of *Adiantum aleuticum* (Fig. 5.58) has a similar discontinuous distribution (Kruckeberg 1964). In Europe, examples abound, especially in the genus *Asplenium. Asplenium adulterinum* is an obligate serpentinophyte, widely distributed in the mountains of south central Europe, with a major disjunction in Fennoscandia (Tutin 1980). Two other European rock ferns, *A. cuneifolium* and *Cheilanthes marantae,* faithfully crop up on serpentines in several European countries; their discontinuous occurrences in Italy have been recorded by Vernano-Gambi (1992). Surely ferns aptly epitomize Beijerinck's law: "Everything is everywhere, but the environment selects" (Sauer 1988).

Cuba's "Serpentine Syndrome." Special places in the world reveal the overriding effects of serpentine geology on plant life. The Pacific Coast of North America, the Balkan Peninsula, and New Caledonia have the outstanding ultramafic displays. But there is one more locale—Cuba—where ultramafic soils are supreme arbiters over where a plant can grow or not grow. The Cuban serpentine story is here awarded special treatment not only for its spectacular attributes but also because it is the most recent to be given a thorough, scholarly treatment. I refer to the

5.60 Ferns occur on ultramafic soils in many places throughout the world. In the Pacific Northwest, three ferns are common on ultramafic areas from sea level to the subalpine. *Polystichum lemmonii* is endemic to mesic serpentine sites from northern California to British Columbia. See also Figs. 5.57 and 5.58. Photo by author.

major monograph *Phytogeography and Vegetation Ecology of Cuba* by the Hungarian geobotanist A. Borhidi (1991). Until Borhidi's book appeared, the vegetation of Cuba had been known only in rather inaccessible and scattered sources. Now this exhaustive opus opens the door to Cuba's botanical riches, with special attention given to the geoedaphic influences on the flora and in particular the consequences of major serpentine occurrences for Cuba's plant life.

Cuba, the largest of the West Indies or Greater Antillean islands, displays a high degree of environmental heterogeneity. Its habitat diversity is the combined result of climate, geology, and vegetation with all their manifold networkings. The island is a microcosm of habitat variety yielding diverse vegetation from xeric cactus scrub to tropical rain forest.

Prominent among the many habitat types are serpentine landscapes, from sea level to montane. Seven percent of the Cuban land surface is serpentinitic: 7500 square kilometers within a 105,007 square kilometer total island area. The island's major mountain systems transform a bioregional tropical climate into sharply contrasting climatic regimes. Borhidi enumerates five different bioclimatic zones for Cuba: (1) hot semidesert with cactus scrub or shrub forest on karst and evergreen shrub woodlands on serpentine; (2) winter-dry tropical zone, the most common, with an average mean temperature range of 20–27°C and average mean precipitation of 800–1800 mm; (3) tropical climate zone with two dry periods; (4) tropical climate, wet all year; and (5) tropical montane zone, wet all year, a regime that fits the *tierra templada* pattern in Latin America. The island's orographic diversity is the overriding influence that yields these several bioclimatic regions.

Cuba is an unsurpassed case history of geoedaphic influences on vegetation and flora (Figs. 5.41–5.44). Diversity of landforms, lithology, and soils conspires to promote a high degree of floristic variety. In Borhidi's words: "The richness of the flora and the variability of the vegetation in Cuba are explained by the varied edaphic conditions. The diversity of soils is attributable to the wide variability of rocks on the island (limestones of various ages, serpentine, dolomite, basalt, granite, diorite, gabbro, sandstone and slate) on which, as a consequence of the varied geological past, soil-developing processes of different duration proceeded." Cuba's serpentines occur as "three large serpentine ranges and nine smaller or larger separate lowland colline serpentine territories of 7500 km^2 area altogether." The high proportion of the endemic flora of Cuba that is obligately endemic to serpentine is remarkable and is reminiscent of the high endemism on New Caledonia's ultramafic regions. "[O]ne-third of the endemic flora of Cuba (920 species, 31.2%) has developed on serpentine areas covering not more than 7% of the whole country. [And] 14.6% of the total flora is endemic exclusively, to serpentine." A manifestation of the singular nature of the serpentine endemic flora is the high incidence of endemics at the generic level. "Of the 72 endemic phanerogamous genera, 24, i.e., 33.3%, live in serpentine areas" (p. 128). The singular serpentine flora occurs in several vegetation types. These are in most of Cuba's bioclimatic zones, from rain forest to dry scrub habitats. Under rain-forest conditions, two serpentine types are found: semi-arid montane serpentine rain forests (*Podocarpus* and *Sloanea* dominants) and the semi-arid montane serpentine shrubwoods (*Clusia* and *Ilex* dominants). In drier areas, Borhidi distinguishes two serpentine shrubwoods, one dominated by *Phyllanthus* and *Neobracea,* the other by *Ariadne* and *Phyllanthus.* Serpentinitic phases of coniferous forest are notable for their rich endemism; the dominant pines are *Pinus caribaea* and *P. cubensis.* Savanna and grassland landscapes have their share of serpentine vegetation types. Most notable is the dwarf palm savanna, with palm species of *Copernicia* and *Coccothrinax*. The herb layer of these communities is a rather rich mix of dwarf grasses and forbs. Substantial areas of savanna and grassland have been converted to agriculture, even on serpentinitic soils.

With serpentine forming such a significant component of Cuba's floristic diversity, it is not surprising to find the ecologist deriving a sense of pattern for the vegetation. Borhidi in fact develops a detailed review of the factors creating the "serpentine syndrome" as it has developed in Cuba. He calls the syndrome the "serpentine combination" and enumerates the various soil factors that constitute it. They are broadly applicable to other serpentine areas as well: (1) calcium magnesium ratios less than 1.0; (2) nutrient deficiency, especially in nitrogen and phosphorus; (3) calcium deficiency; (4) magnesium toxicity; (5) molybdenum deficiency; (6) high iron content; (7) nickel toxicity; (8) slow rock weathering (the dysgeogenous effect); (9) various physical properties of rock and soil; and (10) reduced competition. Some of these factors have been validated for Cuban serpentines. For the calcium/magnesium ratio, it is known that certain Cuban serpentinophytes (e.g., *Leucocroton havanensis*) can absorb calcium far in excess of the low calcium levels in the soil; this same species also accumulates nickel, while other serpentine endemics do not. Borhidi also points out that many Cuban serpentinophytes are accumulators of iron and that dwarfing of plants appears to be associated with high soil iron. It is not surprising that nickel hyperaccumulation occurs in Cuba, with its rich and lavish displays of serpentines. Borhidi et al. (1992) made qualitative tests for high nickel levels on 164 Cuban serpentine taxa: 38 species proved to be hyperaccumulators (>1000 ppm nickel in tissue) and 19 others were moderate nickel accumulators.

Substantiating Borhidi's field tests of high nickel accumulators in Cuba are the quantitative laboratory analyses by Reeves et al. (1996, 1999). They found a grand total of 130 species in many genera belonging to ten angiosperm families in Cuba that accumulate nickel in amounts well above 1000 parts per million: "The number of hyperaccumulators is greatest on the oldest serpentine soils, which are believed to have been available for colonization for the last 10–20 million years. Both Ni hyperaccumulators and serpentine endemic species generally are much more frequent on these old soils, occurring in the eastern and western extremities of Cuba, than on those developed within the last million years in the central part of the country" (Reeves et al. 1999). Hence Cuba's serpentines must hold the world's record for the largest number of hyperaccumulator species, even surpassing New Caledonia.

If serpentine soils possess a set of operational factors, summated as the "serpentine syndrome," then the vegetation on serpentine is expected to display several interrelated attributes. For the Cuba serpentine vegetation, the most evident physiognomic trait is its xeromorphy. In Cuba the serpentine vegetation consists predominantly of sclerophyllous shrubs and small trees; 40 percent of the endemic flora of Cuba is made up of these two life-forms. Borhidi (1991) proposes that xeromorphy is a general response to drought and nutrient stress; it can occur in other "stressed" vegetation types besides serpentine. Other idiosyncrasies of serpentine vegetation are noted: reduced stature and density, reduced productivity, and predominance of certain growth forms (pines, evergreen shrubs, and grasses); serpentine species form their own successional sequences, and often exhibit shifts in altitudinal zonation. These attributes have been noticed elsewhere on serpentines.

It should be apparent from this account of Cuban serpentine phytogeography and ecology that this Caribbean island's flora and vegetation hold a special fascination for the geobotanist. Further, the serpentine story for the island is only part of the geoedaphic fabric. Serpentine contacts other lithologies on the island: karst and other limestone formations, granite, gneiss, diorite, gabbro, dolomite, basalt, sandstone, and slate. And this diverse lithology is placed in a setting of diverse landforms. I can do no better than repeat Borhidi's view that "the richness of the flora and the variability of the vegetation in Cuba are explained by the varied edaphic conditions." But the best way to get the full impact of the influences of Cuban geology on Cuban vegetation is to immerse oneself in Borhidi's remarkable book.

Serpentine Geoedaphics—A Reprise

I confess to a special attachment to the world of serpentines and their plant life, an attachment that is both sentimental and scientific. The bias is acknowledged, yet perhaps defensible. The sentimental side is the consequence of having spent most of my professional career being captivated by the serpentine-plant connection. There is also a more objective impetus. One mode of discovery in science is to take a comparative approach—distinguishing the normal from the abnormal or one state from another. The comparative approach has served biological science well, all the way from the molecular genetic level to the organism, the species, and the ecosystem. By making comparisons—one state or phenomenon with another—we can discern and tabulate differences (and similarities), and then draw inferences or derive tests as to how the differences arose. When the contrasted states are dramatically different, as in vestigial versus wild-type wings in fruit flies, or with serpentine barrens versus a rich conifer-hardwood forest, the comparative approach is quick to yield results. The present account of serpentine botany tries to hew to that approach; normal, fertile substrates are contrasted with the infertile serpentine habitat. The comparisons have yielded a set of attributes—the serpentine syndrome—that defines the habitat and distinguishes it from nonserpentine ones. Thus far the comparison has been mostly ecological and physiological. Comparing flora on normal versus serpentine substrates inevitably leads to questions of relationships—the genetic and evolutionary connection between the two classes of biota. I defer discussion of these matters to Chapter 6.

For now, let me end this section with an aesthetic flavor. Beholders of the stark serpentine landscape have captured its austere essences in word and picture. In Olaf Rune's monograph on Scandinavian serpentines (Rune 1953), his frontispiece portrays a bleak Lappmark scene, captioned with a Viking poem (Havemal): "The pine tree wastes . . . which is perched on the hill, nor bark nor needles shelter it."

William Brewer, botanist with the California Geological Survey team in the 1860s, described the sere, stark nature of serpentine as he stood atop the New Idria serpentine barrens of San Benito County, California: "The view from the summit is extensive and peculiar . . . chain after chain of mountains, most barren and desolate. No words can describe one chain, at the foot of which we had passed on our way—gray and dry rocks or soil, furrowed by ancient streams into mountain canyons, now perfectly dry, without a tree, scarcely a shrub or other vegetation—*none,* absolutely, could be seen. It was a scene of unmixed desolation, more terrible for a stranger to be lost in than even the snows and glaciers of the alps" (Brewer 1949, pp. 139–140).

A modern nature writer is just as struck by the barren serpentine vistas as were the ancient Viking and Brewer. David Raines Wallace writes of his first encounter with serpentine in the Siskiyou Mountains athwart the California-Oregon border. In his charming book, *The Klamath Knot* (1983), he describes his astonishment when he emerged onto a serpentine "waste land" from a lush mesic forest. He devotes a whole chapter, "The Red Rock Forest," to this encounter. Snatches of his prose capture the character of the site: "The red-rock forest may seem hellish to us, but it is a refuge to its flora." His impressions of the serpentine scene continue: "I'd entered a landscape outside normal human ideas of life on earth. The smoky grayish foliage of Jeffrey pine and rigid, jadelike leaves of manzanita abetted the impression. It was as though growing on rock from the earth's deepest recesses had set the plants outside our sanguine historical view." But his final thought seeks out the heart of the conundrum of origins: "It is the obdurate physical adversity of things such as peridotite bedrock which often drives life to its most surprising transformations."

Rocks into Soils: Rock Weathering and Soil Formation

The conversion of fresh rock into soil (Fig. 5.61)—the processes of weathering—has been described as "a hopelessly complicated subject, with a multitude of processes operating on an endless range of rocks and minerals under a great variety of climatic and hydrological conditions" (Ollier 1984). Should this naive geobotanist dare venture into such a thicket and have any hope of coming out with a simplified view of this all-important activity at the surface of the Earth's crust? With the help of geomorphologists and pedologists (Ollier 1984; Jenny 1980; Carroll 1970; and Gerrard 1990), I am emboldened to try. After all, the theme of this book spans the realms of the purely physical (geology) to the crucially biological (surface phenomena in soils and especially vegetation). So let us seek some basic truths, or at least some first approximations.

The fresh rock beneath the Earth's surface (regolith) came into being and resides in place under one set of processes and conditions, far removed from those at the Earth's surface. Magmas are born—or transformed—at high temperatures and usually in the absence of an ambient gaseous aerobic atmosphere. In contrast, rock weathering to yield soils occurs at low temperatures (mostly in the range of biologically tolerant temperatures!) and in the presence of water, oxygen, and other gases of the Earth's atmosphere (Fig. 5.29). This set of conditions encountered by fresh rocks newly exposed at land surfaces creates the context for a definition of weathering: "Weathering is the breakdown and alteration of materials near the earth's surface to products that are more in equilibrium with newly imposed physico-chemical conditions" (Ollier 1984); see Table 5.7 and Fig. 5.62. The subject of weathering is hopelessly complicated, not only for all of the intersecting processes and ingredients, but because one process or product grades into another. Yet this is not a strange milieu for a biologist who is comfortable with, or at least well aware of the John Muir aphorism: "Everything is hitched to everything else." So when we say that weathering consists of three kinds of processes—physical, chemical, and biological—we know in our hearts that all three are intertangled. Yet it behooves us to look at each as though they are separated and sequential, for purposes of a simplifying first approximation. (As an aside, it took up to the Seventh Approximation for American soil scientists to create a classification of soils!)

PHYSICAL WEATHERING

This is the purely mechanical breakdown of rock caused by a variety of forces, within or external to the materials so transformed. Exfoliation or unloading is one

5.61 Talus and scree epitomize early stages of rock weathering in mountains, often harboring unique plant associations: coarse-textured talus of metadiabase rock (foregound) and fine-textured slate scree (middle ground), the latter with a richly diverse herbaceous flora. Wenatchee Mountains, Washington State. Photo by author.

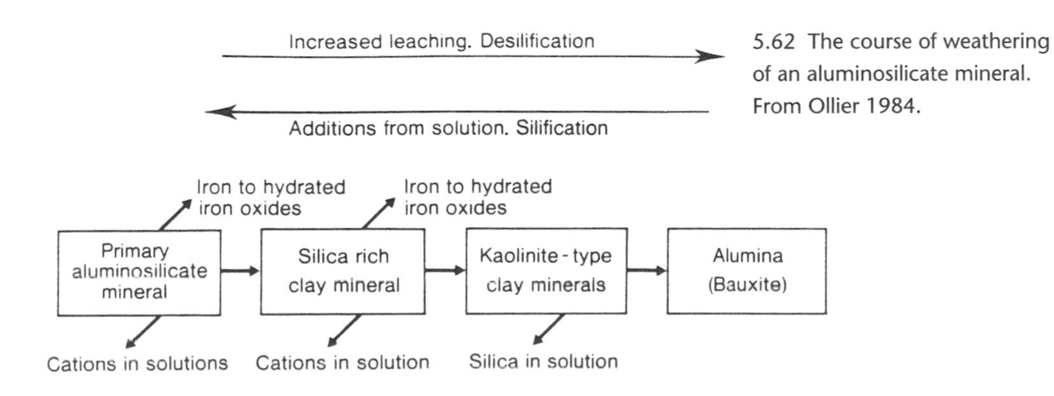

5.62 The course of weathering of an aluminosilicate mineral. From Ollier 1984.

such transformation wherein solid rock is fractured or comminuted by internal expansion or release of external load (e.g., retreat of glaciers) into fragments, often laminar. The granite domes around the world best show this initial weathering, although other massive outcrops can show sheeting or unloading (Ollier 1984). Ollier describes a number of variant forms of purely mechanical fracturing of rock. A type of physical weathering most evident to the temperate zone pedologist is frost weathering. The change in volume of water upon freezing induces great diagenetic effects. Working along planes of fracture (fissility), the expanding ice pries apart angular fragments of rock. A similar fragmentation may be caused by the growth and expansion of salt crystals (salt weathering, *fide* Ollier). If frost action is a major cause of weathering in temperate and cold regions, then changes in temperature above freezing can also weather rocks (insolation weathering). Rocks and their constituent minerals respond differently to temperature changes; so, even in the Sahara and other desert regions, temperature changes can induce physical weathering.

Once fresh rock is subjected to physical weathering, it simultaneously becomes "rotten" rock (saprolite): Physical and chemical changes associated with the surface environment creates saprolite. The changes from fresh to weathered or rotten rock are visible as color or textural changes. Chemical alteration may also have occurred, mostly due to the new oxidizing and hydrating environment. Chemical weathering is largely responsible for the creation of saprolite. It should be pointed out that the nature of the original fresh rock is still evident. Granite still looks like granite; only its surface properties have changed.

CHEMICAL WEATHERING

Parent materials, the weatherable rocks, are ultimately composed of elements arranged in particular configurations—the mineral species. When exposed to certain environments, the equilibrium states of minerals are disturbed, to yield partial breakdown (or transformed) products, mostly as ions or changed into other minerals or compounds usually with less free energy (Ollier 1984). Such reactions proceed only if the weathered products are removed from the site of the reaction. "Chemical weathering results essentially from chemical reactions of minerals with air and water" (Ollier). Solution is the most common first stage in weathering and is the primary means of carrying off the products of chemical weathering. Although water is the most common solvent, acids (e.g., carbonic and sulfuric) are also effective solvents. Other agents of chemical weathering include oxidation, carbonation, hydration, and chelation.

Oxidation of minerals occurs most often in the milieu of water, with its dissolved gaseous oxygen. Silicate minerals, like $FeSiO_3$, are oxidized, for example, to Fe_2O_3 and SiO_3. Since weathering in nature, by oxidation or any other process, takes place at low ambient temperatures, the reactions are slow. Some oxidations-reductions are carried on by microorganisms; reduction in anaerobic environments by bacteria is one such form of chemical weathering.

Carbonation, the reaction of carbonate or bicarbonate ions with minerals, is really a kind of acid weathering, with carbonic acid. Though carbonic acid is a weak acid, geologists now consider it a significant and potent mineral solvent that also assists in the base (cation) exchange process, so important in the interface between plant roots and the clay fraction of the soil.

Hydration, the addition of water to a mineral, plays a major role in clay mineral formation. Hydration also facilitates physical weathering by expanding the crystal lattice of the minerals; exfoliation and granular disintegration will result.

Chelation is a form of chemical weathering that involves organic molecules or complexes. The organic constituent can extract mineral ions from solution and then bind them within the organic molecule matrix. Organic molecules, as chelating agents in weathering, can come from decomposing plant remains (humus, etc.), or may originate as leachates from plant tissue (foliar leachates).

BIOTIC WEATHERING

While weathering can take place in the absence of organisms, their presence significantly enhances the weathering of rocks and rock debris. Plants, animals, and microorganisms all participate in the breakdown of rock. Organisms facilitate both the physical and the chemical weathering processes. The physical mode is aided by organisms exerting pressure and movement, thus exfoliating and transporting rock and rock fragments. Penetration of roots into fissile rock causes splitting or fragmentation. Organisms produce metabolites that facilitate chemical weathering: carbonic and sulfuric acids are notable weathering agents commonly of biological origin. Foliar leachates (a variety of organic metabolites released from leaves) can be effective chemical agents of weathering.

CLIMATE AND WEATHERING

I am not unmindful of the geoedaphic bias of this book. To redress the hiatus in much current ecological thought—the preoccupation with organism interactions and the scant attention given to the physical world in molding floras and vegetation—I have emphasized the manifold effects of geologic processes and products. This is not to discount the other two major environmental influences. Climate and organisms, particularly in the manifold ways organisms interact with each other and with climate, are not to be underestimated as determiners of why plants grow where they grow. And all three—geology, climate, and organisms—demonstrate an intricate functional interconnectedness. I acknowledge once more the holocoenotic view of the biotic and abiotic universes. It is particularly fitting to do so in the context of the processes of weathering.

The very word, weathering, is rooted in a climatic context. Climate, the sum of average weathers, determines the rate and course of weathering, in concert with geology and organisms. It is local to regional annual regimes of temperature and precipitation that contribute to the quality and quantity of weathering. In hot and dry (or cold and dry) climates, weathering is impaired; in hot and wet climates, weathering flourishes. The crucial role of climate in soil formation is succinctly portrayed in Hans Jenny's treatment of soil-forming factors (Jenny 1941, 1980). Climate (*cl*) is one of the five independent variables in the Jenny equation, $S = f(cl, o, r, p, t)$. Holding all but one of the variables constant provides a quantitative method for testing the effects of one of the variables. The emphasis in this book is on testing the effects of the two geoedaphic variables, *p* (parent material) and *r* (topography). A geomorphologist or a pedologist would want to put climate to a similar test, holding all the other factors but climate constant. That testing regime is the subject of many another book. A good start is the chapter on climate and weathering in Ollier (1984).

Two other aspects of weathering merit mention, as their physical and chemical attributes are germane to my geoedaphic bias. The way a particular mineral is constructed (e.g., its crystal size, shape, and the access it provides to weathering agents and to the removal of weathering products) can specify a particular rate of weathering. Thus two minerals of the same chemical composition can have different structures and thus be subject to different rates of *mineral weathering*. Calcite is more stable than aragonite, yet both are $CaCO_3$. In like manner, the architecture of rocks can condition the rates of weathering. Differing porosities and permeabilities of rocks are the variables in *rock weathering*. Such physical properties of rocks as bedding, jointing, brittleness, and response to pressure all determine the ease with which weathering agents gain access to the rock fabric. The three common extrusive, volcanic rocks have a range of weatherability: Basalt > Andesite > Rhyolite. In examples such as these, *p* (parent material) as the variable becomes a subtle arbiter in the outcome of the weathering process.

CLAY MINERALS, THE ULTIMATE PRODUCTS OF WEATHERING

The geology-plant connection is nowhere more intimate than in the link between plant roots and the clay mineral fraction of soil. It is in this milieu that the essential inorganic nutrient exchanges occur. Clay minerals in the colloidal state serve as the sites of cation exchange. And, of course the lithological connection is intimate: clay minerals come from the original unweathered rocks and minerals, the parent materials of soils. Clay materials are of fine particle (colloidal) size range ($<2\ \mu m$), and are mostly crystalline in structure (layered lattices). They may be tiny fragments of the parent minerals but most often are distinct crystalline structures of clay mineral type. "The common rock-forming minerals all weather to clay minerals "(Ollier, 1984, p. 74). Fig. 5.62 portrays the sequence of this transformation.

A particular species of clay mineral need not be in a stable state. Desilicification and silicification are the two major modes of converting one clay mineral into another. Desilicification, common in the tropics, leads to the formation of laterite (ferralite) or to bauxite (gibbsite), the extreme stage. A given clay mineral can be associated with specific zonal soil types. Kaolin predominates in lateritic soils; montmorillonite in chernozems, black earths, prairie soils, and humic gleys; and illite in podsols, brown forest soils, and tundra soils (Ollier, p. 75). Azonal soils like limestone and serpentine may be expected to have different clay minerals. Clay minerals of serpentine soils (Fig. 5.29) may be the same minerals as those of the unweathered parent material; or they may be a distinct clay mineral species, such as smectite or chlorite (Alexander, Wildman, and Lynn, 1985). Clays of limestones are more heterogeneous (diverse): in terra rossa soils the clay is kaolin, and the primary minerals mica and quartz predominate. But a rare clay mineral, attapulgite, serves as the limestone indicator. A dolomite in California weathers to yield a montmorillonite clay, while the limestones of eastern North America can yield illite and glauconite clay minerals (Jenny 1980).

CLAY MINERALS AND PLANT NUTRITION

The fitness of the environment for life, and life's fitness for the environment—a classic duality in biology put into words by L. J. Henderson (1913)—is exemplified in the unique match between the special properties of clay minerals and the nutritional requirements of plants. The ability to exchange essential cations for H^+ ions via the clay colloids must go back to the earliest occurrence of plants inhabiting the land. Weathering to produce clay minerals, coupled with the *cation exchange capacity* (CEC) of the colloids, has provided a ready source of inorganic nutrients, presumably ever since the early Paleozoic. Was this a fortuitous fit—plants preadapted to take advantage of the adsorbed cations? Or did the match of CEC and plant nutrition have to evolve? Most likely the ion exchange capability has been fine-tuned over time; selection for increased efficiency of ion exchange and uptake yielded a genetically determined, efficient mechanism. Both anatomical and physiological adaptations must have taken place. The transition from the simple rootless early land plants to the contemporary vascular land plants with their prodigious masses of roots and root hairs betokens a major adaptive trend. Similarly an adaptive shift in physiological efficiency undoubtedly has occurred over time and in response to a variety of ionic environments. Yet the actual evolutionary pathway of achieving accommodation to the CEC of clay minerals is unknown. We simply put faith in the uniformitarian principle: processes of the present (genetic fixation of adaptedness) took place in the geologic past as well.

The essence of CEC by clay (aluminosilicate) minerals is the bonding of cations electrostatically to the negatively charged surfaces of clay particles. This can be easily demonstrated by leaching a clayey soil with a neutral salt (KCl or NH_4 acetate). The cations of the clay are recovered in the leachate and the cations of the leaching solution are retained in the clay fractions of the soil. Indeed, this is the classic analytical method for determining the CEC of a soil. Different clay mineral species have different CECs: montmorillonite has a much higher CEC than does kaolinite or illite. The reservoir of adsorbed cations is substantial, but not unlimited; the exchange of adsorbed calcium ions for exchangeable hydrogen ions, for example, universally occurs and is facilitated by the secretion of H^+ ions by roots. Acidification of the soil can result.

The nature of the parent rock is an important determiner of the kind of clay mineral that emerges upon weathering. Basalt and other basic rocks usually produce the base-rich montmorillonite (smectite) clay minerals. More siliceous rocks (e.g., granites and arkosic sandstones) are likely to produce kaolinite minerals with a lower CEC. Yet any parent material may give rise to more than one clay species, depending on climate, microenvironment within the soil profile, and vegetation cover. Further, the kind of clay mineral may vary with depth in the soil profile (Ollier 1984).

A cautionary thought: The relationships among rock types, modes of weathering, kind of clay mineral, and particular soil properties are complex and beset by a host of alternative pathways and end products, and these often are not stable. It is reasonable to adapt the soil-forming factor equation of Jenny (1941) to any one of the steps from rock to soil. For example, the particular amount or kind of clay mineral can be a product of climate, topography, time, and organisms—not just parent material. The following references should be consulted for a fuller account of the origins and functions of clay minerals in soils: Greenland and Hayes (1978); Hunt (1972); Jenny (1941, 1980); Birkeland (1974); Gieseking (1975); Jeffrey (1987); and Etherington (1982).

SOIL CLASSIFICATION

Operational, effective environments for land plants are the intimate places where they grow and reproduce. All well and good to recognize the influences of regional climate and geology; they do set the bioregional stage for processes of soil formation and plant survival. Yet where an individual plant makes its home tells us that it is successfully coping with a very local milieu—its microenvironment. Of the several local influences, the nature of the soil is a crucial determiner of a plant's survival. Thus a plant senses a local geoedaphic state through the medium of the soil where it is rooted. Where the parent rock and/or its topographic setting are overriding influences on soil quality, then the geology—soil—plant interface is intimate. Azonal soils fit this model well: the effects of parent material or local topography override the effects of climate and organisms in forming the soil. The geoedaphic bias of this book easily extends to soil formation and the soil-plant connection.

Azonal soils, like those derived from limestones and serpentines, strongly reflect, in physical and chemical properties, their parent lithologies. Zonal soils, on the other hand, are primarily the products of a region's climate; any of a variety of igneous, metamorphic, or sedimentary rocks will yield nearly the same soil type—zonal soils—the product primarily of the prevailing climate.

The urge to classify objects in nature tempts all fields of scientific endeavor. Soil science is no exception. Classification of soils has gone through several phases, culminating in the current United States system, the so-called Seventh Approximation. A fine account of soil classification can be found in C. B. Hunt's (1972) text, *The Geology of Soils*. The classic of soil classification system consists of three Orders (zonal, intrazonal, and azonal), nine suborders, and 37 Great Soil Groups. Zonal soils develop soil profiles (vertically stratified layers) that result from the actions of regional climates. The six zonal suborders comprise soils from cold regions, soils from cool temperate areas (forested and nonforested habitats), and soils of arid regions to soils of warm temperate and wet tropical regions. Many of the Great Soil Groups of the zonal order have old familiar or cognate names, such as tundra soils, desert soils, chernozems, podsols, and laterite soils. Intrazonal soils result from local conditions rather than regional climate; under conditions of poor drainage or with exceptional makeup of parent material, intrazonal soils result. The last soil order, azonal soils, is a group of young soils (less than 2000 years old) that have a poorly developed or no profile. Those azonal soils, often called skeletal soils or lithosols, are intimate products of their parent material and topography. Much rock debris is contained in samples of lithosols. Many of the case histories I have described, where lithology looms large in influencing the resultant soil and plant cover, involve azonal soils. Table 5.23 (from Hunt 1972) and Hunt's accompanying text amplify this old established system.

The "new" *Soil Taxonomy* developed by the Soil Conservation Service of the U.S. Department of Agriculture is the seventh of the approximations leading to its publication. Like the older system, it is hierarchical: Ten Orders of highest rank are subdivided into suborders with Great Groups, Subgroups, and Families. The lowest rank is the soil series, that familiar taxonomic unit of the U.S. Soil Surveys and the field pedologist. The "soil series" is akin to the "species" of the plant taxonomist. This Seventh Approximation has met with mixed favor. Jenny (1980) calls it "a vigorous hybrid that aims to satisfy agricultural demands and pedogenic criteria." But others balk at the system's unwieldy, well-nigh incomprehensible nomenclature. Of the new system, Hunt (1972) is severely critical: "[The system's] merit is lost in its incredibly horrendous nomenclature, which bars the use of any simple English word." This is especially true of some of the names for ranks below the Order. Hunt expands on his critique of the cryptic jargon with some choice quotations. Just one quote here should make the point: "The Umbraqueptic Cryaquents are comparable to the Cryaqueptic Cryaquents in color values, and to the Orthic Cryaquent in chromas" (Soil Survey Staff 1975, p. 107, from p. 180 in Hunt 1972). Both Jenny and Hunt give detailed summaries of the New Taxonomy, at the level of Order. For further elaboration, especially for categories and their definitions below the Order, the reader should consult the "bible" itself: *Soil Classification, Seventh Approximation* (Soil Survey Staff, 1960).

It is instructive to determine how soil classification copes with soils in regions of high lithological diversity, all within a given regional climate. I have chosen Napa County, California, to portray this linkage. Napa County is located in cismontane California, the southernmost portion of the North Coast Ranges, where a Mediterranean type climate prevails. It is highly diverse both topographically and lithologically. With an area of 758 square miles and elevation ranges from near sea level to Mount St. Helena at 4000 feet (1231 m), Napa County is rich in land forms (valleys, hills, mountains, with drainage basins of marked slope and aspect heterogeneity). And all manner of rock types can be found. The Jurassic Franciscan sedimentary rocks (sandstones, shales, and chert) have been intruded everywhere by a variety of igneous and metamorphic rocks—notably gabbro, diabase, and peridotite/serpentinite—as well as Pliocene volcanics (rhyolite, andesite, basalt, and pyroclastics) (Soil Survey of Napa Co. 1978).

The unit of classification at the level of the Soil Survey is the soil series, the lowest yet most operational level in the hierarchy. The Soil Survey for Napa County recognizes 33 different soil series, most of which occur as two or three variant

Table 5.23. Zonal, intrazonal, and azonal soils (from Hunt 1972)

Order	*Suborder*	*Great soil group*
Zonal soils	1. Soils of the cold zone	Tundra soils
	2. Soils of arid regions	Sierozem Brown soils Reddish-brown soils Desert soils Red desert soils
	3. Soils of semi-arid, subhumid, and humid grasslands	Chestnut soils Reddish chestnut soils Chernozem soils Prairie or Brunizem soils Reddish prairie soils
	4. Soils of the forest-grassland transition	Degraded chernozem Noncalcic brown soils
	5. Podzolized soils of the timbered regions	Podzol soils Gray wooded, or gray podzolic soils Brown podzol soils Gray-brown podzolic soils Sol Brun acide Red-yellow podzolic soils
	6. Lateritic soils of forested warm-temperate and tropical regions	Reddish-brown lateritic soils Yellowish-brown lateritic soils Laterite soils
Intrazonal soils	1. Halomorphic (saline and alkali) soils of imperfectly drained arid regions and littoral deposits	Solonchak, or saline soils Solonetz soils (partly leached solonchak) Soloth soils
	2. Hydromorphic soils of marshes, swamps, seep areas, and flats	Humic gley soils Alpine meadow soils Bog and half-bog soils Low-humic gley soils Planosols Groundwater podzol soils Groundwater laterite soils
	3. Calcimorphic soils	Brown forest soils Rendzina soils
Azonal soils		Lithosols Regosols Alluvial soils

textural/slope phases; this brings the total of types recognized to 85. The series are separated into two broad topographic classes. The soils of the first class develop on nearly level terrain and are mostly alluvial in origin. They make up only 16 percent of the county's land surface. The other 84 percent are well-drained upland soils, for the most part primary or residual in origin; that is, they develop in place over parent rock. It is in the latter group where we find the strong impress of lithology and topography on soil attributes. For instance, soils over ultramafic rocks (mostly serpentinites) are mapped as two soil series: the Henneke Series is a gravelly loam occurring in two phases depending on slope (5 to 30 percent and 30 to 75 percent); the Montara Series is a clay loam with two phases, each with two slope parameters. These two series make up 6 percent of the county's land area. Often soil surveys in hilly terrain map the more rugged, precipitous terrain simply as "rough, stony ground" or as "rock outcrop." While these landscapes are unfit for agriculture, they are often the prime sites for native flora that is intimately anchored to some specific lithology.

While the soil series is the "species" of the Soil Taxonomy hierarchy and is the working unit in soil surveys, the county surveys since 1975 attempt to assign the series to their respective higher categories (i.e., the soil Family or above). Thus the two serpentine soils in Napa County, the Henneke and the Montara series, are classified as clayey-skeletal, serpentinitic thermic Lithic Argixerolls and as loamy serpentinitic, thermic Lithic Haploxerolls; both are of the order Entisols. It should be noted that both the Henneke and the Montara series are recognized for other counties in California where the parent rock is serpentinite. One may speculate that only another soil scientist familiar with the Seventh Approximation could find meaning in the diagnoses just given. It is no wonder that Hunt is critical of such arcane terminology.

Soil classification is subject to taxonomic revision, just as is any biological classification. A pertinent case is the change proposed for the Serpentinitic Mineralology Class of soils. Alexander et al. (1985) suggested expanding the class to include soils derived from other ultramafic rocks. The new nomenclature is Ultramafic Mineralogy Class, a grouping to include additional diagnostic minerals.

Other Lithologies (Parent Materials) Yielding Azonal Soils

The temptation to be encyclopedic in recounting the many variant lithologies that yield exceptional soils will be resisted here. After all, the prime examples—limestones, karst terrain, ultramafic rocks, and others discussed earlier—make the point tellingly. Lithology can override climate in producing soils with selective or

5.63 An "island" of Sierran conifers (mostly Jeffrey and ponderosa pine) on hydrothermally altered andesite, surrounded by Great Basin sagebrush (*Artemisia tridentata*) steppe on unaltered andesite. On Peavine Mountain near Reno, Nevada. Photo by D. Billings.

demanding effects on plant life. Yet other lithologies yielding aberrant soils should get at least "honorable mention."

1. *Hydrothermal areas*. Hot waters ascending from great depths cause crustal rocks, mostly of volcanic origin, to become chemically and physically altered. Often in hydrothermal belts, parent material rapidly weathers in the presence of sulfuric acid, to make an acid soil. In the Great Basin of western North America (Nevada), local hydrothermal areas support stands of Jeffrey and ponderosa pine, wholly surrounded by sagebrush (*Artemisia tridentata*) on nearby normal andesite rock (Fig. 5.63). The isolated stands are "conifer islands" in a "sea" of sagebrush (Billings 1950; Delucia, Schlesinger, and Billings 1989; Trimble 1989). Hydrothermally altered volcanic rocks are bound to occur in other parts of the world, though I know of none that portrays such a striking response in the vegetation.

Geothermal belts also foster other local "hot spots." Solfataras, fumaroles, and hot springs are known to support distinctive floras, from blue-green algae to flowering plants. High temperatures as steam or water as well as concentrations of toxic chemicals (sulfuric acid, mercury, etc.) are common attributes of such sites. An example is the hotspring panic grass (*Panicum thermale* = *P. acuminatum* var. *acuminatum* in Hickman 1993) that flourishes around hot springs in the Big Geysers area, California (Spellenberg 1968).

2. *Substrates with mineral deficiencies.* Many of the azonal soils discussed earlier have singular effects on plants by virtue of being deficient in essential elements. Serpentines deficient in calcium and phosphorus, and shale barrens with several elemental deficiencies, typify the nutritional poverty of these substrates. But there are still other parent materials that yield nutritionally impoverished soils. Fossil soils or reworked sedimentary rocks that reemerge as parent materials for another round of weathering and soil formation can be poor in nutrients. The Eocene Ione formation of California is one such geoedaphic case. The substrate is derived from an ancient laterite formed under tropical climate. The resulting Ione soils, a mixture of kaolinitic clay, quartz sand, and ironstone, is a sterile milieu that excludes much of the regional flora. Yet it supports a distinctive acid heath vegetation, including the narrow endemic, Ione manzanita (*Arctostaphylos myrtifolia*). Gankin and Major (1964) conclude that the Ione acid heath maintains itself in the absence of competition from the surrounding flora.

A similar acid heath in California, the pygmy pine barrens (called the "pygmy forest ecological staircase," by Jenny 1980), developed on acid, sterile soils. The pygmy forest soils got their anomalous character from a unique topography (Jenny, Arkley, and Schultz 1969; Westman 1979). Four terraces underlain by sandstone were elevated during the Pleistocene and then remained in place (Fig. 5.64). The slopes bordering the terraces are stocked with a mesic, mixed coast redwood (*Sequoia sempervirens*) forest; the windward edges of the terraces, aeolian in origin, support Bishop pine (*Pinus muricata*) and *Rhododendron* heath. The terraces proper, though, are the startling feature: pygmy conifers and ericaceous heath on acid (highly podzolized) soils underlain with an impervious hardpan. The pygmy conifers, *Pinus contorta* subsp. *bolanderi* and *Cupressus goveniana* subsp. *pigmaea,* both endemics, coexist with a heath vegetation, including the endemic *Arctostaphylos nummularia* (Kruckeberg 1991b).

Ancient soils can serve as parent materials for modern soils (Hunt 1972, chap. 9). The older (fossil) soil overlies the original parent rock; yet the modern soil develops from the fossil soils, not the original parent material. In such cases the nutritional status of the modern soil must depend on the nutrients of the fossil soils. If the latter has been depleted nutritionally, the derived soil will also be deficient in nutrients. A common variant of ancient soils giving rise to a modern soil is the superposed profile, each stratum younger than the one below it. Such profile

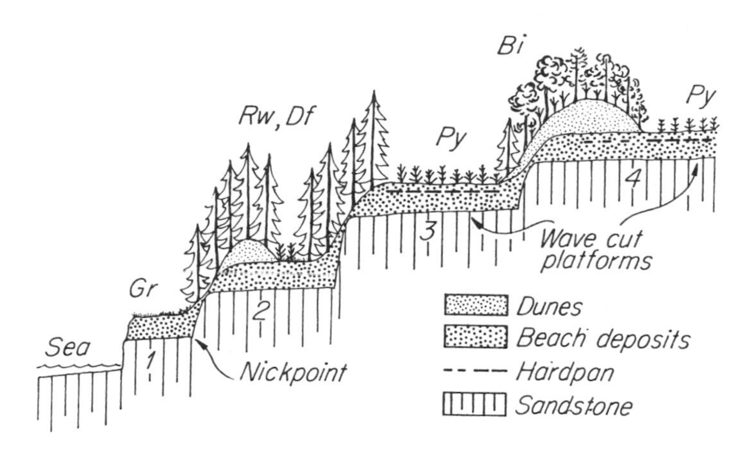

5.64 Sequence of four marine terraces, Mendocino County, California; young dune on second and very old dune on fourth terrace. Gr = grassland. Rw, Df = redwood and Douglas fir forest. Bi = Bishop pine forest. Py = pygmy forest. Horizontal distance is 3 miles; vertical distance is 500 feet above sea level. From Jenny et al. 1969.

sequences are found in places where Pleistocene and post-Pleistocene drift has buried one or more earlier profiles (Hunt, chap. 9). The buried profiles can be the parent materials of the younger soils as well as the drift itself.

3. *Substrates with trace elements* (*toxic, neutral, or micronutrient*). Given the rich elemental variety in the Earth's crust, plant life is exposed to a great many different elements. Besides the essential macronutrients, a multitude of elements exist in trace amounts (<0.1 percent) in soils and parent materials. Nearly every metallic element of the periodic table occurs in trace amounts (or greater) somewhere in the Earth's crust. Trace elements initially in the parent rock become displaced through weathering and mobility into the soil profile.

The most remarkable interactions between trace elements and plants are those instances where an element toxic to most plants is accumulated with impunity by species indigenous to the soil. Plants that take up concentrations exceeding 1000 ppm are called hyperaccumulators (Brooks 1987). The so-called heavy metals like copper, lead, nickel, and zinc are well known as elements (cations) taken up in excessive concentrations by plants tolerant to the mineral. See Brooks (1998), Shaw (1989), and Kabata-Pendias and Pendias (1984) for recent reviews of heavy metals and trace elements in the realm of plant ecophysiology. Some of these toxic elements occur as contaminants (in mine spoils, as pollutants of air and water, etc.). The same elements can also occur in nature spontaneously, both in parent materials and in soil profiles. Thus, nickel is a common element in ultramafic rocks. Copper is associated with a variety of rocks, but is most concentrated in metalliferous rocks, as in the orogenic belts throughout much of south central Africa (Brooks and Malaisse 1985). Lead is a notorious contaminant of soils from a variety of human activities (lead emissions from internal combustion engines, mine tailings, etc.); lead also can occur in substantial amounts in acidic magmatic rocks

(Kabatas-Pendias and Pendias 1984). Besides the widespread toxicity of lead to intolerant plants, genetically fixed racial tolerance to lead is a well-known microevolutionary phenomenon. Hyperaccumulation of lead and tolerance to lead are known to occur in Europe and elsewhere. A notable example is *Thlaspi rotundifolium* var. *cepaeifolium* (Brassicaceae) of northern Italy and adjacent Austria. It accumulates up to 8200 μg/g (dry weight) lead (Brooks 1998). Unique among hyperaccumulators, this *Thlaspi* has the capacity to accumulate both lead and zinc. Zinc occurs widely in both igneous and sedimentary rocks, and is also a frequent soil contaminant of anthropogenic origin. Plant responses on zinc-contaminated soils range from tolerant races of widespread species (e.g., races of *Silene cucubalus, Minuartia verna*) to species endemic to zinc substrates (e.g., the zinc violet, *Viola calaminaria* and *Thlaspi alpestre* subsp. *calaminare*). I have been unable to verify the occurrence of hyperaccumulators of zinc.

The element selenium, although not in the "heavy metal" group of elements, has long been known to be taken up in high amounts by plants. Further, selenium-accumulating plants are highly toxic to animals, and therefore accumulator plants are of economic significance. Selenium mimics, and replaces, sulfur in organic compounds essential in metabolic pathways. Perhaps the most well known among cases of selenium hyperaccumulation are various species of *Astragalus* (Fabaceae) in western North America; because of their adverse effects on livestock, they bear the common name loco weeds. See Brooks (1998) for more on selenium accumulation by plants and the use of these plants in bioremediation of seleniferous soils.

Several of the heavy metals are essential elements in trace amounts, only to become toxic at high concentrations. The ability to tolerate excessive concentrations of (and even to accumulate) various heavy metals has come to be regarded as a model system in evolutionary accommodation to a habitat—all the way from tolerant populations to tolerant endemic species. Heavy metal tolerance now has a rich literature (Baker 1987; Antonovics, Bradshaw, and Turner 1971; Shaw 1989). We will return to this intriguing link between heavy metal habitats and evolutionary accommodation in Chapter 6.

Wetlands Caused by Geologic Events

Surface heterogeneity—that most basic ingredient of landscapes—is a prime cause of conditions creating most wetlands. Uneven terrain includes depressions and drainage channels, occluded (impervious) in varying degrees, and occurs in myriad sizes and shapes. If such occluded landforms have impaired drainage, then water collects and a wetland environment comes into being. Topography conducive to the genesis of wetlands results from the actions of internal, external, and even extraterrestrial processes and unique events. Some authors (e.g., Hunt 1972) conceive of these as geological in origin, either directly or once removed. Thus the internal forces of volcanism and tectonism directly promote concavities (basins and depressions) in which wetlands can develop. The external forces of water, ice, wind, gravity, and organisms may act in concert or separately to create wetland conditions. The extraterrestrial source, meteorite impacts, may be a significant cause of depression topography in unpredictable places on the planet. Another method of partitioning causes of wetlands is to recognize physical versus biogenic causes. Four types of physical causes are evident:(1) geomorphic (tectonic and volcanic), as well as many degradational phenomena; (2) lithological, properties of rock and rock formations; (3) pedological, various degradational changes to soils; and (4) hydrological and climatic. In contrast to these mainly physical causes, biogenic influences include the actions of microorganisms, plants, animals, and humans. When any of the above create the kind of surface heterogeneity that impedes drainage, wetlands can develop.

Earlier in this century, the word wetland was largely unknown. Instead, names for various types of wetlands dominated the aquatic biology literature: marshes, mires, bogs, fens, swamps, and moors. The contemporary term, wetland, is thus generic—embracing all such wet habitats. Further, it became the buzzword of the environmental movement in the waning years of the twentieth century. Disappearance of wetlands at the hand of man has caused global concern for their preservation and critical awareness of their unique ecological roles. Efforts to preserve remaining wetlands are a common focus of preservation activities.

But what is a wetland? The U.S. Army Corps of Engineers' formal definition is a good start (Weinmann et al. 1984): "Those areas that are inundated or saturated by surface or ground water at a frequency and duration sufficient to support and that under normal circumstances do support, a prevalence of vegetation typically adapted for life in saturated soil conditions. Wetlands generally include swamps, marshes, bogs, and similar areas." Wetlands embrace bodies of water of all sizes and include both inland and coastal (tidal and estuarine) partially submerged land. Their configuration, location, and other physical attributes receive a strong causal impetus from geological processes.

TOPOGRAPHIC ORIGINS OF WETLANDS

Montane topography richly fosters conditions for wetlands all the way from alpine lake margins down to the lowland base of mountains, with its deltas, alluvial fans, and the like. Impeded drainages are the most common sources of wetlands; and such conditions are usually caused by geomorphic processes such as the making of depressions and declivities. Both alpine and continental glaciation—as well as deglaciation—produce a rich variety of landforms where water can accumulate: cirques, kettles, at the bases of kames, moraines, and eskers. While the root cause of glaciers is presumably climatic, the most immediate results for yielding wetlands are geomorphic. Glaciers make, and leave behind, surface heterogeneity as well as impervious substrates, some of which can be the makings of wetland topography.

Tectonic movements resulting in the deformation of the lithosphere can create landforms with local to vast areas of impaired drainage. Earthquakes are well-known causes of depressions and dammed terrain that can give rise to wetlands. Recent studies along the North Pacific Coast of North America reveal the creation of wetland conditions caused by slumping due to earthquakes (Atwater 1992). Land slippages may create water impoundments by means other than earthquakes. Unstable slopes can slump by mere gravity. Colluvial materials from mountain slopes can form impoundment basins. All the way from local soil creep to major mass wastages, or rock avalanches—each may provide the dam for impounding water. Fens at the base of serpentine slopes are often formed by slumping of rock and soil from upslope. The famed *Darlingtonia* fens of southwestern Oregon, with their remarkable wetland flora (e.g., *Darlingtonia californica, Cypripedium californicum, Hastingsia* spp., etc.) portray well the biological consequences of this type of wetland (Becking 1986; Bowen et al. 1982; Coleman and Kruckeberg 1999; Kruckeberg 1992).

Surface heterogeneity on a grand scale comes with volcanism—both the volcano-producing type and the lava-emitting fissure type. Volcanic materials can create landscapes with impaired drainages. Surface lava flows on flat or sloping terrain can make dams for wetland genesis. The great lava fields of western North America are pockmarked with depressions, variously irrigated and accumulating wetland vegetation. But the caldera lakes are the most spectacular basins of volcanic origin. Wetlands in caldera lakes are mostly confined to shoreline strips at the water's edge. Such basins, while of geologic origin, succumb to biogenic succession. The lake fills in to become a bog or swamp. The famous wetland of Alakai Swamp in Hawaii has just such a history; cap rock filled the crater, which in turn formed an organic layer of bog vegetation (Carlquist 1970).

WETLANDS CAUSED BY LITHOLOGY (BEDROCK)

The emplacement and specific weatherabilities of rock formations can create catchments for water, leading to wetland vegetation. The disintegration of solid rock by purely physical weathering strips away slabs of rock along bedding or fracture planes. If this occurs irregularly—some rock fragments shifting away from their positions while intact rock segments remain in place—then depressions can form. Deep vertical fissures in bedrock may tap a groundwater source to become, at the surface, springs or seeps. This kind of wetlands is in fact a common phenomenon in ultramafic rock outcrops. Here and there along montane slopes of exposed serpentine rock (or its igneous forms, peridotite and dunite), one inevitably encounters "islands" of wetland vegetation wholly surrounded by the more xeric-appearing serpentine flora (Hardham 1962; Kruckeberg 1985). The astoundingly spectacular *Darlingtonia* fens of northwestern California and southwestern Oregon, mentioned earlier as forming by slumping of slopes, may also originate in the manner just described (Coleman and Kruckeberg 1999).

Perhaps the most spectacular of all wetland habitats produced under unique rock formations are those created by the limestones of karst topography. Rapid differential weathering of limestone creates a host of catchment basins, variously known as dolines, sinkholes, cockpit landforms, compound or valley sinks, and the largest of depressions, called poljes (White 1988). All such depressions, if they have impaired drainage, foster wetland vegetation. Good examples are described by Poldini (1989) for karstic regions near Trieste (Italy and the upper Balkan Peninsula), for the basal sectors of mogotes (karstic towers) in Cuba (Borhidi 1991), and for impeded drainages in the cockpit karst of Jamaica (Asprey and Robbins 1953).

OTHER SOURCES OF WETLANDS

Soils with properties that impede drainage can cause wetland conditions. Level, undulating, and sloping terrains are all sites for edaphically induced wetlands. Impeded drainage occurs anywhere down through a soil profile; any horizon that prevents migration of water can yield wetland conditions. Impervious clay horizons, hardpan layers, or impervious subsoil and parent materials provide the physical states for impaired drainage. Such edaphic conditions can exist in salt, brackish, or freshwater locales. Fine-textured materials contributing to impaired drainage arise mostly from sedimentation; deposits of such materials are common in estuaries, lakes, and river meanders.

Many other geomorphic configurations and processes yield the potential for wetlands. Periglacial, permafrost, and glacial recessional environments (kettles, morainic dams, cirques, etc.) can promote wetland conditions, and any tectonic, volcanic, or geomorphic phenomena that foster impeded drainage in the concavities of heterogeneous surfaces will provide the hydrologic setting for standing water and resultant wetland vegetation. These preconditions are geoedaphic, but the complementary component is excess moisture; only in humid regions where water is in surplus—where runoff exceeds all other environmental demands—do wetlands develop (Garner 1974, p. 306). Hence the linkage of geologic processes with climate is imperative in accounting for the genesis of wetland conditions.

A special type of ephemeral wetland—the vernal pool—is frequently encountered in the Mediterranean climates of California and Chile. Their geomorphic source is the "hogwallow," or Mima type microrelief—arrays of hillocks and depressions, the latter with impaired drainage. The terrain on which vernal pools develop in the Central Valley of California had its origin in a geosynclinal basin. No net gain in the sediments occurs, and the valley soils are mature. They consist of profiles with pedogenic hardpans and/or claypans, often a meter or more thick; they serve as barriers to downward movement of winter rains. This results in water tables perched above the well-developed B horizon. On such poorly drained soils, vernal pools develop where the land surface is hummocky, of the Mima mound type (Holland and Jain 1981; Zedler 1987). But since rainfall is restricted to fall and winter, the pools are transient (vernal), gradually drying up as spring gives way to summer. This progression from pool to dry depression fosters a wave-like progression of different species in flower, some narrowly restricted to the vernal pool habitat.

If wetlands epitomize the synergism of geomorphology and soils with climate, then those wetlands appearing to be solely of biogenic origin add the third dimension—biota, especially vegetation. Late stages of peat bog development appear to be wholly biological in their configurations, makeup, and even their ontogeny. Yet the vegetation of bogs, mires, swamps, and marshes is dependent for its realization on the two physical parameters—concavities and impeded drainage—as well as uneven terrain and rainfall to exceed runoff. So we can invoke the geoedaphic influence on the biogenic wetlands only indirectly.

Plant Responses to Variations in "Normal" Lithologies

It has been easy to find and portray strong linkages between exceptional lithologies and floras that control the quality of plant life. Recall from earlier in this chapter the striking effects of ultramafic and carbonate rocks on plants. Such lithologies and their botanical consequences form much of what the soil scientist calls azonal soil ecosystems. But in many parts of the world, lithologies leading to azonal soils do not exist. Rather, the bedrock displays normal chemical and physical properties; these parent materials weather in close response to regional climates to become zonal soils.

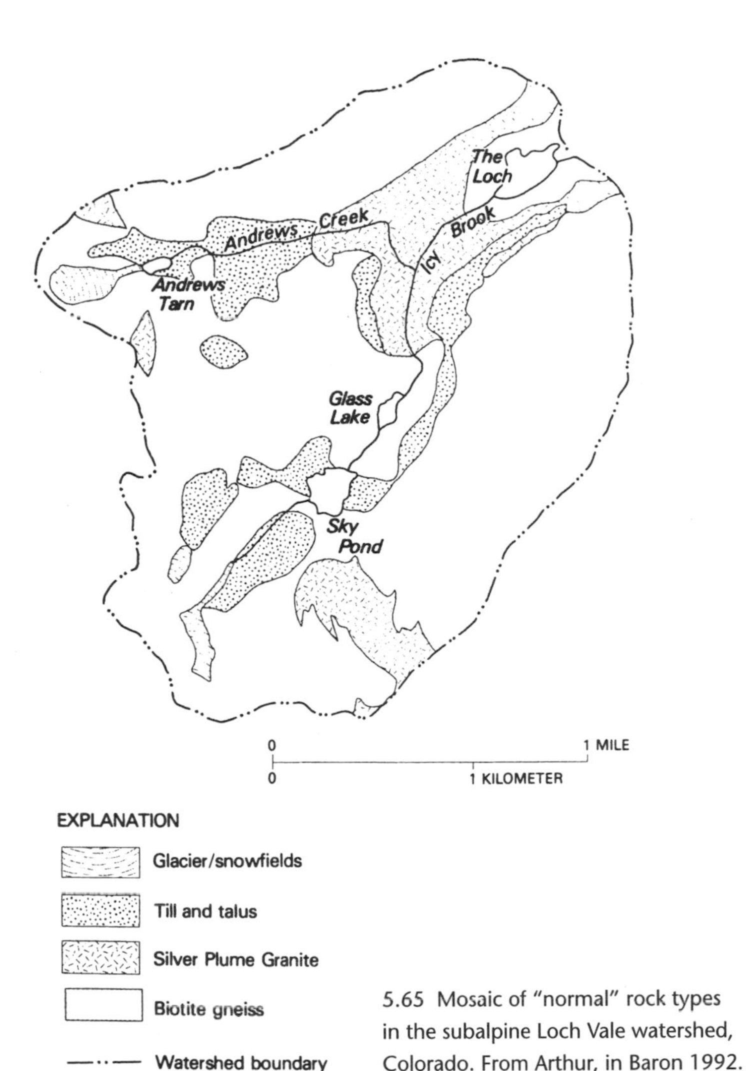

5.65 Mosaic of "normal" rock types in the subalpine Loch Vale watershed, Colorado. From Arthur, in Baron 1992.

Yet the perusal of most geological maps, especially of montane regions, usually reveals a complex mosaic of "normal" lithologies—granites abutting volcanics or sandstones and other sedimentary rocks, and these contacting metamorphic rocks like schist and gneiss. Figure 5.65 portrays just such a mosaic. Yet all the lithological variants in such mosaics of crustal geology are "normal" in the sense of weathering to make zonal soils—soils compounded of the typical elements and attributes required for normal (balanced) plant nutrition. Two early papers (Fernald 1907; Lutz 1958) eloquently addressed this matter of plant responses to different but "normal" soils and lithologies; they are discussed in Chapter 2.

So, what is the geoedaphic question posed by such mosaics of normal lithology?

The question is easy to frame, but it has a vexing paucity of answers: Are there vegetational and floristic differences detectable as one goes from one normal lithology to another—stepping, let us say, from a granodiorite to an arkosic sandstone, or from a rhyolite onto a schist? The expectations from such encounters are several: (1) There is no detectable difference in plant composition from one parent material to the next. (2) Differences exist but are subtle, detectable only quantitatively—variations in frequency, density, gregariousness, and so forth—the usual variables of community ecology, with no change in species composition, just small differences in community statistics. (3) Differences are detectable not only in community statistics but in the makeup of the floras on the adjoining normal lithologies. (4) Differences are pronounced in both species composition and vegetational attributes on neighboring lithologies, presumably due to differences in microtopography, soil physics, mineral nutrition, or to other pedological causes.

Case histories abound when the subject is a lithosequence that includes sharp contrasts in rock type and vegetation—zonal abutting azonal lithologies. We have already paid tribute to the now classic studies by Robert Whittaker (1960) where along the Oregon-California border the lithologies of the Siskiyou Mountains trend from diorite and gabbro to serpentine. The vegetational response to this lithosequence is dramatic, both in physiognomy of the plant cover and the species composition. A similar study by Whittaker and Niering (1968) revealed distinct contrasts along transects from acid soils and rocks to limestone substrates. Jenny (1980) cites two cases where the lithosequence consists of what I am calling "normal" rock types—with no sharp contrasts in physical or chemical properties.

Jenny's first cited study was carried out by Welch and Klemmendson (1973). They found that nutrient and biomass for ponderosa pine (*Pinus ponderosa*) and grassland in northern Arizona varied along a lithosequence involving several volcanic parent materials (basalt, andesite, rhyolite) and a sedimentary rock (limestone). Biomass on basalt contained 1.5 to 2 times as much nitrogen and carbon as the rhyolite soil did. No mention is made of any significant floristic differences for the several parent materials. It would appear that the changes in parent material do induce changes in the pine/grass ecosystem, without materially affecting the plant cover. As predicted, the biotic consequence of changes in normal parent materials along a lithosequence will be subtle, and in this case revealed as a nutritional consequence of substrate differences, interacting with the biotic factor, but without detectable floristic changes.

The other study cited by Jenny (1980) involves a forest ecosystem in the Siskiyou region of southwestern Oregon (Parsons and Harriman 1975), ranging across a rich variation of parent materials (granite to diorite, schist and gneiss, and pyroclastics). The biotic response to this lithosequence of normal rock types was found to be a change in biomass production, with the pyroclastic soils having the highest productivity. The forest dominants, Douglas fir (*Pseudotsuga menziesii*) and grand fir (*Abies grandis*), remained the same along the sequence of substrates. Brief mention is made of differences in species composition of the subordinate plant cover of the woodland, from one lithology to the next: "Salal (*Gaultheria shallon*) and oceanspray (*Holodiscus discolor*) are the most abundant shrubs on the granitic sites, whereas Cascade Oregon grape (*Berberis nervosa*) is common on the sites underlain by schist and pyroclastics. Vine maple (*Acer circinatum*) comprises 50% of the ground cover on the gently sloping site underlain by pyroclastic rocks. Western fescue (*Festuca occidentalis*) occurs on each site" (Parsons and Harriman 1975, p. 944).

Whittaker's (1960) discoveries of vegetation changes across lithological discontinuities also was based in the Oregon Siskiyous. But in contrast to the Parsons and Harriman study, the Whittaker analyses focused on the striking contrasts of diorite to serpentine. However, he did observe floristic change among various rock types other than the mafic rocks (gabbro and serpentine), especially in the Kerby quadrangle and the Grayback Mountain area.

Given the paucity of ecological studies of normal lithosequences, the least one can do is to predict the possible effects of vegetation and flora—in the absence of hard data. One expects the particular properties of rocks and soils to vary along a lithosequence. Variations in the physical properties of adjoining rocks could promote subtle to overt vegetational responses. Differences in weathering of adjacent rocks make them prime candidates for consequent biotic responses. The many different attributes of weathering (hardness of rock, bedding planes, angle of repose, etc.; see Ollier 1984) can affect microrelief, soil depth, texture, and porosity. Differences in weatherability may occur as a mutual effect, where either two rock types meet or different rocks are embedded in a common matrix (e.g., conglomerate or breccia).

Transitions from solid rock to talus and alluvial fans can evoke differences in vegetation. Such catenas may foster plant composition differences even within the same parent material (see Fig. 5.14), and also show differences where different but adjacent lithologies form parallel catenas. In such instances, angle of slope and physical weathering join to create distinct catenas (Jenny 1980). In terms of the vegetation response to such toposequences, one can imagine the following: On a steep face of a mountain slope there may exist two or more contrasting lithologies, say a granite adjoining a chert. Their neighboring catenas could display consolidated rock upslope (ledges, cliffs, rock walls, crevices) with the weathered rock debris below. The granite talus and alluvium might consist of smaller rocks, fragments, and sandy soil in the alluvial fan, since granite (especially decomposed granite) is a softer rock. In contrast, the very hard chert may yield downslope much larger rock fragments and a thin clayey soil in the alluvial end of the catena. These contrasting states should yield floristic and vegetational differences all the way from the upper rock walls to the lowermost alluvial fans.

A transect along changes in "normal" rock types can be expected to exhibit variations in chemical content, from one parent material and its soil to the next. The basis for this assertion is, of course, mineralogical. The constituent rocks of a lithosequence are bound to differ in mineral composition in both quality and quantity. In turn, the different mineral species of the rock sequence will release different elements upon weathering; so adjacent soils in a lithosequence should have both different amounts of the same elements and, likely, different elements. Table 5.4 gives the elemental composition of rocks likely to occur together. A further chemical variable along a sequence of rock types is likely to be changes in the clay fraction and the derived soil—variations in both species of clay colloids and in their absolute amounts (percentage of the clay fraction in the soils). Included in Table 5.18 are examples of variations in the clay minerals, depending on both the mineral composition of the parent rock and the nature of the weathering process. Different clay minerals and their proportionate amounts in soils will directly affect the nutrient status, mostly via exchangeable cations on the base-exchange complex of the soil. From such variant chemistry along a lithosequence, the expectation is to discover that the plant life is affected in some form. Case histories to examine this thesis are hard to come by; most often geoecologists and pedologists have preoccupied themselves with those lithosequences where the contrasts in rock types are sharp. Given below are a few examples illustrating the effects of chemical change along a lithosequence.

The ideal case study of geoedaphic/plant variation along a lithosequence should have the following ingredients: (1) The different rock types must be adjacent, sequential along a given contour level, and with the same slope and exposure. Ideally this could be found in a single drainage basin. (2) Differences in the chemical composition and weatherability of the adjacent rocks are detectable. (3) Transects along the lithosequence reveal quantitative and/or qualitative differences in flora and vegetation.

A study that comes close to meeting these specifications was carried out on a single subalpine drainage basin in the Colorado Front Range, Rocky Mountain National Park (Baron 1992). The aspect of their total ecosystem approach that is germane to the issue of lithosequences is the geology of the Loch Vale watershed.

Rock outcrop dominates the subalpine landscape, with biotite gneiss interfacing Silver Plume granite (Fig. 5.65). Glacial till and talus from the third parent material lies adjacent to the two bedrock types. Chemical analysis of the two outcrop types reveals close similarities, yet with subtle differences: calcium, magnesium, and iron levels are higher in the gneiss while values for potassium are greater in the granite (Mast 1992). Mast examined the consequences of postmetamorphic alteration of the rocks where they contact at faults and fractures. Hydrothermal effects changed the mineralogy of the rocks where, for example, epidote and sericite replace plagioclase and chlorite, which in turn replace biotite. Differences in weathering effects were also noted, both as to rates and end products (differences in clay species). Given this lithological and pedological variety, one looks for mention of biological consequences. At least in the published record of this intensive study, there is no indication that the rock/soil catena affects the flora in any detectable manner. Only 18 percent of the watershed is vegetated, and this is partitioned into subalpine forest (Engelmann spruce, subalpine fir, and lodgepole pine) and wet meadow (Arthur 1992). Possibly the soil/rock differences make no significant (detectable) impact on the vegetation (J. Baron, pers. comm.). Or, has the investigation simply not looked at the possibility of any botanical consequences of the lithosequence?

Concluding Remarks and Summary—Rocks, Soils, and Plants

It should be evident by this, the longest chapter in the book, that lithology makes many and significant liaisons with the plant world. For plant ecologists and geobotanists, it is the parent material (rocks, saprolite, and rock debris) that makes the geology-plant connection most apparent. And the connection is most telling when the rock has some exceptional physical and/or chemical properties. Thus the bulk of the chapter is devoted to the lithologies that give rise to azonal soils and distinct floras—serpentines and other ultramafic rocks, limestones and dolomite, shale, gypsum deposits, et cetera. The first two, the ferromagnesian silicate and the carbonate rocks—and their effects on plant life—have provoked a massive output of published research and speculation in edaphic plant ecology. These two rock types are truly the heart and soul of what I have called *geoedaphics* throughout the book. For these widely occurring azonal lithologies, I have dealt with both their geological and botanical attributes. The petrology, mineralogy, and rock weathering, as well as the chemical constituents of rock and soil, tell the geological side of the azonal story. The plant response to these substrates has been told here in terms of vegetational and floristic attributes: physiognomy, adaptation, endemism, edaphic races, physiological and community ecology, plant nutrition, and so forth.

But there is more to the Earth's crustal lithology encountered by plants than the unusual (azonal) rocks. The last part of the chapter looks at the possible biotic effects of discontinuities in normal lithologies. It is curious that much less is known about the plant responses along "normal" lithosequences than for the azonal cases. Though very limited, the case histories for such arrays of outcrops have shown detectable differences in mineral nutrition and in vegetation from one rock type to another. The other possibilities—either subtle or no detectable differences—simply do not get reported. Yet they are part of the geoedaphic syndrome too. Most often, it is soil scientists who study lithosequences and their consequent soil catenas, but rarely do they take note of the plant response along sequences of rock types.

It is fitting to end this chapter with a prescient quote: "[P]arent rocks are far from being a white sheet of paper on which climate may write anything it desires" (Neustruev 1927, quoted in Lutz 1958). The paper by H. J. Lutz, long-time student of forest soils, presents in masterful detail the case for considering the important influence of geology and soil on forest vegetation. Lutz reviews the waxing and waning in acceptance of two opposing views: Is it climate or geology that determines the nature of forest vegetation? It was a "stampede to the Russian point of view" that climate is primary that led forest ecologists to downplay the role of geology (Marbut 1928, quoted in Lutz 1958). But Lutz makes a strong case for the primacy of geology in influencing the character of forest ecosystems. His essay, a classic in geoedaphics, should be "must reading" for all plant ecologists.

6 Implications of Geoedaphics for Systematics and Evolution

The publication of "new or otherwise noteworthy" additions—new taxa or range extensions—to regional floras continues unabated. And the novelties are usually found in habitats with some exceptional or local edaphic attributes. One has only to leaf through regional journals anywhere in the world to find them. In North America, journals like *Madroño, Castanea,* and the *Canadian Naturalist* frequently publish reports of such new finds. This process by taxonomists of describing unusual entities in azonal habitats has been going on for many years. In this chapter we will take a close look at the nature of the taxonomic recognition of edaphic specialists. In turn, we are led to pose questions about the evolutionary mechanisms whereby edaphic specialists come into being.

Taxonomic Recognition of Geoedaphic Specialists

Recognition of edaphic novelties has been a venerable preoccupation of taxonomists for centuries. Even in pre-Linnaean times, plants of distinctive character growing on exceptional substrates were given names. Cesalpino, in 1583, described a "Lunaria quarta alias Alysson" on serpentine ("black rock") in Italy. Now known as *Alyssum bertolonii,* this serpentine endemic is one of several in the mustard-family genus *Alyssum,* found to be accumulators of nickel (Vernano-Gambi 1992; Reeves, Brooks, and Dudley 1983). With the advent of binomials in post-Linnaean times, two taxonomic consequences can be noted. First, newly discovered edaphic specialists are often recognized as distinct at the species or infraspecific level, either as members of a genus with substantial edaphic restrictions (e.g., *Streptanthus, Leavenworthia, Hesperolinon,* etc.) or in genera of a more generalist nature. Second, if the new entity is found to extend its normal range onto some unusual substrate, this will be noted as simply a range extension or given infraspecific taxonomic status.

The taxonomic rank of geoedaphic specialists ranges from subspecies, variety, and species up to genus and even family. My familiarity with serpentine floras prompts an enumeration of some serpentinicolous examples. The taxa most obviously restricted to serpentine on the basis of their names are those bearing the epithets *serpentina, serpentinicola, ophiticola,* and *ophitidis*. At the infraspecific level, there are such variants as var. *serpentinum* of *Melandrium rubrum* (Scandinavia) and var. *serpentinicola* of *Jasione crispa* (Portugal). Many species bear telltale epithets: *Sedum serpentinum, Verbascum serpentinicum,* and *Anthyllis serpentinicola* (all of the Balkans); then there are the ophiolitic taxa like *Bucida ophiticola* (Cuba) and *Calamagrostis ophitidis* (California). Although many additional serpentine endemics have names that obscure their geoedaphic restriction, by simply scanning the lists of endemics in the reviews of serpentine botany one cannot miss the substantial recognition taxonomists give to the restriction to serpentine (e.g., Brooks 1987; Roberts and Proctor 1992; and Baker, Proctor, and Reeves 1992).

Genera endemic to serpentine occur in many parts of the world. Most are monotypic: *Halacsya sendtneri* (Boraginaceae, the Balkans), *Japanolirion osense* (Liliaceae, Japan), *Shafera platyphylla* (Asteraceae, Cuba), *Neocallitropis pancheri* (Cupressaceae, New Caledonia). The serpentine floras of Cuba and New Caledonia, so rich in endemics, could easily augment this list of mostly monotypic serpentine genera.

At the family level, I turn again to New Caledonia for the only examples where entire, though usually small, families are restricted to serpentine: Amborellaceae (one species), Oncothecaceae (one species), Paracryphiaceae (one species), Phelliniaceae (ten species), and Strasburgeraceae (one species).

While serpentines support an impressive number of named endemic taxa, certain other azonal parent materials may support large numbers of endemics as well. Species epithets that connote substrates include *calcicola, calcarea,* and *dolomitica* for limestone and dolomite endemics; *gypsicola* or *gypsophila* for species on gypsum; and *argillicola* for shale inhabitants. For taxa tolerant or even restricted to metalliferous substrates, taxonomists coined the following epithets: *calaminare* (zinc-rich habitats); *cupriphilus* and *cupricola* for copper substrates; *metallifera* or *metallorum; niccolifera* (nickel habitats); *cobalticola* (cobalt habitats); and so forth. Beyond the telltale namesakes for particular substrates are the many substrate-specific taxa bearing names that do not reveal their edaphic restriction.

When the geoedaphic influence is other than lithological or chemical, there is still nomenclatural recognition of habitat. For instance, topography (elevation, slope, and exposure) and the physical nature of the parent material all have evoked Linnaean epithets as descriptors. For elevation, we find *collina, monticola, montana, nivale, alpina, alpicola,* and others. For rock outcrop habitats, there are the commonly encountered epithets like *saxosa, saxatilis, rupicola, petrophila,* and *lapidicola*—not to mention *rimicola, cinicola,* and *cratericola.*

In formal taxonomic listing of a geoedaphic specialist, publication of a new entity *must* be based on more than a habitat attribute. Taxonomic recognition requires that discrete entities be identified by morphological attributes. However slight these differences may be (often to the dismay of the user of taxonomic keys), the newly described entity must be distinguishable from its nearest relative. In the case of serpentine taxa, often the morphological markers are one or more of the so-called serpentinomorphoses (Ritter-Studnicka 1968; Kruckeberg 1985): (1) xeromorphic foliage (sclerophylly), glaucousness, size reduction, reduced or increased pubescence, anthocyanous coloration, et cetera; (2) reduction in stature (shrubbiness of arborescent species, dwarfing and plagiotropism [prostrate habit] of herbaceous species); and (3) increase in root system. In fact, some form of xeromorphy is likely to be the hallmark of any edaphic novelty—a response by stems and leaves to stress. The linkage of distinctive morphological characters and geoedaphically unique habitat justifies taxonomic recognition. By these criteria—singular character and habitat—"good" species (those readily told from their nearest relatives) abound worldwide, serving as indicators of some unique site.

Geoedaphic specialists at taxonomic levels from infraspecific variants (variety or subspecies) to genus occur in the floras of any region where geological diversity appears. They are often the majority of endemics within a local or regional flora. Numerous examples have been given in previous chapters, especially Chapter 5. But what kind of taxa are they? The possibilities can be categorized by evolutionary age (old or new taxa), range (distribution as narrow endemics to regional edaphic specialists), mode of origin (gradual to abrupt), cytogenetic attributes (ploidal or other chromosomal differences and/or oligogenic to polygenic heredities), degree of affinity with presumed relatives, and so forth. These several possibilities merit amplification; see also Kruckeberg and Rabinowitz (1985) for an extended treatment of endemism in higher plants.

Age of a geoedaphic specialist can rarely be determined with confidence. Old taxa (paleoendemics, *sensu* Stebbins and Major 1965; Raven and Axelrod 1978) may be recognized by either their relictual nature or their high polyploid level, or both. Geoedaphic endemics are likely to be relictual if they occur in more than one disjunct population and if they have no close relatives nearby. That is, they are systematically (taxonomically) isolated. In their extended treatment of kinds of endemics in the California Floristic Province, Raven and Axelrod (1978) cite the example of the serpentine endemic *Phacelia dalesiana* (Hydrophyllaceae) as a paleoendemic; it is not closely related to any others of its genus, and occurs in the geologically old region of the Klamath-Siskiyou Mountains of northwestern California. The singular, monotypic *Darlingtonia californica* (Fig. 5.49), mostly confined to boggy sites on serpentine, is probably another paleoendemic. For the criterion of ploidal level in related taxonomic groups, some monotypic diploids are older than low polyploids, while rare taxa of high polyploid levels as well as monotypic diploids are also paleoendemics (Kruckeberg and Rabinowitz 1985).

Where azonal soils have persisted over long periods of geological time, one expects large numbers of paleoendemics. The serpentine and metalliferous habitats of the Great Dyke in Zimbabwe have been exposed and open to colonization since before (Cambrian) the dawn of the angiosperms in the early Cretaceous (Wild and Bradshaw 1977). While a few endemics occur on these ancient, anomalous substrates, there is no clear indication of a rich paleoendemic flora. Wild and Bradshaw believe that climatic and other changes through time have transformed the flora; some species have gone extinct and others have replaced them.

Neoendemics, taxa of relatively recent origin, usually can be recognized by their close affinity to other species of their genus, often appearing near one another regionally, and by their occurrence in one or a very few neighboring sites. Probably many geoedaphic endemics are youthful by these criteria. The five species of the Hesperides group in *Streptanthus* (Brassicaceae) are wholly confined to California serpentines; they appear to be neoendemics by the above criteria (Kruckeberg and Morrison 1983).

The geographic range of geoedaphic endemics can be a distinctive attribute. Some are very local, confined to only one or a few sites (the narrow endemics of Mason 1946a, b). Wherever substrates are local or highly disjunct, one expects to encounter taxa narrowly confined to them. Serpentine examples occur in many places worldwide. In California, *Streptanthus batrachopus* is confined to serpentine outcrops only on the upper slopes Mount Tamalpais, Marin County. In Japan, *Viola yubariensis* is found only on serpentine barrens at high elevations of Mount Yubari, Hokkaido. *Myosotis monroi* is only on the dunite of Dun Mountain, South Island of New Zealand. Narrow endemics are also known for limestone, dolomite, gypsum, and those highly disjunct granitic balds in southeastern United States and Western Australia (Ornduff 1987).

Not all edaphic endemics are narrowly restricted to a single or a few habitats.

Often they are faithful to an azonal habitat wherever it may occur within a region's flora. Several serpentine endemics in California faithfully track the substrate regionally: *Quercus durata* and *Ceanothus jepsonii* are near-constant indicators of serpentine in chaparral communities throughout the Coast Ranges; and the former even occurs frequently on the Sierra Nevada serpentines. *Streptanthus breweri* tracks serpentine in the Coast Ranges from New Idria in the south, north to Lake and Mendocino Counties; *S. polygaloides* (Fig. 5.50) can be expected on most of the larger serpentine outcrops in the western Sierra Nevada, from Tulare County north to Placer County. These are regional edaphic endemics; their counterparts can be identified on serpentines or other azonal substrates in many other places in the world. Thus *Halacsya sendtneri* (Boraginaceae) is a regional serpentine endemic in the Balkans; *Kobresia myosuroides* (Cyperaceae) is a faithful calcicole throughout the Alps of Europe.

Azonal habitats may support floras with more than edaphic endemics. It is usual to find azonal environments with floras consisting of a mix of species with varying fidelities to a particular substrate. Besides strict endemics, there are two other kinds of entities showing some degree of geoedaphic partiality. Each type is constituted of species with rather wide habitat tolerance and of wide regional distribution. One class of species can be called local or regional indicators. These may be common on a variety of "normal" substrates in one part of their range but confined to an azonal soil elsewhere, often at the limit of their total range. Of the several serpentine examples cited in Kruckeberg (1985), those like incense cedar (*Calocedrus decurrens*) and Jeffrey pine (*Pinus jeffreyi*) epitomize the pattern. Incense cedar and Jeffrey pine are widespread in the Sierra Nevada and Transverse Ranges of California on a variety of soils, often granitic. But in the North Coast Ranges, both are highly faithful to serpentine.

The other type of species that occurs on both zonal and azonal soils has been called indifferent, bodenvag, or ubiquist (Kruckeberg 1985). All three terms describe taxa that occur in both zonal and azonal habitats. In any locally delimited edaphic site, its flora will usually contain representatives of this type. The taxa can be of any life-form (herbaceous to woody) and from many different plant families, even a sampling of the region's flora. For western North American serpentine, such bodenvag species are numerous. Conifers like Douglas fir (*Pseudotsuga menziesii*) and lodgepole pine (*Pinus contorta* subsp. *latifolia*), shrubs like *Ceanothus cuneatus* (Rhamnaceae) and *Adenostoma fasciculatum* (Rosaceae), and herbs like *Achillea millefolium* subsp. *lanulosa* and *Anaphalis margaritacea* (both Asteraceae) grow on serpentine substrates as well as on a variety of other soils. Bodenvag taxa are well represented on other special substrates, such as carbonates, gypsum, and shale.

Does this apparent indifference to substrate suggest that tolerance to the conditions of an azonal part of their total range is simply a manifestation of phenotypic plasticity (or broadly tolerant genotypes)? For native species, it has been amply demonstrated that tolerance has a genotypic basis. Wherever tested experimentally, indifferent species are mostly found to be made up of races some of which are tolerant and others intolerant to a local azonal habitat. Ecotypic differentiation is the model system for such taxa. Ecotypic variation has been identified for many such taxa; where they occur on an azonal substrate, they are likely to be genetically distinct, tolerant races (Kruckeberg 1951, 1967, 1995). Such edaphic races rarely receive taxonomic recognition, either as species or as infraspecific variants (varieties or subspecies). Yet, as John Harper points out (1981), such unnamed races are distinct entities adapted to a local azonal site. As local endemic races, they, like rare and endangered species, merit preservation as rarities.

Not all bodenvag or edaphically indifferent species have undergone ecotypic differentiation to form substrate-tolerant races. The alternative mode of achieving tolerance involves what Herbert Baker (1965) heuristically calls the "general purpose genotype." In a recent paper, Baker (1995) cites several examples of edaphic tolerance that could be the consequence of phenotypically plastic genotypes.

Taxonomists seem not to have acknowledged the antithesis of geoedaphic specialists and edaphic tolerance; avoidance of a substrate rarely has evoked any taxonomic recognition. Yet the opposite of edaphic tolerance does exist as an ecophysiological phenomenon; perhaps the majority of species in a regional flora avoid edaphic "islands." Avoidance of an azonal substrate does get its share of terminology. But it is ecologists who invent the descriptors. For taxa that avoid carbonate rocks, there are the familiar terms, calciphobe and calcifuge. It is curious that similar terms to express avoidance of serpentine, gypsum, heavy metal soils, et cetera, have not been coined.

How "good" are the taxa described for exceptional edaphic habitats? Does a named limestone or serpentine endemic have unqualified singular quality and taxonomic acceptance? The distinctness of such taxa ranges from slight to very distinct. Naming edaphic variants, usually rare and unusual ones, may reflect the universal human propensity to single out and give names to novelties, rare and unusual artifacts, and organisms. (Kruckeberg and Rabinowitz 1985):

> Rarity of all sorts piques the curiosity of humans. Rare objects, whether discoveries of natural origin or artifacts of cultures, are avidly sought by collectors and are treasured, housed, and exhibited for the benefit of others. This fascination with that which is rare may partially explain the natural-

ists' and systematists' time-honored preoccupation with rare and endemic taxa. To be sure, there are sound justifications for acquiring and studying rare organisms. But we suspect that the rarity-seeking syndrome in humans has fostered much of the biologist's preoccupation with rare and endemic plants and animals.

When the rarity is a plant found on some local or unusual substrate, it is likely to be recognized at the species (or even genus) level even when its distinctness from its nearest relatives is slight. Taxonomists are fascinated by the rare and unusual. Once a region's flora is described—written up as a published flora—the quest for the unusual continues; and the rarities usually get named. But a binomial that gets into print need not be eternal. To paraphrase an old adage, "One person's species is another one's variety"! Taxonomic inflation may be the result of initial discovery, but a later judgment may be deflationary. For edaphic endemics, this deflation (or its obverse) commonly occurs. There are numerous examples among the serpentine endemics of western North America. E. L. Greene named the Sierran endemic jewel flower as *Microsemia polygaloides*—a monotypic genus. Other botanists, from Watson and Gray to Hickman (1993), place this serpentine endemic in *Streptanthus* (*S. polygaloides,* Fig. 5.50), even though it is the only known serpentine *Streptanthus* that is a hyperaccumulator of nickel (Kruckeberg and Reeves 1995). As a species, *S. polygaloides* has all the hallmarks of a "good" species—singular vegetative and floral morphology, as well as absolute fidelity to serpentine. In fact, a case has been made (Reeves et al. 1981) to restore it to the monotypic *Microsemia* because of its singular capacity for concentrating nickel. Other serpentine jewel flowers have been reduced from species to infraspecific status. Thus in the new *Jepson Manual* for California (Hickman 1993), *S. hesperidis* is now *S. breweri,* var. *hesperidis; S. lyonii* is now *S. insignis* subsp. *lyonii;* and *S. tortuosus* var. *optatus* has been merged with var. *suffrutescens* of *S. tortuosus.* Yet other serpentine jewel flowers have withstood the scrutiny of the compilers of the new *Jepson* flora. The local serpentine endemics *S. brachiatus, S. drepanoides, S. niger,* and *S. morrisonii* are retained at the species level. Most recently a change in status of a serpentine endemic went in the other direction: *Acanthomintha obovata* var. *duttonii* is now elevated to *A. duttonii* (Jokerst 1991).

These changes in taxonomic rank can be found in other floras: inflation, deflation, and retention of status quo for edaphic endemics are noted in the serpentine floras of Japan, the Balkans, and New Caledonia.

Scanning current taxonomic literature is a good way to detect the significance of geoedaphic novelties in the systematics of native (and even introduced) flora.

Recent regional publications are much better in their recording of habitat attributes than their predecessors. *Flora Europea* (Tutin 1964 to 1980), the new *Jepson Manual* for California (Hickman 1993), and the ongoing *Intermountain Flora* (*Vascular Plants of the Intermountain West,* Cronquist et al. 1972 to 1984), are examples of this closer attention to the substrate or other geoedaphic attributes of a plant's habitat. Current numbers of regional botanical journals are also sources of newly described geoedaphic novelties. Many new taxa have unique edaphic attributes. Also, published range extensions or new occurrences of taxa often are based on a geological feature. I have scanned recent numbers of *Madroño: A West American Journal of Botany,* for occurrences of this sort. One or more of such "new or otherwise noteworthy" additions to a flora can be found in nearly every issue of that journal. Similar findings are to be expected in other places where geology generates a variety of unique habitats.

Evolution of Edaphic Specialists

The many kinds of accommodation by plants to geologically conditioned environments provoke questions about their evolutionary origins. Conventional wisdom would consider such accommodation adaptive; plants growing on talus, limestone, serpentine, or any other geoedaphically created environment are presumed to have been selected for and thereby adapted to the particular habitat types. The match of an organism to its environment is commonly accounted for in terms of the neo-Darwinian paradigm. Its basic assumptions are: (1) a reservoir of genetic variability; (2) selection by the environment of the matching genetic potential for meeting the challenge; and (3) genotypic accommodation to the particular habitat factors—that is, genetically determined adaptation. These first three steps may be the final evolutionary end product: a population adapted to a given geoedaphic challenge. In ecotypic or racial terms, such a microevolutionary result is variously called an edaphic ecotype, an edaphic race, or (as named by a taxonomist) as a variety or subspecies. Edaphic variants are recognized as part of a species' array of populations, each having specific adaptive properties. The neo-Darwinian model embraces yet other evolutionary outcomes. If a race or infraspecific variant becomes isolated genetically and/or ecogeographically, it may further diverge from its nearest relative. Such isolates, having attained habitat specialization coupled with distinctness in morphology as well as ecophysiological attributes, become species. In short, microevolution (race formation) may also be a stepping-stone to macroevolution (speciation).

We will provisionally use this conventional neo-Darwinian explanation to account for genotypic adaptation to particular habitats and to assess the origins of geoedaphic novelties. Yet we are bound to wonder if this model gives good answers to all the questions that surface upon encountering the variety of edaphically unique sites occupied by local, distinctive floras. Does a particular physical environment elicit a specific genotypic response? Or is the response the product of some earlier adaptation that happens to work in the new setting? For example, is inherited heavy metal tolerance a specific or "generic" response? Applying the word adaptation to any unique fit of organism to environment is a handy "explain-all" device. But can a presumed adaptation be verified? We are bound to deal with these questions as we now take up, first, the microevolutionary response to geoedaphic factors. The same questions will surface when we later examine the macroevolutionary origin of distinct isolates: species formation and beyond.

MICROEVOLUTION AND GEOEDAPHICS

Infraspecific variation in response to geoedaphic selection lies in the realm of population biology, or what is sometimes called genecology. The key concept in genecology is the ecotype—"the product arising as a result of the genotypical response of an ecospecies [i.e., a Linnaean species] to a particular habitat" (Turesson in Clausen, Keck, and Hiesey 1940). In modern terms, such ecotypic variants are locally adapted populations or races. Racial differentiation within a species can come about as the result of matching suitable genotypes with habitats. Edaphic races or ecotypes are common geoedaphic responses to a habitat. Racial differences in response to such substrates as serpentine, limestone, heavy metal sites, and other clearly delimited azonal substrates are well known; see Shaw (1989) and Kruckeberg, Walker, and Leviton (1995) for major reviews. Tests for edaphic races include reciprocal transplants of cloned ramets, and seedlings or growth responses of cultured excised roots to particular substrates or to levels of a given inorganic element. Accessions of seed or ramets of a species are taken from both normal and azonal sites and are grown reciprocally on the normal and azonal soils (or artificial simulations of the contrasting substrates). If differences in tolerance are detected, it is concluded that the population tolerant to the azonal substrate is a genetic race. While most of the proven cases of racial tolerance have come from studies on limestone, serpentine, or heavy metal sites, it is expected that racial variants have been selected by other geoedaphic challenges. Talus races, rock-crevice races, and elevational or other topographically defined races are likely outcomes as well. The local or regional geoedaphic mosaic of environments should evoke a variety of genotypic responses.

What do we know about the genetics of edaphic races? Is tolerance the result of polygenic differences or a simple oligogenic difference? Is tolerance a dominant trait? Does tolerance to one edaphic factor confer tolerance to other edaphic stresses? Again, we must turn to case histories of heavy metal tolerance for answers—the only substrates for which genetic analysis is available. The most comprehensive review of the genetics of heavy metal tolerance is by MacNair (1993). MacNair is critical of earlier reports of polygenic control of tolerance. Interpreting the tests for tolerance can be confounded by phenotypic variance, resulting in misinterpretation. One case study by MacNair (1983), tolerance to copper by *Mimulus guttatus,* decisively revealed that a single major gene was involved, though modifier genes were also implicated.

Not all gene pools have the genetic resources (preadaptation) that permit the evolution of tolerant races (edaphic ecotypes). Many species on normal sites are unable to grow on nearby more demanding, or merely different, edaphic environments. Though the term is perhaps a bit anthropomorphic, such exclusion has been called "avoidance." Some species of a regional flora do not exhibit the bodenvag (soil-wandering) ability, genetic or otherwise. Yet other species are well-known bodenvag (indifferent, ubiquist) types. For example, in California many bodenvags, on and off serpentine soils, are known: tree species like *Pinus ponderosa, P. jeffreyi, P. sabiniana, Umbellularia californica, Calocedrus decurrens;* shrubs like *Ceanothus cuneatus, Quercus vacciniifolia, Adenostoma fasciculatum;* and herbs like *Achillea millefolium lanulosa, Gilia capitata,* and *Salvia columbariae* (Kruckeberg 1985). Either the bodenvags are edaphic ecotypic variants or possess general-purpose genotypes (Baker 1965), and thus have the capacity for tolerance to a demanding substrate. But how do these types arise? Evolutionary theory has a facile answer to the question—preadaptation, as the initial stage. By preadaptation is meant the capacity of a gene pool to possess genotypes to meet new challenges. If the gene pool is highly polymorphic, then some of its genotypes may be preadapted to fit a new or more demanding habitat.

Preadaptation may represent the first stage in the evolution of edaphic specialization. Hence we need to elaborate on its place in evolutionary theory, especially since it can be interpreted in more than one way. Preadaptation is a current and acceptable concept in the major texts on evolution (e.g., Dobzhansky et al. 1977; Futuyma 1979). An eloquent statement in its defense merits quoting (Dobzhansky et al., p. 125):

> A genotype only passably successful now may flourish in some future environments. Such a genotype is often said to be "preadapted" to what it will meet in the future. The concept of preadaptation is a fallacy if it is taken to

mean that evolution somehow contrives adaptations for use in as yet nonexistent (or untried) environments. Nevertheless, in building adaptedness for present environments, natural selection can only operate with what is inherited from the past.

This simple theoretical proposition of preadaptation is amenable to testing by putting samples of a parent gene pool in a new edaphic situation and then screening for adapted, tolerant individuals. The potential for tolerance to biocides (pesticides, herbicides, pollutants, etc.) has been demonstrated in a variety of organisms (e.g., insects, weedy plants). So it is not surprising that tolerance to stressful substrates can originate rapidly via preadaptation. The best examples come from the studies on heavy metal tolerance (e.g. Antonovics, Bradshaw, and Turner 1971; Antonovics 1975; Bradshaw 1976). Bradshaw and colleagues have tested the hypothesis that some constituents of a flora that have not as yet encountered soils contaminated with heavy metals may be preadapted and thus able to grow on such soils. Several authors have reported on substrate tests of "intolerant" populations and found that some populations indeed contain preadapted genotypes. For example, Gartside and McNeilly (1974) tested several British presumed nontolerant (normal) pasture plants for degree of tolerance to copper. Two grasses, *Agrostis tenuis* and *Dactylis glomerata,* yielded a low frequency (0.08%) of fully tolerant survivors. Other species yielded a few individuals with either low tolerance (*Poa trivialis, Lolium perenne, Arrhenatherium elatius, Cynosurus cristatus*) or no copper tolerance at all (*Plantago lanceolata, Anthoxanthum odoratum, Trifolium repens*). From this and similar studies (Antonovics, Bradshaw, and Turner 1971; Antonovics 1975; Wu 1989), we draw the following conclusion. "Normal" populations may have the genetic potential for occupying edaphically stressful sites. But not all such nontolerant species are so preadapted. Those not preadapted will be excluded from the edaphic site absolutely or may ultimately acquire genetic resources for tolerance.

The acquisition of edaphic tolerance raises several questions at the level of racial tolerance and in relation to the potential for further evolutionary divergence. The heavy metal studies raise the issue of cross-tolerance. Is a copper-tolerant race also tolerant to other toxic heavy metals? Most of the evidence suggests that tolerance to one metal does not confer tolerance to another: "[T]he mechanisms promoting tolerance to different metals are independent from one another both genetically and physiologically" (Walley, Khan, and Bradshaw 1974). This simple answer leaves the question unanswered for species not yet tested. For other edaphically restrictive sites (e.g., limestone or serpentine), do we know to what physical and chemical features of the substrate a population is tolerant? In the case of the ser-

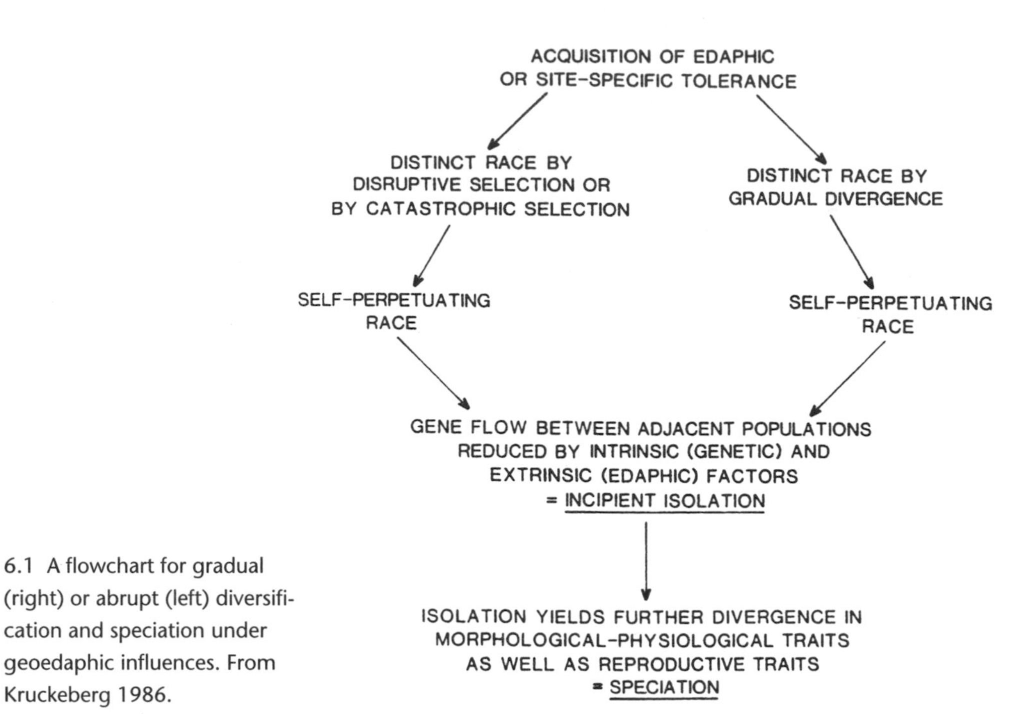

6.1 A flowchart for gradual (right) or abrupt (left) diversification and speciation under geoedaphic influences. From Kruckeberg 1986.

pentine syndrome, tolerance may be to more than one physical or chemical factor. Or it may arise in response to some unexpected physical factor (drought, soil property, etc.) rather than to an expected chemical one.

Having acquired tolerance, say, to heavy metals, does a tolerant population give up (lose) some attribute for survival on normal soils? Again, the Bradshaw school has an answer. Tolerant races may have reduced growth on normal soils or reduced competitive ability with nontolerant (normal soil) races. Nicholls and McNeilly (1985) state: "Correlated effects between resistance to toxic substances and low fitness in non-toxic situations are known from several plant and animal species." They cite examples of pesticide resistance where the resistant forms have reduced fitness in nontoxic situations.

The preceding discussion of preadaptation suggests that it is idiosyncratically expressed. Some organisms have the capacity for edaphic preadaptation; others do not. Further, the acquisition of edaphic tolerance may result in loss of competitive ability or some other survival attribute on normal soils.

We now examine preadaptation—the acquisition of edaphic tolerance—as the likely initial step in the evolution of edaphic specialization. Scenarios can be crafted for the sequence of stages in evolution from preadapted genotype and race formation to edaphically restricted species. The scenarios can be visualized as a flowchart (Fig. 6.1). How evolution might proceed from preadaptation within a normal population to a full-fledged edaphic endemic species can be rendered in theoretical terms. This has been done for the serpentine environment (Kruckeberg 1986). Here it will be applied to the more general cases of any edaphic specialization.

The sequence of stages in the genesis of a geoedaphic specialist at the species level can be looked upon either as a continuum or as a series of discrete steps. Both paths invoke known evolutionary processes. The selective agents in the geoedaphic milieu operate on gene pools and populations to achieve spatial and edaphic isolation and ultimately genetic isolation. Thus the model given here could apply equally to cases where the selective agents are biotic, not geoedaphic. Hence the evolution of a pollination or herbivore specialist could follow a similar pathway. In other words, the major tenets of contemporary Darwinian evolution apply to the origin of both kinds of end products—species that fit either a special biotic niche or a geoedaphic one. The neo-Darwinian model is described in all the major texts on evolution (e.g., Dobzhansky et al. 1977; Futuyma 1979; Mayr 1970; and Stebbins 1950).

Is this evolutionary pathway smoothly continuous, or marked with small to major discontinuities? Debate persists on this question, with views ranging from the continuum (or gradualist) camp (Dobzhansky, et al. 1977; Mayr 1970) to the abrupt discontinuity, or punctuated equilibrium proponents of Eldredge and Gould (1972). I favor an intermediate position: some gradualism, preceded or followed by an event marking a discontinuity (Kruckeberg 1986). The possible evolutionary pathways are presented next, to be followed by evidence for their support.

The initial steps in divergence, from an ancestral or parental population to one venturing into a new habitat, require appropriate resources in the parental gene pool. Some degree of genetically based tolerance to a new geoedaphic challenge via preadaptation must be present. One or a few genotypes from a polymorphic (heterozygous) gene pool gain a toehold on a new substrate or geomorphically distinct site. The initial preadaptation is most likely a simple genetic trait, perhaps controlled by a single locus. The initial tolerance probably will be less than optimally fit. So, if they are to survive and increase, the incipient "founder individuals" will undergo further genetic repatterning. A single major gene may acquire modifiers, or the initial tolerant genotype may be reinforced by still other genes by mutation, recombination, and selection to perfect the geoedaphic adaptation. Most populations at this stage are thus differentiated from an immediate ancestral population by several to many genes; the local population thus attains fitness to a geoedaphic habitat with polygenic control. This level of distinctness is the familiar status of ecotypic differentiation. A single polymorphic species now has one or more edaphic races (ecotypes). This level of racial divergence may persist without further evolutionary divergence. The right-hand pathway in the flowchart (Fig. 6.1) illustrates this gradual divergence to race formation.

A more abrupt genesis of a distinct geoedaphic race is shown as the left-hand pathway of the flowchart. Raven (1964) drew upon Harlan Lewis's (1962) model of catastrophic selection to posit a rapid fixation of an edaphic specialist. The model proposes that a marginal population or one decimated by a transitory severe environmental stress may consist of a lone genotype that has been drastically altered cytogenetically (via chromosomal inversions or translocations). The surviving fragment of the population is not only fit ecologically but reproductively isolated by a cytogenetic barrier from its nearest relatives. Such rapid (catastrophic) selection produces a new population isolated spatially and genetically from its decimated predecessor. A new species is born. Raven and Lewis call this rapid succession of events *saltational speciation.* Since both the ancestral and the derived populations are sympatric (occurring in the same place), the saltation model appears to be a special case of sympatric speciation, presumably rare in animals but not uncommon in plants, as we shall now see.

In fact, another mode of sympatric speciation in plants, allopolyploidy, may also yield edaphic specialists. Interspecific hybridization followed by chromosome doubling is well known in ferns and flowering plants (Stebbins 1950, 1971). The process is initiated when two diploid species (genomes AA and BB) hybridize to form a sterile hybrid (AB). Fertility is restored in the hybrid progeny by chromosome doubling (AABB). The morphology of the AABB tetraploid is distinct from that of the parents and is genetically isolated from them. This rapid merging of genomes of two species to produce a third can initiate the occupancy of a new habitat. The allopolyploid derivative commonly occupies a geologically more recently available terrain (Stebbins 1950). Post-Pleistocene recessional habitats are frequently sites for the new polyploid species (e.g., the *Iris* example of Anderson 1936). Edaphic habitat selection can occur at a local microscale. In Ogotoruk Creek Valley, Alaska, Johnson and Packer (1965) found that polyploids of several herbaceous perennials were most frequent on edaphically more stressful sites.

MACROEVOLUTION AND GEOEDAPHICS: SPECIATION AND BEYOND

The key attribute to divergence beyond the acquisition of racial tolerance is *isolation.* In the cases where the geoedaphic habitats are the primary discontinuities, isolation is both spatial and ecological (edaphic). Such physical isolation can permit the isolated gene pool to further diverge. In addition, the isolation may be enhanced by the genesis of reproductive isolating mechanisms. When ecogeographic isolation is combined with reproductive barriers, the discrete genetic assemblage has the attributes of a species. During the process of isolation, the incipient species may acquire morphological features that further distinguish it from its nearest ancestor. This sequence of events is called speciation. The many edaphic specialist species chosen as examples throughout this book are postulated as likely to have originated via one or the other of the pathways just outlined. But can we marshal any evidence to support this hypothetical sequence?

A species is not only a name in a regional flora: it is a testable hypothesis. Though this idea rarely appears in the conventional taxonomic literature, it deserves recognition (*fide* Paul Illg, invertebrate taxonomist). The tests for validating a species' integrity include morphological markers, and ecophysiological fitness for a particular geoedaphic niche. These attributes will be based on significant genetic differences, often polygenically controlled. Further, the distinct phenotypic hallmarks of the species should be maintained by reproductive barriers. Case studies that have met the above tests are rare. Most taxonomists will assume that the "good" species is its own outcome of just such a testing. Yet as I will now demonstrate, there are examples for each major stage in the speciation process.

The stages prior to genetic and spatial isolation—preadaptation and ecotypic variation—were dealt with earlier in this chapter. There is good evidence for their having taken place (Kruckeberg 1986; Kruckeberg, Walker, and Leviton 1995). The later stages of isolation and speciation are less commonly demonstrable. Case histories are wholly confined to the literature on tolerance to toxic heavy metals. Antonovics (1975) reviewed the evidence for "putative speciation" of races tolerant to mine spoils. Attributes of incipient speciation—isolating mechanisms and some degree of cross incompatibility with intolerant relatives—have been detected. A more recent case history has put the hypothesis of a speciation event to a full test. In a series of papers, Mark Macnair (references to follow) supports his contention that *Mimulus cupriphilus* is a newly evolved copper-tolerant species. It and its putative parental species, *M. guttatus,* occur on mine tailings in the Copperopolis area of the Sierra Nevada foothills, California. That *M. cupriphilus* merits separate species-level recognition from *M. guttatus* is supported by several lines of evidence. First, the two species more or less pass the test of sympatry. That is, they seldom hybridize, due to differences in pollination mechanisms (Macnair 1989). Differences in floral and vegetative morphology further support the distinctness of *M. cupriphilus*. Even though hybridizations between the species are rare in the wild, the two can be intentionally crossed to yield fertile progeny. Thus it is possible to determine the genetic basis of the species differences. The morphological differences are controlled by four polygenic genetic systems. In contrast, a single dominant gene for copper tolerance is shared by *M. guttatus* and *M. cupriphilus*

(Macnair and Cumbes 1989). Here is a clear case of speciation in response to a geoedaphic stimulus; the case is supported by tolerance tests, morphological attributes, reproductive behavior, and genetic analysis. Referring to the flowchart of Figure 6.1, we are curious to know if *M. cupriphilus* went through any of the earlier steps prior to speciation. Undoubtedly its ancestors were preadapted genotypes, but there may be no way of telling if there was an edaphic ecotype (race) stage. Perhaps the copper-tolerant *M. guttatus* is the edaphic race?

Macnair holds that the speciation event is of recent origin, since the copper mine sites are less than 150 years old. The newly arisen copper mine species, *M. cupriphilus,* must have resulted from rapid speciation. Yet it does not quite fit the Lewis-Raven model of saltational speciation. The rapid evolution under the saltation model requires extinction of the ancestral population and cytogenetic repatterning of the survivor. Neither condition seems to apply to the *M. cupriphilus* case. In fact, the hypothesis of catastrophic selection has been applied only to studies of *Clarkia* species (Lewis 1962).

Other than direct genetic evidence for edaphic speciation, we can find indirect evidence for the evolution of edaphic specialists by examining comparative systematic studies. The mustard genus *Streptanthus* illustrates this approach. The subgenus Euclisia, with sixteen species, consists of eleven taxa showing varying degrees of restriction to serpentine, as well as a substantial range of uniqueness in morphological attributes.

Populations of the *Streptanthus glandulosus* species complex portray the initial stages of edaphic specialization. Local populations of this wide-ranging species complex occur on and off serpentine soils in the Coast Ranges of California. Progeny testing of plants from both substrates on a standard serpentine soil revealed that the nonserpentine (NS) accessions failed to grow on serpentine (S). I concluded (Kruckeberg 1951) that the species displays ecotypic variation; S and NS races were present. In terms of reproductive isolation, populations of the *S. glandulosus* complex revealed, from intentional crosses, a range of interfertility from populations fully interfertile to those completely sterile (Kruckeberg 1957). Some populations were distinct in morphology and were reproductively isolated via intersterility from other members of the group. They have been recognized as distinct species: for example, *S. niger* and *S. albidus* (Kruckeberg 1958; Hickman 1993). *Streptanthus glandulosus* and its subspecies showed a unique kind of reproductive isolation, one correlated with distance. Populations from nearby sites were fully fertile; as distance between populations increased, interfertility of artificial hybrids between the more remote populations decreased. Crosses between populations at the opposite ends of the range of *S. glandulosus* were completely sterile (Kruckeberg 1957).

Partial reproductive isolation, following racial differentiation, is illustrated by the named variants in several species of *Streptanthus*. *Streptanthus insignis* subsp. *insignis* and *S. insignis* subsp. *lyonii,* both on serpentine of the South Coast Ranges of California, are separated by only small spatial discontinuities in the serpentine lithology. They are genetically distinct morphs, but are highly interfertile (Kruckeberg and Morrison 1983). *Streptanthus drepanoides,* in the section *Hesperides,* achieves species recognition for its several singular morphological features and its reduced fertility with its nearest congeners (Kruckeberg and Morrison 1983). Still a later stage in species divergence appears exemplified by *S. barbatus* and *S. howellii.* These distinct species are endemic to Klamath-Siskiyou serpentines (northwestern California); yet both retain features showing their kinship to the wide-ranging, arid-land, nonserpentine species, *S. cordatus* (subgenus Pleiocardia). A final stage has been reached in subgenus Euclisia by *S. polygaloides,* endemic to Sierra Nevada serpentines. It is so distinct a species that it defies alignment with any other Euclisian taxon. Indeed, *S. polygaloides* was placed in its own genus, *Microsemia,* by E. L. Greene (1904), and in recent times that status has been supported on the basis of its being the only hyperaccumulator of nickel in *Streptanthus* (Reeves Brooks, and MacFarlane 1981; Kruckeberg 1986; Kruckeberg and Reeves 1995).

I have phrased the processes of speciation in the conventional terms of the neo-Darwinian paradigm: Variation + Selection + Isolation. That is, mostly by gradualism writ large! Such a putative evolutionary sequence would be acceptable to most biologists. Yet I have detected chinks in the neo-Darwinian armor, coming from disparate biological disciplines, ranging from molecular development to paleobiology. Can we make other evolutionary paradigms work to account for the macroevolutionary emergence of unique geoedaphic specialists, as new species or taxa of higher rank? Two recent examples of evolutionary models departing from the neo-Darwinian paradigm open new doors to finding answers to the big question: How does genuine newness—radical departures via major quantum leaps—occur? One hypothesis is from George Miklos (1993), a molecular embryologist/morphologist. Miklos seeks to discover how major body plans (e.g., new phyla of animals) come into being. He rejects the neo-Darwinian model as not effectively going beyond "allelic bookkeeping" (microevolution) to account for the genesis of wholly new lineages (phyla). Miklos proposes an alternative approach—the union of molecular embryology, genetic engineering, and paleontology. He argues that out of this union can come an understanding of the root causes of not only commonality but new and radical departures in the milieu of molecules and body form-and-function. Miklos thus offers a new venue for seeking answers to macroevolutionary questions. I am not aware that Miklos's counterparts in plant biology are "rethinking the evolutionary process for autotrophs" (Miklos, pers. comm.).

The other iconoclasm—questioning the utility of the gradualist neo-Darwinian model—comes from paleobotanists Richard Bateman and William DiMichele (1994). They have restored to respectability and verification the concept of saltational evolution, advanced by Richard Goldschmidt (1940). They define saltational evolution as "a genetic modification that is expressed as a profound phenotypic change across a single generation and results in a potentially independent evolutionary lineage (prospecies = the 'hopeful monster' of Richard Goldschmidt)." Rejecting Goldschmidt's single systemic mutation mechanism, they invoke the possibility of the occurrence of many "hopeful monsters" by mutation of key homeotic genes (genes yielding a phenotype radically different from the ancestral type); most such homeotic mutants fail to meet the fitness challenge of highly selective, competitive niches. However, some homeotic mutants can survive under "the temporary release from selection in unoccupied niches." Such hopeful monsters (prospecies) may persist by yielding to normal neo-Darwinian processes as the niche becomes competitive-selectional. It is tempting to adopt the neo-Goldschmidtian saltation model for the origin of distinctive geoedaphic specialists at the species, genus, or family level. What is required to fit the model to the origin of edaphic specialists? First would be the detection of homeotic mutants that depart radically in structure and function from their zonal precursors. This would be followed by adaptive accommodation of the prospecies to the new geoedaphic challenge. Thus, one could view the origin of such physiological-morphological breakthroughs as the cactoid habit or the metallophyte-hyperaccumulator syndrome by saltational evolution. To those intrigued with macroevolutionary questions (and dissatisfied with the gradualist model), I commend the two papers just cited. My brief summary barely does them justice.

ORIGIN OF HIGHER CATEGORIES AS EDAPHIC SPECIALISTS

Can the models for the origin of geoedaphic specialist species be extended to higher categories—genus and even family? Before we explore this question, it is useful to examine those genera and families with a high frequency of edaphic restriction. For north temperate floras there are examples at the family level: Asteraceae (Compositae), Brassicaceae (Cruciferae), Caryophyllaceae, Liliaceae, and Poaceae (Gramineae). Each has a number of species in various genera that are edaphic specialists. Of course, such large families can hardly be called edaphic specialist groups, since each has many species that occupy habitats not distinguished for their edaphic uniqueness. Yet there does seem to be a tendency in these families for a "constitutional tolerance" to special edaphic sites. Alan Baker (1987) applied the term "constitutional tolerance" to those taxa that have attained broad tolerance to a variety of edaphic sites, especially to heavy metal substrates. I would extend the idea of constitutional tolerance to the capacity for edaphic tolerance via preadaptation. The families in question, then, would be seen to have a facility for meeting an edaphic challenge with their ability to generate preadapted genotypes. With this capability, the pathway leading to edaphic endemism is begun.

It is to the small families of vascular plants that we must turn for examples of unique evolutionary responses to geoedaphics. The ultramafic terrain of New Caledonia is home to two monotypic families, Oncothecaceae and Strasburgeraceae (Jaffré 1980). Other examples might surface if taxonomists dealing with higher categories would include in their family synopses some mention of ecological niches. Cronquist (1981), in his monumental work on classification of flowering plants, is very thorough on morphological and chemical attributes of families, but rarely gives any indication of ecological preference. To be sure, many of the small families may not occupy unique habitats; rather they may be relictual (paleoendemics) of ancient zonal habitats, especially tropical rain forests.

At the level of genus, the number of geoedaphic endemics substantially rises. Again, it is on the ultramafic substrates where most are found. Temperate examples are few: *Solanoa* (Asclepiadaceae, California) and *Halacsya* (Boraginaceae, the Balkans). From the mustard genus *Streptanthus* could be carved out the monotypic *Microsemia* (*M. polygaloides*). As mentioned earlier, it is so distinct in its morphology and its singular status as a hyperaccumulator of nickel that it may yet merit recognition as a monotypic genus.

Geoedaphically unique genera abound in the tropics—all ultramafic endemics. In New Caledonia (Jaffré 1980) and Cuba (Borhidi 1991) there are numerous genera restricted to serpentine and peridotite. Often they are monotypic or have only a few species.

This generic endemism to ultramafic substrates merits comment. The endemic genera in New Caledonia are most likely paleoendemics. Jaffré et al. (1987) contend that they may have had their beginnings in the Eocene, when even more of the land surface of New Caledonia was ultramafic. These authors ascribe the diversification over the stretches of Tertiary time to "the phenomena of extinction, selection, diversification and preservation, induced by the nature and extent of the geological formation, the diversity of the biotopes involved, and the fragmentation of the original vegetation cover."

The number and variety of serpentine endemic genera in Cuba are just as spectacular as the New Caledonia roster. The monumental monograph on the vegetation of Cuba by Borhidi (1991) devotes extensive text to the Cuban serpentines (see the earlier discussion of this work). Of the 72 genera endemic to Cuba, 23 are restricted to serpentine (33.3 percent). Several are monotypic, and they all range across the taxonomic "landscape": monocots and dicots in diverse families. For example, the monotypic *Gastrococos crispa* (Fig. 5.42) is a striking palm of the secondary savanna, and *Shafera platyphylla* is a distinctive woody composite (Asteraceae) of montane serpentine pine forests.

Turning to the floristically rich and unique metalliferous soils (nickel and copper) of south central Africa (Brooks and Malaisse 1985), one learns of a region rich in edaphic endemics, mostly tied to soils with high levels of copper, nickel, and cobalt. Yet all are species of genera that range beyond the metalliferous substrates. As far as I can discover, there are no endemic genera (or families) on these azonal soils of Zimbabwe, Zambia, or Congo (formerly Zaire). Two possible explanations surface to deal with this absence of endemic genera. First of all, one can hypothesize that edaphic diversification has taken place within genera; taxonomists recognize them as members of existing genera. A second thought invokes the notion of taxonomic inflation (or deflation). One taxonomist's endemic species is another's endemic monotypic genus. In Africa, perhaps, taxonomists have been more conservative than in New Caledonia or Cuba. I suspect that both explanations are correct. Some endemics are clearly monotypic genera, like the Cuban palm, while others simply may be singular endemic species of larger genera.

Epilogue

For plants with a geoedaphic preference, neither the search for their evolutionary origins nor the question of their taxonomic affinities is a closed issue. I predict that most of the new or otherwise noteworthy taxa that are described in print will have some singular edaphic or geological attribute. But the more perplexing and challenging unfinished business is in the arena of evolution. Notwithstanding the scenarios elaborated earlier in this chapter, testing them as hypotheses is largely on the agenda for the future. What are some of the more critical unsolved problems concerning evolution from edaphically normal plants?

MICROEVOLUTIONARY QUESTIONS

Is preadaptation, the presumed first stage in the evolution of edaphic novelties, widespread and detectable? Tests for the presence of edaphic preadaptedness should be done first in those genera and families that already have species now fully adapted. Their normal-soil counterpart habitats are most likely the ones to harbor preadapted genotypes. Any indication of preadaptedness, as well as no potential for edaphic accommodation, will be useful information. A related question is the crucial genetic one: What is the hereditary basis for incipient preadaptation, as well as for phenotypes in those populations well established as geoedaphic specialists? The essential first step in probing the genetic question is to make intentional hybrids between tolerant and intolerant races of the same species or crosses of related species with contrasting tolerances. The latter test depends on interfertility of the parents. Desiderata for carrying out such a crossing program are: (1) the plants are self-compatible, so that their F–1 and later generation progeny can be selfed; (2) a well-defined phenotype has been identified in the tolerant parent (e.g., tolerance to low calcium, high copper, etc.); and (3) the test species should be easily cultivated, should flower and fruit within a season, and have easily manipulable flowers for crossing. My choice of plants would be contrasting edaphic races or species that occur on and off serpentine, of crucifers like *Alyssum, Streptanthus,* and *Thlaspi*. Given the increasing accessibility of molecular genetic techniques, it should be possible to identify the genes for edaphic tolerance. A corollary probe to the genetic one is the detection of the physiological basis of tolerance. How are the genes for tolerance translated into physiological tolerance, at levels from the whole plant to the cellular and molecular? The same question arises for hyperaccumulation of heavy metals. What mechanisms, presumably cellular-molecular, account for concentrating (or alternatively, excluding) copper or nickel?

MACROEVOLUTIONARY QUESTIONS

Going from edaphically tolerant populations and races (microevolution) to evolving distinct and unique geoedaphic taxa (species and higher categories) requires not only major evolutionary shifts but also a leap of faith. There is yet no clear proof for how macroevolution may work. In this chapter, I have examined various models for this higher level of evolution. It will be instructive—and exciting—to watch how these contrasting models weather future testing. Will the gradualist model of the neo-Darwinian camp be replaced by saltation? Or will the neo-Darwinian model merely be retooled, given its broad explanatory powers?

The sharp contrasts in phenotypes, both physiological and morphological, that we have been discussing throughout this book offer unparalleled model systems for studying fundamental problems in ecophysiology and evolution. They could help answer the eternal ecological question: Why do plants grow where they grow?

7 Geoedaphics and Biogeography (Geology and the Distribution of Plants)

"Why do plants grow where they grow?" That simple, yet profound, question turns out to have a host of answers, depending on the plant and the biases of the observers. But surely, biogeography should be a fruitful synthesis of all those biases and contributory disciplines that seek answers to the elusive "why?" question. In my preoccupation with the role of geology in controlling plant habitats, we have visited many domains in natural science. A synthesis of these biological and physical approaches is the fecund field of plant geography. It, like taxonomy and evolution, draws upon a host of disciplines. But unlike those other fields of synthesis, plant geography is concerned with the what, how, and why of plant distributions in space and time. From the simple cataloging of the geographic ranges of plants and animals to the search for causes, biogeography has an eclectic bent—choosing all manner of data to explain the distributions of organisms.

I now employ biogeography to incorporate geoedaphics into the context of distributions in space and time. Events, processes, and phenomena of a geological nature have profound effects on the distribution of species, floras, and biotic communities. An underlying proposition in this linkage concerns the causes and effects of *discontinuity* in space and time. No terrain is homogeneously continuous; gaps of varying sizes promote the isolation that yields discrete biotas. The time component is the domain of the historical biogeographer. Variation in patterns of distribution over time has a dynamic history, linked to continental drift and accompanying orogenies, as well as to the chance or directed dispersal of disseminules. Two time-bounded approaches have, in fact, their respective champions: the vicariance biogeographers and the long-distance dispersal aficionados. Though they may be mutually exclusive approaches, vicariance and long-distance dispersal can operate together, and both have geological preconditions. This linkage will be examined shortly.

Another, more contemporary, aspect of biogeography is the quest to understand extant distributions. Since its biospheric stage is spatially and biologically heterogeneous and thus ecologically diverse, it is called ecological biogeography (Myers and Giller 1988). We will also probe biogeography's substantial connections to geoedaphics.

The linkages between geology and distributions of plants in space and time are fundamental and far-reaching, extending from the global and continental realms to the provincial and local context. Each linkage has both a historical and an ecological context to account for the distribution of biota. Each is also characterized by gaps in distribution, so that discontinuous rather than continuous distributions are the rule. This range of linkages will serve as the primary focus for most of what follows later in this chapter. But first, we examine concepts, or governing principles, in plant geography.

Some Guiding Principles of Phytogeography

The most pervasive "rule" in biology is that there are exceptions to most every rule (Stebbins 1982). This is especially true for biological systems above the individual organism level. The so-called laws of plant geography are especially pliant—subject to exceptions. So it is well to consider each such "law" simply as a first approximation. Each of the following dicta will have relevance to our theme of geoedaphics; geology has shaped the spaces over and on which distributions occur. Stanley Cain (1944) assembled an impressive list of principles in his classic *Foundations of Plant Geography;* they appear here in Table 7.1 (p. 10 of Cain). I will expand on certain of these and then add to the Cain roster other guiding dicta. All but the principle "Biotic factors are also of importance" of Cain's list of dicta relating to the environment have a geoedaphic context.

1. "Climatic control is primary." The broad belts of zonal temperature and precipitation determine the nature of vegetation globally. The four primary "biochores" (environments for organisms)—forest, savanna, grassland, and desert—are the products of major climatic belts. How does this dictum square with my championing the broadly encompassing idea of geoedaphics? I have contended that global to

Table 7.1. Some principles of phytogeography (from Cain 1944)

Principles concerning the environment:

1. Climatic control is primary.
2. Climate has varied in the past.
3. The relations of land and sea have varied in the past.
4. Edaphic control is secondary.
5. Biotic factors are also of importance.
6. The environment is holocoenotic.

Principles concerning plant responses:

7. Ranges of plants are limited by tolerances.
8. Tolerances have a genetic basis.
9. Different ontogenetic phases have different tolerances.

Principles concerning the migration of floras and climaxes:

10. Great migrations have taken place.
11. Migrations result from transport and establishment.

Principles concerning the perpetuation and evolution of floras and climaxes:

12. Perpetuation depends upon migration and evolution.
13. Evolution of floras depends upon migration, evolution, and environmental selection.

regional and local climates are shaped by some attributes of geology, especially landforms. Global climates "need" a round globe with its dynamics of diurnal and seasonal changes; the round Earth is a product of geophysical forces, so its climates have origins in geologic phenomena. Regional climates, even more, reflect the influence of landforms. Whenever geologic forces have created surface heterogeneity, the broad climates of the four biochores are modified. The major north-south cordilleras in North America exemplify this influence on climates by landforms. Indeed, mountains are the most dramatic sources of surface heterogeneity worldwide that render major climates subject to geology (Gerrard 1990; Barry 1992). In my view, though "climatic controls are primary," they are ultimately set in motion by global and regional geology.

Two corollaries to Cain's "climate-is-primary" law are principles 2 and 3 of Table 7.1: Climates have varied in the past, as have configurations of land and sea. Both these geohistorical attributes are ultimately conditioned by geological events and trends. Dansereau (1957) describes these major global changes as oscillating between "normal" and "revolutionary" (Fig. 7.1). For long stretches of time, landmasses were relatively small, had low relief, and thus had uniform

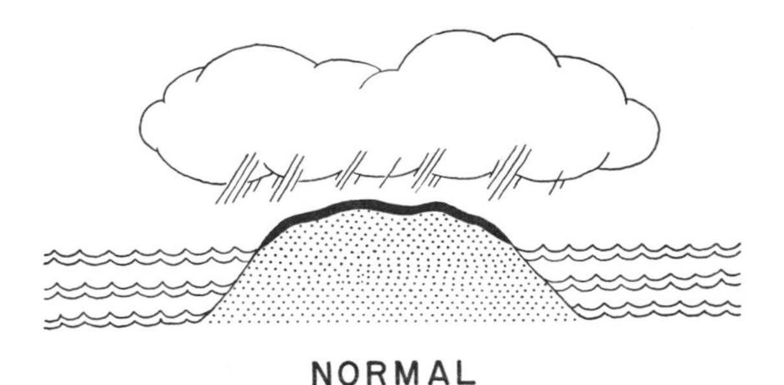

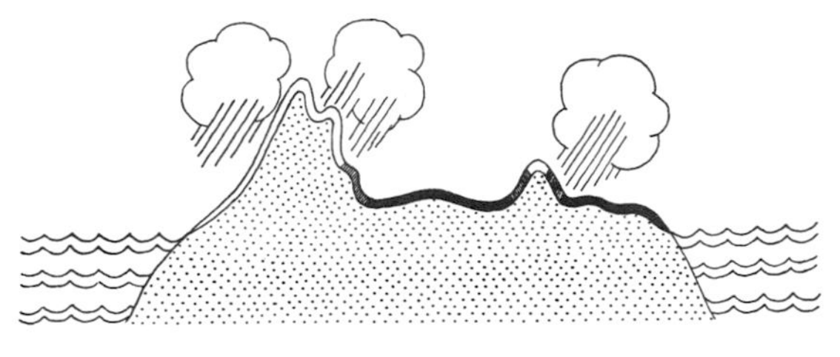

7.1 Stylized view of contrasting landform-climate features yielding normal and revolutionary climates. The latter have larger continents, more relief, and thus greater climatic differentiation. From Dansereau 1957.

climates (high temperatures and humidity, but low precipitation). Then for briefer periods—"moments of terror" à la Eldredge and Gould (1972) and John Le Carré—continents became larger, the seas were diminished, and relief became more pronounced. Revolutionary climates thus prevailed in which temperature and moisture differed markedly from place to place. Both normal and revolutionary climates, especially the latter, are the products of tectonic geology.

2. "Edaphic control is secondary." Here Cain takes the traditional view that soils, especially zonal, are products of their climates. For zonal soils, it is conventional to say that climate is the primary determinant of soil characteristics. Following this conventional view, even with azonal soils (e.g., limestones and serpentines), the impress of regional climate still holds sway, despite the influence of unique parent materials. To rephrase this "law" to read "Geoedaphic controls are secondary" would be wrong! *Geoedaphic controls* (in the broad sense) *are primary.* Climate and soils are ultimately the products of geology. Cain's remaining principles are self-explanatory. Suffice it to say that principles 7 to 13 can be connected to the "geoedaphic-control-is-primary" dictum, as I have done elsewhere in this book. Highly relevant to the geoedaphic theme is the pair of linked generalizations concerning surface heterogeneity and discontinuity. Disjunct distributions and ecogeographic isolation are the universal consequences of spatial, topographic, and lithological discontinuity. Diversity at the levels of species and the community is

primarily fostered by these two linked attributes of the Earth's crust and its biosphere. The Laws of Distribution formulated by Pierre Dansereau (1957, pp. 54–55) merit quoting in full:

> 1. Law of Geological Alternation. Normal and revolutionary periods do not have the same selective force upon the biota. 2. Law of Availability. The geographical distribution of plants and animals is limited in the first instance by their place and time of origin. 3. Law of Migration. Migration is determined by population pressure and/or environmental change. 4. Law of Differential Evolution. Geographic and ecological barriers favor independent evolution, but vicariant pairs are not necessarily proportionate in their divergence to the gravity of the barrier or the duration of isolation. 5. Law of Phylogenetic Traces. The relative geographical positions, within species (but more often genera and families), of primitive and advanced phylogenetic features are good indicators of the trends of migration.

At the level of biotic community, organizing principles center around the events of establishment, replacement, and equilibrium. Dansereau (p. 203) formulated these as the "Laws of Community Adjustment." Most relevant to the distribution of geoedaphic specialists are the concepts of niche, succession, and climax. Although green plants share universally in the common niche function of photosynthesis, a more unique niche definition for plants is derived from the "law" of tolerance ranges. Genetically fixed spans of tolerance vary widely both within and between floristic regions. Thus, edaphic specialists, like serpentine or limestone endemics, could be considered niche specialists, able to tolerate the exceptional properties of their substrates *and* thereby avoid competition with zonal species (Gankin and Major 1964).

Are there successional species in edaphically demanding habitats? In the most extreme sites, like serpentine barrens, their sparse species composition (low density and diversity) probably remains the same from initial colonization to their attaining equilibrium in density and species composition (the steady state of the climax). In these instances, there is essentially no succession (apart from disturbance), only climax (Whittaker 1960). Furthermore, the recruitment for and the resultant "steady state" composition of an edaphic community is best viewed as "Gleasonian"; individual populations or species come together as a result of the overlapping of their individual tolerance ranges (Gleason 1939).

There remain still other generalizations significant for plant distribution, especially those geoedaphically circumscribed. Dansereau (1957) discusses the cornering phenomenon: Abrupt changes in topography may truncate or eliminate specific

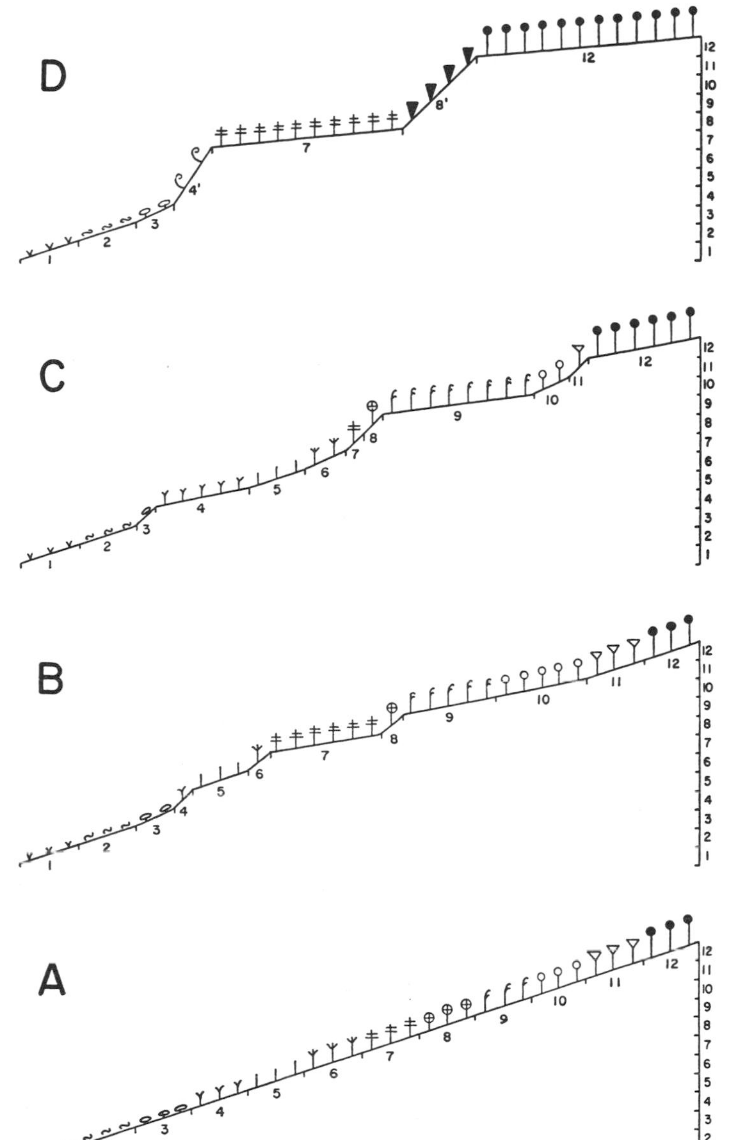

7.2 "Cornering" as shown in four hypothetical riparian transects. A shows regular zonation where all belts (1-12) are equally developed because of the regularity of the slope. B and C, with an uneven slope, show a certain amount of cornering and expansion as the levels that favor some belts are compressed or increased. In D, at 4 feet and 8 feet, such drastic changes in slope have occurred that new belts are introduced into the zonation. From Dansereau 1957.

communities or species assemblages that would be found on more gradual topographic gradients. "Such topographic inhibitions are not rare; in fact they are an outstanding feature in almost all landscapes where cliffs, moving sand, and permanently or periodically submerged land are of considerable spatial importance" (pp. 169–171); an idealized representation of this phenomenon is shown in Figure 7.2.

The dictum, "Climatic control is primary" (Cain 1944) has a corollary that can

7.3 Lithology and topography exert striking contrasts between vegetation types in the White Mountains of eastern California. On the left are dolomite flats and slopes with *Pinus aristata* but no *Artemisia*; on the right, sandstone flats and slopes support a sagebrush (*Artemisia tridentata*) community. Photo by D. Henderson

be applied to geoedaphically controlled distributions. It is Mason's so-called law of the extremes: "In any given region the extremes of these factors may be more significant than the means" (Mason 1936). Not only do climates generate extreme deviations from the normal; topography and lithology have extreme manifestations, which may be the primary arbiters of distributions. Most elements of a zonal flora cannot transcend such extremes as topographic or lithological barriers (Fig. 7.3).

A special biogeographic concept has to do with insularity in distributions. Ever since the classic MacArthur and Wilson book *Island Biogeography* (1967) was published, attempts to test the validity of the "equilibrium theory of island biogeography" have flourished. Recently the theory has been applied to conservation efforts where limited size and fragmentation of preserves lead to insularity. The theory's connections to geoedaphics is significant; special edaphic sites are insular, whether on oceanic islands or on mainlands (Kruckeberg 1991b).

Major Global and Regional Effects of Geology on Plant Distribution

CONTINENTAL DRIFT (PLATE TECTONICS)

"[C]hanging physical environments governed by plate tectonics have had a major role in evolutionary history." This epigraphic statement by Raven and Axelrod (1974) acknowledges the central role that global geologic processes have played in the distribution of angiosperm floras. Long before the theory of plate tectonics transformed physical and historical geology, biogeographers welcomed Alfred Wegener's "continental drift" hypothesis as a mechanism to account for many intercontinental disjunctions in related biota. In modern times, the fruitful connection between biogeography and plate tectonics has been exhaustively explored, both in concept and in specific case histories (e.g., Raven and Axelrod 1974; Davidse 1983; Briggs 1987). Rather than try to summarize much of this literature, I will examine the linkage between phytogeography and plate tectonics in the context of geoedaphic controls of plant distribution. Of all those geologic influences on plant life, plate tectonics is the grandest arbiter of major floristic patterns—both for connections and for discontinuities of major floras. Entire regional and even continental biotas (floras as well as faunas) have been, and are being, shifted in position by plate movements. And at the risk of redundancy, we must acknowledge that these realignments of the Earth's biota are geologic in origin.

For biologists who deal largely in contemporary time rather than deep or geotime, there are questions provoked by the biological responses to plate movements. How do floras and faunas respond to the movements of continental plates? Assuming plate movements of from 2 to 10 centimeters per year, the short-term displacement should not affect the composition of a biota moving on the gliding crust. But in time, organisms riding the plates will become displaced—separated in space—and are likely to be thrust into new climatic and topographic environments. This gradual transformation assumes that the biota has not been submerged by marine waters during periods of isostatic depression/rebound. Assuming continuous availability of land, how do terrestrial organisms respond eventually to the movements of continents and terranes? Three far-reaching responses to this geologic disturbance seem evident. First, extinctions of some taxa may befall biotas. Or evolutionary changes—shifts in adaptedness as well as in nonadaptive traits—are certain to have occurred with the attendant isolation. Still a third biological consequence will be both stepwise and long-distance migration. So, when the landmasses reach various stages of disjunction and foster new environments, biotas will be affected by selective extinctions, evolutionary change, or migration. Conceptually, this is all quite plausible; it must have taken place. Yet it is hard to visualize, even with millennial vision, the transformations in motion as well as the particular effects on the distributions of organisms.

BIOTA OF ACCRETED TERRANES

If drifting continents have displaced segments of continental biotas, then what effect has regional shifting of terranes (microcontinents) had on floras? The phenomenon of terranes is a further embellishment to plate tectonic theory. Exotic (alien) fragments (terranes or microcontinents) are rafted often at great distances, to be accreted to continental margins, far from their places of origin. The western margin of North America is a showplace—and jigsaw puzzle—of accreted terranes, both verified and suspected. When not subducted under continental plate, the terranes persist as telltale fragments of crust plastered helter-skelter along the margins of a continent (Skinner and Porter 1992, p. 456). Their role in transforming distributions of plants is unclear. At the least, they provoke speculation: Should a portion of a biota persist on a moving terrane, (1) it will be displaced and isolated from its parent biota; or (2) it will evolve under a new environmental regime; or (3) it will retain the character (species composition) of its former contiguous biota; or (4) it will become extinct. The northern suite of terranes in western North America probably had "docked" before the Pleistocene; hence any "rafted" biota would have been obliterated or displaced southward by the continental ice sheets. But the more southerly accreted terranes could have carried a persisting biota to its final docking site, thus achieving a significant disjunct distribution. Although some terranes originated underwater (seamounts or limestone platforms), others may have been terrestrial throughout the time span of their drifting prior to accretion. Even if there are questions about rafted biota on terranes, there is no doubt about their lithologies and their distances of displacement. Hence the isolated lithologies could be foci for edaphic endemism or other singular plant response. Thus verified terranes, as well as suspect terranes, may be hosts to floras whose attributes may be tied either to origins and displacements of terranes or to their final isolated lithologies. Terranes stimulate conjecture for the biogeographer—puzzles yet to be resolved. The title of John McPhee's book (1993) on the plate tectonic geology of California captures the essence of migrating terranes: *Assembling California.*

OROGENY AND ASSOCIATED LITHOLOGIES AT PLATE BOUNDARIES

Crustal deformation and uplift at plate boundaries (orogeny) can form major mountain systems. The cordilleras of the New World from Alaska to South America and the Himalayas are spectacular examples of orogenic belts. In Chapter 4, the role of mountains in fostering unique plant distributions and ecological settings was explored. For biogeography, the role of orogenies has been powerful in many ways. Foremost is the genesis of barriers to migration with resultant discontinuities in plant distributions. Then the mountains themselves become a mosaic of surface heterogeneity and diverse habitats in which floras will evolve. Mountain biota are rich in regional to local endemics.

Along with the geomorphic and structural consequences of orogenies (mountain-building), the orogenic belts consist of a welter of differing lithologies. All major rock types—igneous (both intrusive and extrusive), sedimentary, and metamorphic—reside in orogenic belts. Here, the rich mosaic of contrasting rock types serves to fix regional to local plant distributions, when the rocks are transformed into soils, both zonal and azonal. The role of lithological variety for plant ecology was the focus of Chapter 5. For the biogeographic connection, I am simply recasting the association of rock type with plant distribution as one of the major geoedaphic causes of particular patterns of distributions of floras, community types, and species. Two examples from the west coast of North America illustrate the orogeny-lithology-plant distribution connection. Vulcanism is a spectacular consequence of Pacific Coast plate tectonics, extending from Mount Lassen in northern California to Mount Garibaldi in southern British Columbia. A number of endemic species are restricted to one or more of the volcanoes; thus *Pedicularis rainierensis* is only on Mount Rainier while *Hulsea nana* is on each successive volcano from Mount Lassen to Mount Rainier (Kruckeberg 1987a). The other example comes from the orogenically generated ultramafic rocks, from California to central British Columbia. The association of Pacific Coast ultramafic areas with plate tectonics is set forth by Coleman and Jove (1992). While we have illustrated the restriction of plants to ultramafic substrates in Chapters 5 and 6, an additional case history has special biogeographical interest. The Shasta holly fern, *Polystichum lemmonii* (Fig. 5.60), occurs on every major exposure of olivine-derived peridotite and serpentinite from northern California (Mount Eddy and Scott Valley) to northern Washington (the Twin Sisters Mountain dunite). At first, *P. lemmonii* was considered conspecific with *P. mohrioides* of Chile and thus an example of an amphitropic disjunct. But D. Wagner (1979) has convincingly proved that the serpentine endemic and the Chilean fern are merely superficial look-alikes; the Chilean fern differs in significant morphological characters and is *not* found on ultramafic sites. But *P. lemmonii's* fidelity to West Coast ultramafic areas is biogeographically remarkable enough, without the embellishment of amphitropical distribution.

Vicariance or Long-distance Dispersal? Their Geoedaphic Relationships

Puzzlement over causes of major intercontinental disjunctions of floras and faunas has stimulated two contrasting explanations: vicariance versus long-distance dispersal. Both have crucial linkages to geologic phenomena. The gist of the vicariance paradigm is that on a global basis "the general features of modern biotic distribution have been determined by subdivision of ancestral biotas in response to changing geography" (Croizat, Nelson, and Rosen 1974, in Pielou 1979, p. 79). Pielou expands on this definition: "The splitting of populations that is a prerequisite for evolutionary divergence often results from the fragmentation of landmasses and the rafting apart of their biotas. Other causes of splitting are the formation of physiographic or climatic barriers such as mountain ranges or deserts." In this view the ultimate fragmentation is caused by plate movements and orogenies. The isolation thus achieved is the precursor to allopatric speciation and the evolution of new biota, yet with ancestral, phylogenetic linkage to the original floras and faunas. The contrasting model, long-distance dispersal, emphasizes the dissemination of founder biota over water and land by various dispersal mechanisms. While the two modes of creating discontinuity have coexisted as "either/or" explanations of disjunctions and allopatric evolution, there is good reason to consider *both* as causal explanations (Briggs 1987). Briggs harmonizes the two succinctly:

> Dispersal is an everyday occurrence undertaken by succeeding generations of almost all species while vicarianism is an event of much greater rarity since it must involve the creation of a barrier to separate existing populations. A most important point is that when vicariance does take place it appears to offer, at the same time, unusual dispersal opportunities for some groups of species. So dispersion may be looked upon as a continuing, inexorable process while vicariance, when it occurs in one habitat usually stimulates dispersal in another. This is particularly true in regard to continental movement with its making and breaking of land and sea barriers.

A novel linkage between vicariance (i.e., allopatric speciation) and volcanism has been proposed by John Morony (1999). He illustrates his thesis by associating discontinuous distributions of birds and plants between the volcanic and nonvolcanic areas along the Andes of western South America. He suggests that volcanic domains that promote discontinuity and allopatric speciation not only include the centers of volcanism but are affected by the more extensive areas of volcanic ash (tephra) fallout that promoted extinctions leading to vicariance. These ideas are a valuable contribution to biogeography and to geoedaphics. As we pointed out earlier, volcanoes are a special case of mountainous terrain influencing many aspects of their floras and faunas. Here, Morony adds to the uniqueness of volcanism in terms of the evolution and distribution of their biota.

Inescapable is the linkage between biotic disjunctions and physical geology. Discontinuities in terrestrial biotas reflect the present and past gaps in the configurations of landmasses. Moreover, the disjunctions in both biota and their anchoring of land areas are dynamically orchestrated—in a state of flux. The magnitudes of separations do change, owing to geologic change (plate movements and orogenies) and to the biological opportunism of chance dispersal and allopatric evolution.

Biogeography of Landforms Created by Climates

"Climatic control is primary." This widely held view, concisely stated by Stanley Cain (1944), is one I have challenged. Yet that dictum has validity only within the context of global, continental, and regional geologic controls. For plant geographers, especially those dealing with short-term ecological time frames, regional climates foster the broad vegetation zones of the world (Cain 1944; Dansereau 1957; Walter 1979). Moreover, many landforms are the immediate or past results of prevailing climates—either created *de novo* or greatly altered from their initial form by meteorological events. In turn, climatically created landforms, and the prevailing climates in which they form, constrain the nature of the occupancy by organisms of these landforms. How landforms molded by regional climates influence plant distributions merits elaboration.

The genesis or alteration of landforms caused by prevailing or past climates is the domain of climatic geomorphology, simply defined as "the product of climate acting on geology" (Howard and Mitchell 1985). Zonal landforms, like zonal soils, derive their particular features from the actions of major (zonal) climates. Thus zonal landforms in humid regions tend to have smoother (less angular) profiles. Further, the result of weathering in warm humid climates is the extensive accretion of fine sediments in river basins. The dense vegetation of humid lands also plays a part in the formation of zonal landforms. In contrast, arid climatic zones yield landforms with sharper, more angular relief. Weathering tends to be physical rather than chemical or biogenic, and the sparse vegetation cover has little influence on the shapes of landforms.

Topographies that are given their final shapes by the actions of climate abound world wide. Though I maintain that their "time-zero" origins are geological (structural-orogenic and lithological), the end product may be vastly modified by temperature and precipitation. The two crucial environmental modelers are weathering (physical and chemical) and transport by ice, gravity, wind, and water. There are many examples of climatically altered (or created) landforms that set the stage for particular distribution patterns for plants and animals. Some examples follow:

1. Karst landforms regularly display spectacular surface heterogeneity, with such features as dolines, dogtooth surfaces, tower karst, cockpit topography, and so forth (see Chapters 4 and 5). The conventional view has been that prevailing climates from cold to temperate and tropical determine the nature and degree of karstic landforms. This may be true in some instances, but structure (bedding planes, jointing, etc.) of the carbonate rocks can effectively control the kind and amount of weathering the rocks may undergo (Jennings 1985). Temperate karst, like that of the Mediterranean region, abounds in a variety of landforms. Dolines (gorges in all their variant sizes and forms) are discontinuously arrayed and thus foster isolation of biota. This isolation can result in the evolution of endemic taxa as well as fostering singular, habitat-specific plant communities (Poldini 1989). It is in the humid tropics that karst landscapes take on the most bizarre and spectacular forms. Dogtooth karst with its flat but serrated surfaces, cockpit terrain, and the cone and tower karsts of Cuba, subtropical China, and Malaysia are products of high rainfall and mild annual temperatures. The mogotes (inclined cone karst) of Cuba exemplify the potential in the tropics for local endemism and singular plant communities resulting from isolation and habitat uniqueness (Borhidi 1991). With karst, climate and lithology are linked.

2. The intermountain country of western North America is a showplace for landforms modified by weathering and transport. Water, wind, and—to a lesser, more indirect extent—glacial ice have molded the raw materials of sedimentary and basaltic lithology into grand, awesome relief: Grand Canyon, Bryce and Zion National Parks, Cedar Breaks, Canyonlands (Figs. 7.4 and 7.5), Monument Valley, and others. Even in these arid environments, climate has transformed topography. Since the elaborate erosional landforms are discontinuously arrayed in space, they afford the stages for acting out unique distributions of organisms,

ranging from local endemics and habitat-specific communities, to ranges of wide regional extent (Cronquist et al. 1972). It is the Colorado Plateau country that is grandly dissected by erosion. Local endemics confined to the dissected terrain of the Grand Canyon and the Canyonlands exemplify the linkage between landforms and biogeography.

7.4 Canyonlands National Park, Utah, reveals influential features of sedimentary geology (variations in landform and in lithology) that affect the distribution and composition of the flora and vegetation. Pictured here are intricately dissected topographic surfaces of a portion of Canyonlands as viewed from Grandview Point. Photo by B. van Diver.

In the same general region, the intermountain West, several remarkable landforms are associated with a succession of catastrophic meteorological events. The channeled scablands of eastern Washington State consist of a gigantic network of steep-sided coulees and runoff channels, as well as vast areas in Columbian Plateau country denuded of their original loess; all indications are that these features were the result of massive flooding. J. Harlan Bretz (1959) proposed that in

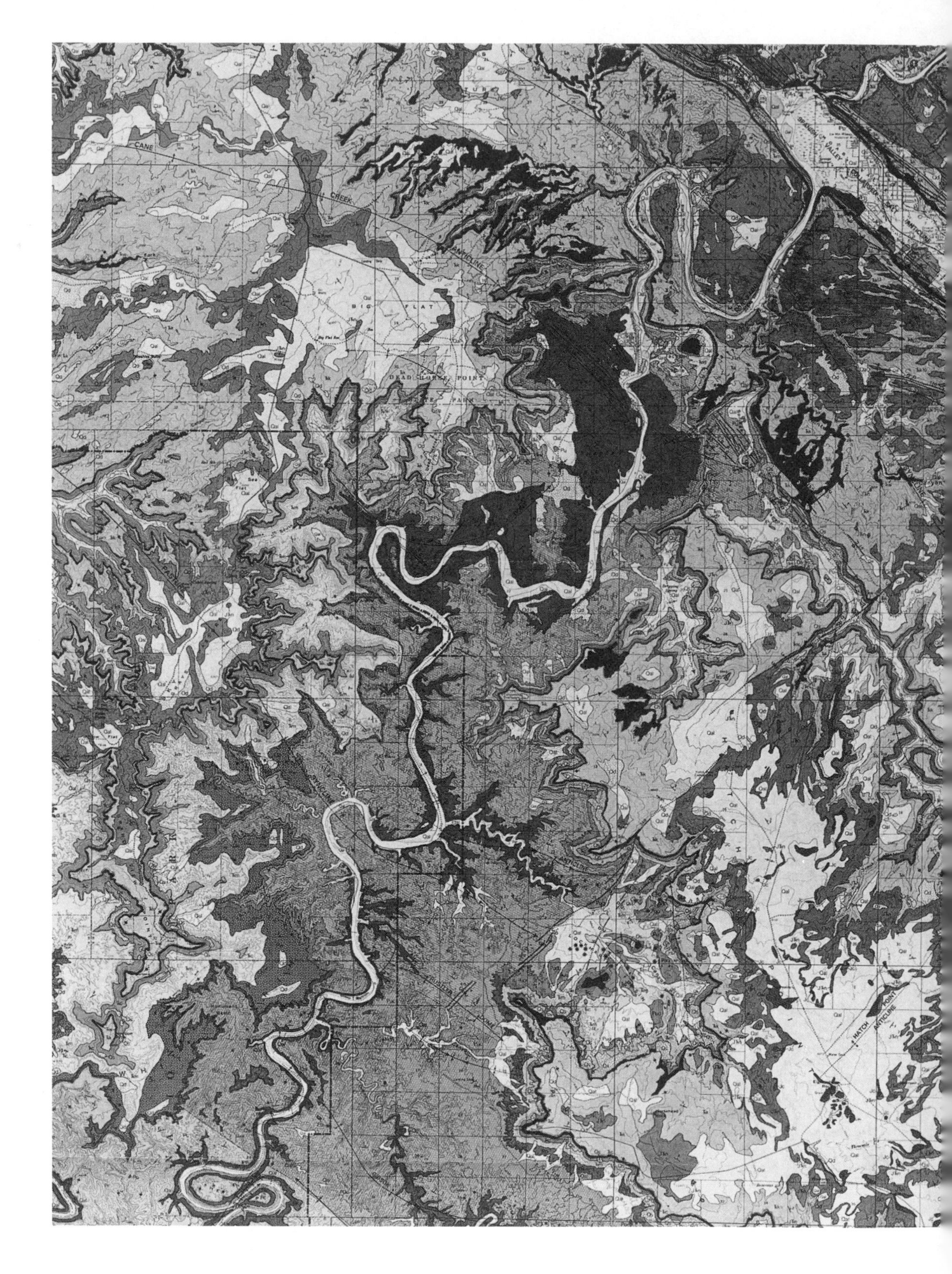
CANE
CREEK
BIG FLAT
DEAD HORSE POINT

a two-week period a series of floods scoured the Columbia Plateau region, leaving a variety of instantaneously created landforms. The cause of the repeated flooding in Pleistocene times was the breakup of the huge ice dam holding back glacial Lake Missoula (Allen and Burns 1986). Can this be interpreted as a sudden transformation of landscapes due to climate? The genesis—along with the waxing and waning of continental ice—is in itself a climate-induced phenomenon. The possibility exists that the breaking of the ice dam was triggered by a brief warming period. Other rapid changes in landscapes triggered by extreme (catastrophic) meterological events are most often the local alteration of slopes. Mudflows, landslides, slumping, and so forth, can dramatically change relief (Starkel 1976).

The biogeographic consequences of such extreme meterologically caused alterations of landscapes appear not to have been addressed. The removal of preflood soils, especially the loess of the Washington Palouse grasslands, has given way to lithosols (rocky, skeletal soils), habitats for a distinctly different set of plant communities (Daubenmire 1970). The cliffs and their colluvial talus, common features of coulees and runoff channels caused by the violent flooding, acquired a characteristic flora as well. Among features of the channeled scablands are numerous sinks that have become ponds, supporting a characteristic vegetation. Elsewhere the sudden formation of bare slopes or similar substrates from flooding can create pioneer habitats for revegetation as recruits from nearby unflooded communities as well as from chance dispersal far away from the denuded site.

Landforms and their exposed lithology, when fashioned by rainfall and other hydrologic influences, may seem to be exceptions to the central theme of this book—the primacy of geologic factors in promoting plant life. After all, such landforms and their bedrock are, in their final shaping, the products of climate—in the geomorphologist's view. The dissected plateau country of the American southwest seems to epitomize the primacy of regional climate as the ultimate shaper of landscapes. All well and good. Yet once fashioned by climate and its hydrologic consequences, the finished (or as yet unfinished) products—the land forms and their bedrock—enter the geoedaphic realm. They become the geologic framework for occupancy by contemporary flora. Hence, however originally formed, a dissected plateau and its bedrock exert their intrinsic influences on the biota.

7.5 This geologic map of a portion of Canyonlands shows a rich variety of sedimentary rock types offering a diverse series of substrates for plants. Taken from a map in Huntoon et al. 1982.

Landforms Caused by Biological Agents (Biogeomorphology)

The primary thrust throughout this book has been to highlight the effects of geology on the living world, especially plants. But there is a reciprocal proposition. It is: "Life is a geological force" (Westbroek 1991). The biosphere is a nearly global skin coating the Earth's crust with organisms and their by-products. Everywhere there is life, organisms are altering the fabric of the crust. The particular aspect of that universality addressed here is biogeomorphology—the fashioning of landforms by organisms. And then we ask: How does the superposition of life on geology affect the distribution of organisms? In her introduction to *Biogeomorphology,* Heather Viles (1988) encapsulates the organism/landform synergism as a kind of reciprocity: "1. The influence of landforms/geomorphology on the distributions and development of plants, animals and microorganisms [is linked to]: 2. The influence of plants, animals and micororganisms on earth surface processes and the development of landforms." It is the latter part of this duality that we now take up.

In concert with climate, organisms alter landscapes by processes causing weathering, erosion, and transport. Any group of organisms can induce these processes—microorganisms, plants, and animals (both invertebrates and vertebrates). The results of such biological activities on geological materials can be detected at all scales, from regional (macrotopographic) to the ultralocal (microtopographic). Of the many and diverse biogenic influences on landforms (and indeed on ocean basins), a selection of those that alter or make topography on a large scale will suffice to illustrate "life as a geological force."

ORGANISMS AND MACROTOPOGRAPHY

Three of the major global biomes—humid tropics, arid lands, and arctic-alpine (periglacial) regions—display landscapes modified or even created by biological agents.

1. The vegetation of the wet tropics, with its massive organic load of plant substance and its year-round continual respiration, strongly influences the inorganic surfaces it densely covers. But peculiar to the tropics is the reciprocal idea that the tropical "landscape and its associated climate may have little effect on the rain forest ecosystem" (C. F. Jordan in Viles 1988). The immediate effect of this immense biological activity is deep weathering. It results in transforming hard parent rock into saprolite, a softer derivative. It also confers a smooth and rounded topography to the land, and contributes to frequent and massive landslides. Since these biogenically created landscapes are so vast and continuous, their effect on

plant distribution is to create more of a continuum than sharp discontinuities. Speciation leading to the immense diversity of the tropics is stimulated more by biologic than geoedaphic causes. Yet we noted earlier the dramatic effects in the tropics on speciation of landforms: the tepuis (isolated high plateaus of Venezuela) and the high volcanic peaks of East Africa, both landforms in the tropics.

2. In arid and semi-arid lands, where annual precipitation is limiting to plant growth, the major role of vegetation (including lichens, algae, and microorganisms) is to limit sediment transport by wind and water (S. G. Thomas in Viles 1988). Stabilization of landforms like dunes, flats and slopes is thus achieved by plant and microbial cover against the incessant aeolian transport of materials.

Biocrusts and desert varnish are two unique soil sediment stabilizers in arid lands. Biocrusts are surface networks composed of lichens and algae; most often they occur between clumps of higher plant vegetation. They prevent erosion by wind and water. When disturbed (e.g., by grazing), the crusts are broken or destroyed, allowing for enhanced sediment transport. An exceptional impact on biocrusts was the blanketing of semi-arid lands by aeolian volcanic ash following the 1980 eruption of Mount St. Helens. Vast areas of cold steppe desert were covered by ash, to the lee of the eruption (Mack 1981, and in Bilderback 1987).

Desert varnish coats rock surfaces as a thin mix of dust cemented by the biological activity of blue-green algae. This coating serves to reduce rock weathering and thus aids in preserving the angular shapes of arid land rock outcrops (Thomas in Viles 1988).

Calcrete (caliche) deposits are widespread in arid regions. They contribute to the geomorphology of a region both by the forms taken by the deposits themselves and by their ability to prevent sediment transport. Calcrete is a calcium carbonate rock produced by the combined actions of lichens and plant roots. Some calcrete produced by lichens are laminar in texture and have been called lichen stromatolites, owing to their resemblance to the algal stromatolites of marine environments (Thomas in Viles 1988).

3. In the relatively harsh environments of arctic and alpine (periglacial) regions, climate dominates but vegetation modifies the geomorphic processes. Low-relief landforms may originate wholly or in part by the actions of plant life. Palsas, formed on peatlands, begin as low hummocks and eventually attain conical shapes up to 10 meters high. The interior of a palsa consists of alternating layers of frozen peat and ice. Patterned ground epitomizes the contest between climate and vegetation. Thus the common frost scars of north latitudes are bordered by low vegetation. The plants may temporarily advance over the frost scar, only to be restrained by the physical actions of ice (R. B. G. Williams in Viles 1988). Mam-

mals can also shape landscapes in boreal regions. The prodigious workings of beavers may result in telltale changes in valley geomorphology. A succession of beaver dams can leave behind a succession of ponds or lakes. Beaver dams also serve to check sediment transport and control runoff. Other mammals contribute mostly to local topographic modifications; caves of bears, tunnels of pocket gophers, and even the migrating effects of lemming populations can directly or indirectly alter relief (Williams in Viles 1988).

ORGANISMS AND KARST

The exceptional landforms created by karst result from the dissolution of rock, mostly calcium carbonate. Primary agents abetting this dissolution are organisms: higher plants, cryptogams, and microorganisms (Viles 1988). When the karstic geomorphology is mainly due to organisms, the product is called "phytokarst" or "biokarst." The most frequent primary agents are algae and lichens, though "root karst" is fashioned by vascular plant roots, especially in the tropics. Dissolution and absorption of calcium can alter limestone surfaces; mangrove roots create a highly dissected root karst (Viles 1988). The shapes and textures of biokarst range from serrated pinnacles up to 3 meters high to the minute pits and grooves (microrelief) caused by lichens and snails. This biogenically caused relief would provide the heterogeneous surfaces for plants, resulting in discontinuous distributions.

MIMA-TYPE RELIEF ("MIMA MOUNDS")

A special kind of microrelief of biogenic origin has been observed in many parts of the world. Mounds of uniform size and shape most often are hemispheric and evenly spaced; large mounds can be up to 2 meters high and 10–20 meters in diameter. The name "Mima microterrain" or Mima mounds is taken from their type locality, at Mima Prairie, Thurston County, Washington State (Fig. 4.27), where they were first described in detail and their biogenic origin espoused (Dahlquest and Scheffer 1942). While various hypotheses, from purely physical to biogenic, have been proposed, the most highly regarded explanation is that territorial fossorial rodents make the mounds. This and alternative hypotheses are discussed in Cox (1984), Kruckeberg (1991a), and Dahlquest and Scheffer (1942). However formed, do Mima mounds foster particular ecologic and biogeographic patterns in vegetation both locally and regionally? The mounds of Mima Prairie display both obvious and subtle patterns. Giles (1969) found that species composition changed from the north-facing to the south-facing sectors of a mound. Del Moral and Deardorff (1976), elaborating on that finding, showed that vegetation patterns varied on at least three scales: the microtopographic level of a single mound, an intermediate scale of variation among mounds depending on exposure, and the elevation of the mounds. At the third level, a successional change in vegetation composition occurs. Where Douglas fir (*Pseudotsuga menziesii*) is invading the mounds, the conifers create a new environment on the mounds—an increase in mesophytic species.

TERMITE-CREATED MICRORELIEF

Termite nests can create spectacular surface heterogeneity; the heaps and mounds are often the conspicuous topographic features of otherwise flat terrain (A. S. Goudie in Viles 1988, pp. 166–192). A remarkable instance of change in floristics due to termite mounds occurs on the serpentines of the Great Dyke in Zimbabwe (Wild 1975). The primary plant responses are an increase in vegetation cover, a change in species composition, and increased palatability of the grassland forage. Wild found that although nickel increased in the mounds, calcium and magnesium also increased; the net effect is to foster a greater biomass and a greater number of serpentine-indifferent (bodenvag) species. See Butler (1995) for a discussion of termites as geomorphic agents.

THE HUMAN FACTOR IN THE GEOEDAPHICS OF PLANT DISTRIBUTION

Changes wrought by humans in the distributions of plant species have come to dominate the kinds of perturbations inflicted on biota, globally to locally. Extinctions of indigenous species, introductions of alien organisms, and wholesale alterations or outright destruction of habitats all have profoundly affected the plant world. It is not my intention to embark on all aspects of this massive set of disturbances. Only in the context of biogeomorphology will we examine the far-reaching human impacts on native floras. We look at two aspects: landform changes and alterations of soils, and the ways these geomorphic transformations have affected the distribution of biota.

Whenever human activity has altered landforms—or created new ones—the consequences for indigenous floras are to drastically reduce or eliminate species, especially rare or local endemics; or the human presence provides opportunities for the spread of natives, usually common and successional species, as well as the spread of aliens on the disturbed sites. In montane areas, the topographic changes are mostly due to human-caused erosion or intentional alteration of relief. Effects of erosion are most severe in the humid tropics and in wet temperate regions (Pitty 1971). The loss of montane habitat due to soil erosion following deforestation has been devastating. Parts of tropical Madagascar, Cuba, New Caledonia, and India have been transformed into biological deserts, especially where deep gullies and sediment fluxes become devoid of any native vegetation. Upland areas in Hawaii, once they were converted from forest to agricultural lands, inevitably suffered losses in species diversity of the indigenous flora. Most of Hawaii's extinct or endangered plant species owe their plight to landscape alteration through human activity.

It is in lowland topography that the pursuits of humans have taken the greatest toll on native floras. Commandeering of water resources in lowland catchments runs the gamut from major dams to canals and drainage ditches. Canyons become lakes, like the transformation of the Colorado River into Lake Mead in western United States. Natural river courses are altered by diking, ditching, and diversion to create new landforms out of old. Canyon, riparian, and estuarine native species are decimated or lost in the wake of these wholesale interventions. And if native flora is reduced or lost by such human activities, then the disturbance invites new flora, mostly aggressive introductions from other lands. Human disturbance also creates habitats for alien weeds: "Weeds are plants (other than crop species) that thrive under human disturbance" (Baker 1965). In mountain areas, road-building, logging, and fire open up pioneer sites for invasive aliens; lowland landscape alteration achieves the same opportunity for introduced species.

The earth of Planet Earth has undergone vast alterations since the dawn of agriculture and civilization. Soil losses have meant the loss of habitats for native species. This alarming circumstance has been the subject of countless publications (e.g., Ehrlich and Ehrlich 1981, 1987). More germane to the geoedaphic/biogeography connection are the human causes of heavy metal exposure. Mine spoils initially are wastelands. But the heaps and hillocks from mine workings eventually develop a rudimentary soil and become sites for exclusion of adjacent biota and for colonizing by tolerant genotypes. Where the mine spoils contain heavy metals like lead, copper, nickel, zinc, and mercury, the genesis of metal-tolerant races is nearly inevitable. Here, human disturbance has provoked an unparalleled microevolutionary response (Bradshaw 1976). The bottom line to be read from the above: "Man is a practicing biogeomorphologist!" The final chapter of this book amplifies on the theme of the human influences on the geology-plant interface.

Island Biogeography and Geoedaphics

"Insularity . . . is a universal feature of biogeography" (MacArthur and Wilson 1967). And such insularity can take many shapes and contexts. As a manifestation of discontinuity in the distribution of habitats and associated biota, insularity has pervasive geoedaphic components. Separation of landmasses, and the formation of discrete topographic units and lithologies, make geography a prime arbiter of insular distributions. There are three broad patterns of insularity. The classic and most obvious is related to oceanic islands. Then there are those islands occurring on the continental shelf, close to mainland terrain (continents), yet still surrounded by water; these are the offshore continental islands. The third mode of insularity has to do with topographic (geomorphic) and edaphic islands in a "sea" of normal (zonal) mainland environments (Kruckeberg 1991b). It is this latter type that I have used to explore the applicability of island biogeographical theory (IBT).

Do topographic and lithologic/edaphic islands in a "sea" of contrasting mainland environments fit the model for oceanic islands? The model and the theory (IBT) as espoused by MacArthur and Wilson (1967) are derived from the premise that the numbers of species on islands cannot be accounted for by *in situ* evolution alone (Pielou 1979). The species-stocking of islands is achieved by striking an equilibrium between extinction and migration. Further, for oceanic and offshore islands, IBT considers that species numbers at equilibrium are determined by the number of species in the source biota (mainland taxa), by the size of the island, and by its distance from the source biota. How do these premises work for geoedaphically distinct "islands" on mainlands?

Mainland islands that display unique topographic or edaphic attributes possess a number of features in common with oceanic islands: distance from normal environments, size of island, and number of species in the surrounding source biota. However, there are significant differences between oceanic and mainland islands that would constrain the wholesale application of the IBT to mainland islands (Kruckeberg 1991b): (1) Proximity of recruits: Mainland geoedaphic islands are inevitably surrounded by floras on normal (zonal) terrain. This is in marked contrast to water-bounded islands that are usually far removed from their source biota. Thus for the mainland islands, ease of migration, often local, is inevitable. All it takes is for preadapted genotypes present on a nearby zonal substrate (granite, sandstone, etc.) to colonize a serpentine outcrop nearby. This proximity of recruits also confers a close taxonomic affinity between the source biota and the plants of the edaphic island. (2) Mainland islands will be stocked with at least three kinds of taxa: (*a*) those species found on nearby normal habitats (the so-called indifferent or bodenvag species); (*b*) species reaching the mainland island as outliers or as extensions of their usual ranges; and (*c*) taxa peculiar to the island, such as local endemics or edaphically tolerant races. Since biotas of oceanic islands originate as recruits from far removed areas, they will often be a mix of endemics as well as species commonly found on oceanic islands. (3) Mainland islands may engender interspecific competition not usually found on oceanic islands. Bodenvag species from adjacent normal substrates may outcompete insular endemics. Only when the island is a geoedaphically stressful habitat, such as serpentine, might the competition be reduced. An island, like a local serpentine outcrop, can effectively exclude most species found on the nearby normal substrates. (4) If extinction of a local endemic occurs on an oceanic island, its replacement is most unlikely, given the remoteness of source biota. But for mainland islands, proximity of recruits should foster replacement of an island extinction.

Even with the above differences between mainland and oceanic islands, IBT is still a heuristic model for determining the origins of insular endemics. The parameters of distance between source and founders, species richness of source biota, and size of island are still valid.

The "Inselberge" phenomenon is a well-known example of topographically created islands (Porembski and Barthlott 2000). Isolated mountain peaks or ranges elevated above the "sea" of lowland terrain epitomize Inselberge relief. The remarkable alpine floras of the East African volcanic peaks typify the Inselberge mode of discontinuous distribution (Hedberg 1970). Though poor in species, the rigorous, diurnally stressful habitats of the Afroalpine have fostered a spectacular Inselberge flora: local endemics and unique life-forms (e.g., the giant pachycaul senecios) are on one or more of the isolated peaks. In western North America, alpine islands in the "ocean" of prairie or desert abound; their floristics show the effect of insularity (Billings 1978). Another illustration of the Inselberge effect is on the high mountains of the southern Appalachians in southeastern United States. White et al. (1984) tested the hypothesis that on these montane islands "vascular plant richness is related to island size" and found that "the species-area relationship of these mountain tops has the steep slope [of the MacArthur-Wilson model]." White and colleagues found species richness to be positively correlated with size of area, number of peaks, maximum elevation, and number of community types present. Here, there is a significant difference between these montane island patterns and that predicted from IBT. The two smallest montane islands have higher richness than what was expected from size alone. The past history of the regional flora can account for this discrepancy. The conclusion reached was that "extirpation has been more important than immigration in shaping the recent floristic richness of the high peaks."

Floristic islands created by edaphic discontinuity would appear to epitomize the conditions for support of IBT. Unusual lithologies that lead to azonal soils are often "insular" in distribution. Islands or archipelagos of serpentine outcrops (see map, Fig. 7.6) reveal much the same insular pattern as the Inselberge phenomenon. Discontinuity of a regional flora, exclusion of many taxa, and significant endemism are the hallmarks of serpentine islands (Kruckeberg 1991b).

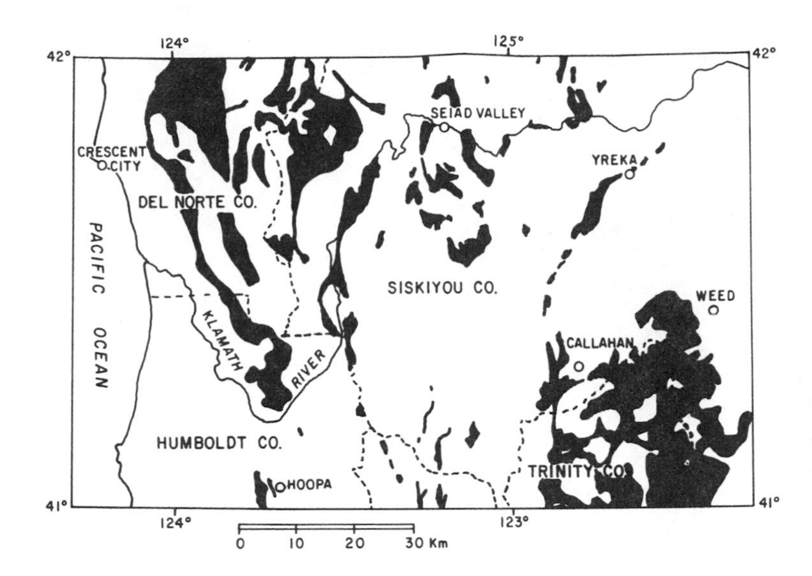

7.6 Discontinuous distribution (insular pattern) of serpentine outcrops in northwestern California. From Kruckeberg 1985.

Granite outcrops, often discontinuous in occurrence, appear many places worldwide. They can harbor distinctive floras and plant communities. The granite "flat rocks" of southeastern United States (Fig. 7.7) are exposed, floristic islands surrounded by mesic forest. Even a single granite outcrop can take the form of an archipelago of microisland communities. Burbanck and Platt (1964) focused on these intra-outcrop islands of vegetation, calling them "island communities." They are in depressions with soil, scattered across the face of the bare rock. Depending on the depth of soils, they support four distinct community types. The prospect for testing island biogeography theory on these island chains is promising. Burbanck and Platt did look at size of island community, but chose to relate it to soil depth rather than to species diversity. Murdy (1968) sees the disjunction of granite outcrops as a propitious condition for both speciation and extinction. Endemic taxa may be found on one outcrop but missing on a nearby outcrop. If extinction is really involved, then this suggests the Lewis (1962) model of catastrophic selection leading to saltational speciation for the survivors.

Wyatt and Fowler (1977) subjected their floristic data on North Carolina granite outcrops to the MacArthur-Wilson model for islands. They found a positive, linear relationship between area and species number. To the west, granitic outcrops in the central mineral region of Texas were found to have fewer endemics than on the southeastern United States outcrops (Walters and Wyatt 1982). They ascribed this difference to the greater geographic isolation and the sharper discontinuity with surrounding vegetation for the southeastern granite outcrop floras.

7.7 One of the many granite outcrops (balds) in southeastern United States, at Overton Rock, Franklin County, North Carolina. Ephemeral annuals, some endemic to the balds, are frequent on these granitic "islands." Photo by C. and J. Baskin.

Though the outcrops of the two areas harbor different floras, they were nearly identical in life-form spectra (a predominance of annuals).

Granite outcrops occur on other continents. Some examples are the "kopjes" of South Africa and the granitic domes of Australia. Ornduff (1987) gives an extensive and fascinating account of the latter edaphic islands in southern Western Australia. The flora of these granite domes is not simply "a random subset of the Western Australia flora." Some families are underrepresented, while others are overrepresented on granitic outcrops. Further, there is a bias in life-form: nearly two-thirds of the native vascular flora on granite outcrops are annuals, while annuals are only one-twelfth of the Western Australian floras as a whole. Geophytes and "resurrection" plants are conspicuous elements of the outcrop floras. Ornduff, using island/area analysis, found that the species number/area prediction did not hold for these granitic outcrops. Even the smallest outcrop (Nettleton Rock) supported the largest number of species.

It is well to point out here, as does Ornduff (1987, p. 19), that MacArthur and Wilson (1967) exempted mainland islands from some of their conclusions for oceanic islands. They argued, as I have, that mainland "habitat islands" are sur-

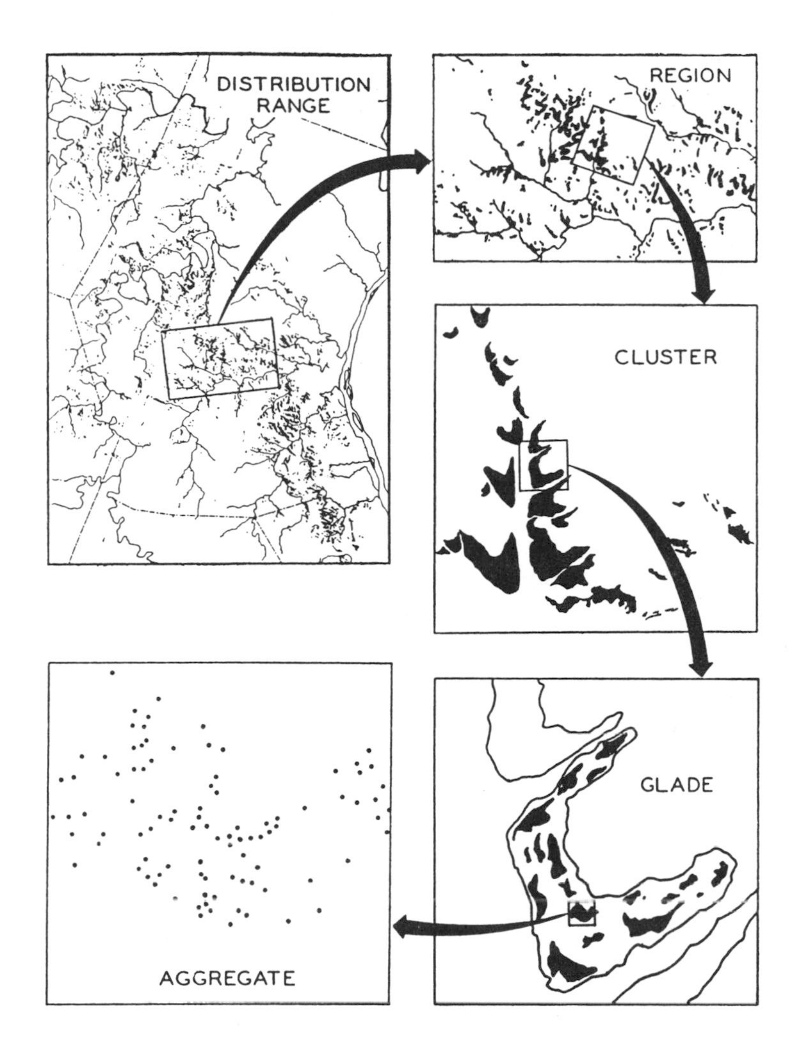

7.8 This series of maps with increasing resolution dramatizes the insular pattern of distribution of *Clematis fremontii* var. *riehlii* on isolated limestone outcrops in the Ozark Mountains, United States. From Erickson 1945.

rounded by land areas with potential immigrants and competitors, unlike oceanic islands.

Similar patterns of insularity are evoked by limestone and dolomite. An elegant account of such a case history is Erickson's (1945) study of the distribution of *Clematis fremontii* var. *riehlii* on limestone glades in the Ozarks of Missouri. The plant is confined to rocky, barren openings (glades) on outcrops of thin-bedded dolomite. The character of the edaphically defined distribution is dramatically portrayed by Erickson's sequence of range maps, from region to local aggregate (Fig. 7.8). The ultimate local habitat is the single aggregate, a colony of *Clematis* at a given glade. The insular character of the distribution includes a hint of island size and species diversity. Erickson compiled records of the plant's absence as well as its presence. At least the number of *Clematis* plants is lower on small glades, or the taxon is absent altogether from some of the smallest glades. This subject of edaphic islands in the context of IBT is amplified in Kruckeberg (1991b).

Historical Biogeography and Geoedaphics

Contemporary distributions of organisms have direct links to the past, both recent and ancient. Thus, where plants grow has been conditioned by events and processes of the past. The principles, or simply the truisms, that arise from this historical view of biogeography merit restating. The principle of uniformitarianism—the present is key to the past—has special significance in our putting past distributions in the context of changing manifestations of geoedaphics. Past landforms, lithologies, and soils, by this principle, were not unlike those existing now and were caused by similar processes. Over geologic time, these geoedaphic factors have varied in space, form, and substance. Likewise, climate has varied in the past.

Although paleoecology and paleobiogeography have respectable histories all their own, it is only in recent times that paleobiologists have begun to put ideas of ecology, biogeography, and evolution to the tests of history. Some recent reviews of this emphasis on causal paleobiology are: Behrensmeyer et al. (1992); DiMichele, Phillips, and Olmstead (1987); Dodd and Stanton (1990); Ross and Allmon (1990).

Hallam (1994) defines the capabilities, and pitfalls, for trying to reconstruct past distributions of organisms. He says: "The problem arises of accommodating inconsistencies with *ad hoc* assumptions, producing on occasion a confusing mass of facts, ideas and circularity, something which has bedeviled biogeography in the past." Yet paleobiogeography can claim a respected niche in the pantheon of natural sciences. Hallam's quote from Ball (1983) says it well: "[H]istorical biogeography must remain a narrative science, with pluralistic methodology and using circumstantial evidence."

What can students of past floras tell us about the physical environments to which the plants were adapted? Is there any evidence to suggest that species and communities in the past were restricted to particular topographies, parent materials, or soils, as they so often are today? Alas, there are only limited answers to those questions. The critical problem impeding answers is the difficulty in reconstructing past physical environments from the limited and often distorted evidence that remains—the fossils and their associated hints of landforms and their lithologies. But paleobiologists are, more than ever, trying to make such reconstructions—for example, DiMichele, Phillips, and Olmstead (1987) and Ross and Allmon (1990).

	EXTRINSIC FACTORS	INTRINSIC FACTORS	TOTAL
BIOTIC FACTORS	80 (38%)	37 (18%)	117 (55%)
ABIOTIC FACTORS	80 (38%)	14 (7%)	104 (45%)
TOTAL	160 (76%)	50 (24%)	211

7.9 "Frequency of sorts of evolutionary factors considered in the journal *Paleobiology* in the years 1975-87. In each cell, the top number is the frequency out of 146 papers classified, the bottom number the percentage of the total number of factors considered (211)" (Allmon and Ross 1990). Extrinsic/abiotic factors loom large. It would be nice to know what portion of the extrinsic/abiotic bias deals with landforms and lithology of the past.

The evolution of past floras and their constituent species has been tied to a combination of major causes, both biotic and abiotic. These in turn are under the influence of intrinsic or extrinsic factors. Four permutations of the biotic/abiotic, extrinsic/intrinsic attributes can be visualized as combinations within a $2 \times 2 \times 2$ table (Fig. 7.9, from Allmon and Ross 1990). One of the four combinations—abiotic (physical) extrinsic—is the set of attributes that includes, besides climate, the geoedaphic influences.

Landform sizes, shapes, and distributions are often reliably reconstructed from the fossil and geological record. We have already elaborated on the effects of plate tectonics in the realignment or "drift" of continents and terranes. This is the most far-reaching and best documented of the geoedaphic influences on past distributions of flora. Invoking continental drift and terrane relocations to explain floristic disjunctions must be done within the appropriate time frame. Since modern floras go back only to the Cenozoic, continents will have already become positioned in roughly their present spatial configurations. See Briggs (1987) for an excellent series of world maps portraying Triassic to Present landmasses. The time constraints on landform configuration are illustrated by comparing Briggs's map of the Miocene world with that of the Present. Major tectonic and sea level changes did occur even between the Miocene and the Present, to significantly affect biotic distribution patterns.

On a more local to regional scale, we can expect that past landform changes also affected floristic distribution patterns. But here the evidence is less substantial than for continental drifting. Two areas of past topographic diversity have been detected: uplands and swamps. Upland topographies are detected by the presence of andesitic volcanism and the identification of colliding plate boundaries. See Tiffney and Niklas (1990). These authors have given a thorough statistical account of geographic variability and plant diversity from the Devonian to the Miocene. They point out that the evidence for the relative importance of land area size and topographic variability is ambiguous: "[T]he correlation of numbers of uplands and land area makes it impossible to distinguish the hypothesis that vascular plant diversity is a function of area from the hypothesis that diversity is a function of terrestrial topographic variability. Both may be important" (p. 93).

Evidence for past diversity of landforms should include indications of insularity, both oceanic and mainland. Island biogeography must have been a significant element in the fabric of arrays of floras and landmasses. Tiffney and Niklas (1990) caution that the evidence is equivocal. It is difficult to distinguish real (oceanic, offshore) islands from peninsulas, as well as isolated uplands from continuous upland terrain (the latter is my conjecture). So the paleobiogeography of topographic and edaphic islands must await further evidence.

Edaphic islands in the past, as today, were, and are, caused by discontinuities in topography (e.g., swamps and bogs) or by discontinuities in lithology (e.g., islands of serpentine, limestone, or other azonal substrates). I find no evidence whatsoever to indicate that lithology was a causal variable in the past. I want to come back to this hiatus in the record a bit later. However, good evidence exists for swamps and bogs and their azonal floras all the way back to the Devonian. See DiMichele, Phillips, and Olmstead (1987). These authors describe the swamp/bog habitat as one under intrinsic abiotic stress. Such a stress is confined to edaphically specialized habitats and is due to the physical and chemical attributes of those habitats. Coal swamps of the middle Paleozoic and bald cypress swamps of the Tertiary hold in common the following features:

> Entry into these habitats is made possible by preadaptation. They generally are low diversity "species sinks" with highly specialized floras, rarely sites of evolutionary innovations that subsequently spread into non-stressful habitats. Due to limited species exchange with surrounding environments, these kinds of habitats tend to become progressively more archaic through time, preserving a flora that changes little for long periods. They are very susceptible to mass extinction and vegetational reorganization, at which point the cycle of increasing archaicism is reset.

In their paper on abiotic environmental stress, DiMichele et al. (1987) cite modern cases of abiotic stress elicited by azonal parent materials. But they apparently have no evidence for azonal lithology as having affected evolution and distribution of plants in the past. This is puzzling, for one would expect that exceptional lithologies would be preserved. But if so, is there evidence for an associated fossil flora? For instance, it seems likely that a Mesozoic serpentine outcrop might be preserved or even persist surficially to the present. In the former possibility, the serpentine flora may or may not be fossilized. And in the latter case—continuous site for plant life—its earliest flora may not be preserved; only the recent to current flora would persist. An intriguing example appears to be the serpentines of the Great Dyke of Zimbabwe which have been available for colonization since the Paleozoic. Wild and Bradshaw (1977) state: "The endemic heavy metal and serpentine species must . . . have begun their evolution in South Central Africa and perhaps in other parts of the tropics soon after the angiosperms emerged as a distinct group" (p. 289). How old the paleoendemics of the Great Dyke may be can only be conjectured. I suspect that during this long exposure of serpentine the flora has changed, yet leaving no fossil evidence of antecedents. I imagine that a unique azonal habitat like serpentine has to be both buried and then rediscovered in order to yield evidence for an ancient edaphic island.

Occurrences of lithological variety must have been present in the distant geological past. Further, surface exposures of unusual rock types (limestone, dolomite, serpentines, etc.) must have elicited inherited responses (from racial differentiation to speciation) from the regional floras, as accommodations to such azonal geoedaphic sites. So far as I can discover, paleoecologists have not yet reported on any such occurrences; neither unusual rock types nor associated flora seem to have been unearthed. Why this hiatus in the fossil record? Should not there be ancient "fossilized" serpentine or limestone outcrops and their associated fossil flora? I see at least two causes for this gap in the geological record. First, such outcrops would have been upland sites and hence their floras probably not preserved in place; rather they would have been transported elsewhere. Vastly more often preserved were floras of depressions—in swamps et cetera (DiMichele et al. 1987). Second, upland lithology of the past would likely have undergone major distortions by tectonic activity, so as to obscure both the original sites and surficial occurrences. At any event, the lack of evidence for past connections between lithological diversity and specialized floras is likely to persist as a significant gap in the fossil record. All we can do is to assume that the uniformitarian principle holds for this aspect of Earth's history.

Geoedaphic Endemics and Plant Distribution

Endemism is a hallmark of specialized edaphic habitats. Biogeographers give much weight to the evidence from endemism in determining the origins, migrations, and contemporary distributions of biota. Stanley Cain (1944) devoted an entire chapter to endemism, and more recently Jack Major (1988), in his valuable synthesis of concepts on "Endemism: A Botanical Perspective," stressed the value of endemics for phytogeography with a classic quote from Braun-Blanquet: "[T]he study and precise interpretation of endemism of a territory constitute the supreme criterion, indispensable for consideration of the origin and age of its plant population" (Braun-Blanquet 1923). Having dealt with the kinds and degrees of endemism earlier (Chapters 4 to 6), in the present context we ask: How do geoedaphically constrained endemics bear on matters of distribution of species, floras, and even of plant communities? Since substrate-specific endemics closely track a particular parent material and soil type, their distribution can coincide with the areal extent of a particular geologic formation. For example, woody endemics like *Quercus durata, Garrya congdonii,* and *Ceanothus jepsonii* are indicators of substrate and also by their high fidelity to California serpentines can indicate the age and areal extent of the edaphic habitat. A most valued contribution of endemics to phytogeography is their manifestation of disjunction and insularity for mainland areas of sharply contrasting lithologies. The distribution patterns on mainland islands of unique topography or substrate permit the testing of island biogeography theory (Kruckeberg 1991b, and earlier in this chapter).

Those endemics restricted by the nature of their terrain (landforms or substrates) illustrate several types of restricted distribution: (1) They can be either paleo- or neoendemics. Those with but a single population and/or without related species nearby are thought to be relictual—paleoendemics. Neoendemics are more likely to have close relatives in nearby normal habitats. (2) Local and regional endemics may differ ecologically. Regional endemics are often the products of historical changes in climate; local or narrow endemics are mostly edaphic specialists (Kruckeberg and Rabinowitz 1985; Raven and Axelrod 1978; Stebbins and Major 1965). (3) The number of endemics in an area may be a product of the size of the azonal locale as well as its geologic age. The larger (or older) the area, the greater will be the number of endemics. (4) Geoedaphic endemics and disjuncts can arise by either vicariance or long-distance dispersal. Major Earth history events can create disjunctions, thus illustrating the vicariance model. Long-distance dispersal as a cause of geoedaphic endemism depends on (*a*) the presence of disjunct edaphic sites and (*b*) their colonization by preadapted migrants followed by evolutionary divergence. The vernal pools of California and Chile illustrate this latter type of disjunction and endemism.

In short, geoedaphic specialists illustrate a wide range of biogeographic verities. Thus it is no wonder that plant geographers have found the terrain-restricted endemics to be good examples for most any pattern, model, or hypothesis; the references cited in the previous paragraphs support this contention (see previously cited Cain, Major, Kruckeberg and Rabinowitz, Raven and Axelrod, Stebbins and Major).

Species Diversity and Its Geoedaphic Connections

Species diversity, as a function of physical and biotic environments, takes on a central position in biogeography (Whittaker 1975). Further, preservation of species diversity has taken on high priority in global to regional conservation efforts (Falk and Holsinger 1991; Orians et al. 1990). Here I pose a purely geoecological question: How do indices of species diversity differ when one compares normal (zonal) geoedaphic environments with those having singular, azonal attributes? Of the several measures of alpha (within habitat) diversity (Whittaker 1975), I will use a simple statistic, the total number of species found in a sample plot. This measure is also called species richness.

There is a vast literature on measures, environmental vectors, and causes of species diversity. Most are associated with spatial (or temporal) changes in regional climates. For instance, the global perspective on species diversity as a function of latitude and attendant shifts in climate yields the generalization: species diversity increases from high to low latitudes. That is, species richness is vastly greater in the tropics than in high latitude regions. Climate, as influenced by landforms, can be the primary arbiter for species diversity on a more regional scale as well. For example, species diversity is higher in the less mesic (rain shadow) east side of the Cascade Range of northwestern North America. Del Moral and Watson (1978) contend that "the shift to a more continental climate opens up the forest canopy to permit greater niche differentiation and hence greater species numbers within a community." So we do recognize that many changes in species diversity are directly linked to climate, which in turn is often a consequence of some topographic barrier.

But what of variation in species richness within a regional climate? Is there a geoedaphic explanation for regional to local variations in species richness? I now examine these questions in the context of surface heterogeneity of landscapes and of lithology and soils.

SURFACE HETEROGENEITY AND SPECIES DIVERSITY

A working hypothesis is that environments with irregular landforms should be richer in species than monotonously flat terrain. Those landforms with surface heterogeneity should yield a greater variety of microsites or niches than those of level lands (Harper 1977). To make the comparison both meaningful and scientifically valid, one should have inventoried the plant cover on adjacent landscapes under the same climatic regimes; ideally the comparison should be between adjoining level and irregular surfaces. I have not found any published case histories that illustrate differences in species diversity for such a contrast of landforms. However, there are some examples that approximate the model.

Species diversity on the microscale of Mima mounds terrain is richer on the mounds than in the intermound (level to concave) areas as well as in the adjoining nonmounded prairie. This difference has been reported for the type locality of Mima mounds (Mima Prairie, Thurston County, Washington) by Giles (1969) and also is true for mounded topography in Kenya (Cox and Gakahu 1985). Similar to Mima topography is the undulating terrain that produced vernal pools in the depressions. Vernal pool microtopography is richly manifest on the coastal mesas of southern California (Zedler 1987) and in the Great Valley of central California (Holland and Jain 1981). More than one grade of species richness occurs in vernal pool landscapes. Richness is greater in the vernal pool depressions than in the elevated interpool terrain. Species composition is markedly different in the two adjacent habitats, since aquatics are restricted to the seasonally drying pools. Species diversity may also vary from pool to pool as well as along the gradient within a pool, from shallow to deep (Holland and Jain 1981; Zedler 1987).

Scaling upward from the intimacy of microtopography to what might be called mesotopography, there is ample opportunity for landform variation to promote species richness. I have in mind those local variations mostly found in mountains: talus and scree slopes, alluvial fans and bajadas, recently deglaciated terrain, and other surface heterogeneity in montane regions. Not all of these local landforms have been examined for their ability to promote species diversity. In his comprehensive review of the ecology of recently deglaciated terrain, Matthews (1992) makes the point that glacial forelands offer the ecologist a host of structure and function for the testing of ecological theory. For species diversity and its link to terrain, Matthews states: "Almost all studies indicate an increase in the number of species [richness] with increasing terrain age, at least on younger ground" (p. 182). Species diversity on deglaciated foreground may peak fairly early in succession, and then declines.

For other mesoscale landforms, we may have to be content with prediction rather than published results. On talus and scree, I predict that (1) species diversity will be greater on fine-textured materials, and on stabilized rock debris; (2) species diversity should peak midway along a scree slope, since the upper reaches are new and unstable while the lowermost have coarse boulder debris unsuited for colonization. The complex of alluvial fan and bajada in semi-arid regions should display a similar range of species richness, depending on position along the catena. I predict that the lower end of the alluvial fan and its adjoining outwash slope or bajada will be richer in species than the more transient and unstable upper part of the alluvial fan.

At the macroscale of single mountains or mountain ranges, the display of topographic diversity is spectacular (Gerrard 1990; Price 1981). Along any spatial vector, change is ever present. The most obvious and well-studied montane gradient is the elevational one. Are there peaks of species diversity along an altitudinal gradient? If substrate could be held constant, I would predict that species richness would peak in the upper montane. This seems to be true in the Pacific Northwest of North America, where the subalpine parkland—a mosaic with "atolls" of forest (tree clumps) in a "sea" of mountain meadow—has a rich inventory of species and community types (Franklin and Dyrness 1973). This high elevation zone derives its rich floristic diversity from climate, community structure, and landform diversity.

Another manifestation of species diversity in montane terrain is the degree of isolation and areal extent of individual high peaks in a mountain range. The high peaks of the southern Appalachians offer a choice example. White, Miller, and Ramseur (1984) framed their study in the context of species richness and island biogeography. Ten high elevation areas above 5500 feet (1680 m) were inventoried; a total of 342 vascular plant species were recorded. "Species richness on the separated areas [was] positively correlated with size of the area, number of peaks, maximum elevation and number of communities present" (p. 47). This case history clearly links landform heterogeneity with species richness. Undoubtedly a similar relationship could be discovered for other high peaks and mountain ranges of the world. It is my notion that these upper montane, species-rich areas gain their highest floristic diversity at elevations where continuous forest has ceased, and where the terrain at a given elevational belt is elaborated into many different landforms: meadows, rock outcrops, scree and talus, avalanche slopes, tree clumps, and others. I would predict similarly high species diversity for the higher volcanoes of the Cascade Range, the treeline areas of the Rockies, the Sierra Nevada, and the Japanese alps; in fact, the pattern should appear in any mountain system in the temperate world.

EDAPHIC VARIATION AND SPECIES DIVERSITY

In earlier chapters, I provided illustrations of the effect of parent material and soil diversity on plant life. Can we now find some connection between edaphic heterogeneity and species diversity? Does species richness increase, decrease, or stay the same as one tallies it along an edaphic gradient, from zonal to azonal sites? All three outcomes can occur, which suggests that the comparisons of zonal with azonal habitats can be confounded by the interplay of several variables. A mosaic of lithologies and soil types promotes habitat diversity, one cause of increase in species diversity (Brown 1988). A countervailing attribute that could reduce species diversity of an azonal habitat is the severity of physical conditions on the azonal sectors of a lithologic mosaic. Brown (p. 170) calls this variable "favorableness": "Environments that support diverse species appear to have mild physical conditions that could be tolerated by many species. . . . In contrast, habitats that are depauperate in species often have harsh and/or unpredictable physical conditions that require special adaptations and would be extremely stressful to most species that occur in other regions." So theory would allow for either increase or decrease of species diversity as one records richness along the sequence from zonal to azonal substrates. Indeed, the real world is complex, with spatial and temporal variables interacting with biotic and geoedaphic variables.

The "serpentine syndrome" (Jenny 1980) illustrates well the indeterminacy of substrate-specific species richness. Some serpentines (e.g., in New Caledonia, Brazil, Cuba) are extremely species-rich habitats. Yet many temperate serpentine outcrops are the "barrens" well known to the field naturalist, with few, a single, or even no species, and sparse plant cover (Kruckeberg 1999). Major (1988) contends that "high species richness accompanies high endemism" (p. 332). For serpentine floras the connection is impressive; apart from the extreme serpentine barren habitat, an increase in species number is often associated with an increase in number of serpentine endemics (Kruckeberg 1985). This contention is supported by data from Whittaker's (1960) elegant study of substrate-specific floristics and ecology for the Siskiyou Mountains of western United States. In his section on species diversity (pp. 319–320), Whittaker observed several trends: (1) Species diversity increases from coastal to inland areas of the Siskiyous. (2) Species diversity is highest on mesic or intermediate sites, for "normal" diorite and gabbro as well as "abnormal" serpentine. (3) "Even more striking is the increase in floristic diversity from the 'normal' diorite to the 'abnormal' serpentine" (Whittaker 1960, p. 319).*

Species richness data for other azonal habitats are likely to show the same patterns as those for serpentines. Floras on limestone and dolomite rocks, as well as gypsum substrates, should have higher—or lower—species diversities than those of adjacent zonal habitats. Again, the major determinant is the severity of the azonal

* See recent papers by Harrison et al., dealing with species diversity on serpentines (Harrison 1997, 1999a, b, c; Wolf, Brodmann, and Harrison 1999).

physical environment. Limestone outcrop habitats figure prominently in the rock outcrop vegetation of southeastern United States (Quarterman et al. 1993). While no direct comparisons of species richness with that of nearby zonal habitats have been published, it is evident that attributes related to species diversity differ significantly. Life-form spectra are strikingly different. Instead of deciduous forest, the limestones (locally called "cedar glades") foster on some outcrops herbaceous perennials, though annuals may dominate the flora. Endemism is also a common attribute of species diversity in the cedar glade habitats (Baskin and Baskin 1986, 1989). The unique floristic composition of the limestone outcrops creates characteristic plant communities. For sandstone and granite outcrops in the same region, similar floristic and vegetation attributes (endemism, life-form shifts, etc.) have been recorded (Quarterman, Burbanck, and Fralish 1993). But in this excellent review, no direct attention was given to comparing species richness of these remarkable azonal sites with the region's zonal vegetation.

The many connections between geoedaphics and the distribution of plants range widely in kind. The key phenomenon is discontinuity—gaps in distribution. These are brought about by gaps in landforms, lithologies, and soils. What is more, the scale of such discontinuous distributions ranges from the micro and local to regional. Published information on the distribution of plants as affected by the geophysical environment is still manifestly sparse. This "unfinished business" can be initiated with questions as yet largely unanswered: (1) Are plant distributions affected by discontinuities from one normal (zonal) substrate to another? (2) How does small-scale (local to micro) topographic variation affect distributions? (3) Which is the better measure of floristic uniqueness—species richness or degree of endemism? These questions and other unrequited issues presented in this chapter offer fruitful research opportunities for the future.

8 Human Influences on the Geology-Plant Interface

Humankind has not been kind to our planet. Assaults, incursions, and perturbations on and in the Earth's crust by members of our species have gone on for millennia. Of course, all life alters its contexts, and living things (especially heterotrophs) are selfish and aggrandizing. By far our greatest global concerns are the result of witnessing humanity's impacts on other forms of life. Losses of biodiversity at the population, species, and community levels are at the core of conservation efforts worldwide. But humans have also massively altered geoedaphic environments, either directly or indirectly, by impacting the biota inhabiting special landforms and substrates. Here, I want to focus on some of the many human activities that have altered or eliminated landforms and lithologies.

Alterations of physical landscapes have ranged from local pinpricks to massive perturbations, mainly through agriculture and forestry, all forms of mineral extraction, warfare, and urbanization and its attendant communication networks. Severity of impact has been in about that order.

Agricultural and Extractive Impacts

Agriculture is intertwined with geology and geomorphology in many ways. I confine this search for connections with geoedaphics to activities where tilling and cropping (= "mining") the soil as well as tending domestic livestock on the land have made major impacts on landforms, lithology, and soils. Modification of landforms by agricultural practices has only rarely involved direct alterations of physiography, rendering the contours of the land into new shapes. Where agriculture is confined to level ground, no change in landform is evident. But where agriculture has attempted to accommodate to uneven or even hilly terrain, its primary impact has been on the soil mantle as a consequence. The ancient rice paddy culture of eastern Asia has adapted to hilly terrain by elaborate contouring and irrigation. These venerable practices mostly had benign effects, but surely have resulted in the extirpation of the floras of those areas. In modern times—from the incursions of the plow and Fresno scraper to the bulldozer—the ability to alter landforms to "suit" agriculture has reached alarming magnitudes. Whole new landscapes are created by massive losses of soil horizons and even parent materials as a result of certain agricultural practices. We witness in many forms worldwide the loss of soil by erosion. Mass wasting and gullying can create denuded slopes and even canyons where once was level or gently contoured terrain. Runoff accumulation from aggravated sedimentation downslope from tilled or grazed lands is also a consequence of erosion.

Salinization is yet another consequence of converting native vegetation to agriculture. In Western Australia salinization followed after removal of native woodlands and the planting of wheat (*fide* R. Ornduff).

While agriculture has been tried on all manner of substrates, it is the great zonal soil groups of the world that have fostered long sustained and productive tillage and farming. Mostly, zonal soils develop on nutritionally balanced parent materials—either bedrock, alluvium, or aeolian deposits. Initially their profiles are deep and well differentiated, with adequate nutrient content. These soils once supported native grassland, savanna, or bottomland forest. They are located on level to gently contoured terrain, mostly at the lower margins of montane foothills and out into major drainage basins. Prior to cultivation, these zonal soils supported a region's climax vegetation. After cultivation begins, zonal soils almost inevitably undergo alterations: changes occur in fertility as well as physical transformations due to erosion and loss of native biota. The ultimate "insult" to zonal soils is to take them out of production; urbanization and road-building are the major abuses.

In contrast, those azonal soils that mirror some unique quality of their parent material have been tried for farming over and over again. While such fitful attempts at tillage meet with dubious results—marginal success to total failure—they have altered the geoedaphic character of the specialized substrates. Agriculture on soils over limestone bedrock may succeed only under intensive management. But in the attempts, limestone terrain and soils are often drastically altered. Their high porosity and solubility means that their shape is continually changing, and tilling can aggravate the changes in landforms. Where karst topography results from solutional transformations of limestone, the effects of human activities can be substantial. Not only do agricultural practices have an impact on karstic terrains, but many other human activities produce adverse effects as well. For example, the nearly flat floors of the closed karst basins (dolines) have been tilled for centuries in the Balkans. Grazing by domestic animals has drastically altered the vegetation on karst in the same region—as has deforestation. Quarrying for limestone and marble from karst formations has gone on ever since Minoan times (Williams 1993). Many of the classic architectural marvels of ancient Mediterranean cultures were fashioned out of karst deposits. Limestone quarrying continues at an accelerated pace in modern times, especially for cement, fertilizer, and building materials. Strip mining for bauxite (aluminum ore) in karst country creates barren wastelands devoid of soil and vegetation; parts of the remarkable cockpit terrain of Jamaica have suffered this drastic alteration (Howard and Proctor 1957). Unique to karst is its remarkable water-supply system, which has been exploited since ancient Greek and Roman times. Depletion, along with pollution, of karstic groundwater has resulted in numerous adverse consequences, including collapse of overlying terrain. These and other human intrusions on karst, as well as issues of its preservation, are the subjects of the Williams (1993) symposium papers.

Agriculture on ultramafic substrates is even more problematic. Where serpentine soils develop on alluvial catchments from nearby serpentine uplands, cultivation is tempting. The deep, heavy-textured (clayey) soils seem likely prospects for farming. But attempts to till them usually fail. Overcoming the adverse magnesium/calcium ratio, heavy metal toxicity, and generally low fertility is too costly to belabor (Martin et al. 1953; Vlamis 1949; Walker 1948). So soils derived from ultramafic rocks only occasionally get altered by agricultural practices. Other uses (mining, grazing, et al.) have far greater impacts on these outcrops. The highest and best uses for soils over ultramafic bedrock—or its alluvial derivative—are still the dual passive uses: for wildlife and as watersheds.

I presume that attempts have been made to grow crops on other azonal substrates. Since such attempts are likely to end in early failure, their impacts on the azonal landscapes should be moderate. I have in mind such azonal substrates as gypsum or heavy-metal impregnated soils (either natural or human-caused, such as mine spoils, etc.); also soils from sedimentary rocks with low fertility (shales, certain sandstones, mudstones, etc.) have been tried, mostly without success.

Forestry, though akin to agriculture in the sense of harvesting a renewable resource, can have even more far-reaching effects on landforms, soils, and even

8.1 Logging impacts on an ultramafic area (dunite), Twin Sisters Mountain, Washington State. Photo by author.

parent materials. In the first place, trees will be harvested wherever they grow (Fig. 8.1). Logging occurs on flat terrain to steep uplands, and trees are harvested on both azonal and zonal soils, either on a one-time basis or repeatedly under sustained-yield practices. Although deliberate changes to landforms are rare, inadvertent change is common, particularly in logging upslope in mountainous terrain. Not only does the removal of trees affect landforms, but road-building for access and tree removal can have major impacts. In addition to local erosion and sedimentation, such harvesting practices on uplands can result in displacement of the regolith, caused by earth and rock slides as well as gullying of the terrain (e.g., Fredriksen 1970). The net effect is largely on the soil mantle, though underlying parent material may also be disturbed by mass wasting (displacement as colluvium) and exposure to weathering.

The ancient cultural practice of pasturage, especially on uplands or hilly terrain, has been notorious for its degradation of landscapes. Livestock of all kinds can effect geomorphic changes. Like upland agriculture and forestry, the impact

8.2 Aerial view (1962) of attempts to control gully erosion by dams, catch basins, and reforestation, with the upper watershed still in grazing use, Mississippi. U.S. Forest Service photo.

of domestic animals is first on the native vegetation. But with severe overgrazing, the regolith (soil and its parent material) is subject to massive change through gullying, erosion, sedimentation, and so forth (Fig. 8.2). The classic case histories of denudation and landform change by domestic animals are in the regions of the world inhabited by humans and their livestock since the dawn of domestication. The Mediterranean basin, the Near East, and the Asian continent all exhibit the horrors and permanent scars of sustained overgrazing. And in modern times, the introduction of livestock to the Americas has had some of the same consequences as were witnessed in Greco-Roman

times. New assaults on terrain, especially by goats and other feral animals, are poised to repeat the geoedaphically ruinous transformations of the Old World. For the role of domestic animals as geomorphic agents, see Butler (1995).

8.3 Massive gravel extraction near Tacoma, Washington, bordering Puget Sound. Photo by L. Delano.

The most direct human assaults on landforms and especially on lithology (soil, loose regolith, and rock formations) have been from mining (Fig. 8.3). Extraction of rock as ore and for other uses has gone on worldwide for millennia. Rock and other crustal materials are quarried for road construction, building stone, et cetera, and crushed to make sand and gravel (Fig. 8.4). Nearly all such extractive enterprises disturb the nearby or overlying soils and vegetation. For various methods of surface mining (strip, placer, and hydraulic) the overlying soil is usually treated like dirt—a "useless" overburden to be disposed of in some way (Figs. 8.5 and 8.6). In the most sophisticated and ecologically accountable mining operations, the soil may be set aside for later reclamation. In recent times, this has been done with spectacular results in North America and elsewhere. For example, in Natal, South Africa, the vast surface mining enterprise to recover precious metals in the sands near Richards Bay routinely sets aside the topsoil and returns it to the site for replanting with native coastal species. In the United States, recla-

8.4 Gravel mining on the remarkable Mima Mound landform, Thurston County, Washington State. Photo by J. McDermott.

8.5 Aerial view (1940) of gold-dredging spoils, Eldorado County, California. U.S. Forest Service photo.

8.6 Reforestation of strip-mined land in Illinois, 1958. U.S. Forest Service photo.

mation with soil and vegetation is mandatory for strip mines. The Homestake Mine in northern California is attempting to restore a serpentine chaparral after surface disturbance.

But these reclamation efforts are relatively new. In the past, vastly more soil was simply washed away, buried, or otherwise rendered nonproductive (Fig. 8.2). The magnitude of rock extraction has varied from trivial "pinpricks" on the Earth's crust to vast areas of removal. Even the simple mine shaft and its adit, although underground, inevitably leaves a disturbed perimeter outside the entrance. But far greater dislocation of the Earth's crust results from great open-pit extractions, massive strip operations, and hydraulic mining. The impact of hydraulic operations is awesomely exemplified by the consequences of the "gold rush" in California from 1848 onward. Menard (1974, pp. 353–354) vividly describes the "by-products" of gold extraction in the Sierra Nevada:

> 1200 million cubic meters of soil and gravel were mined hydraulically to separate them from their gold. The mining itself probably destroyed approximately 120–200 square kilometers of the Sierra Nevada. However, the wastes destroyed an area perhaps ten times as great by burying fertile topsoil under sterile sediment. [The sterile sediment was traced] down the river valleys as a wave that spread over the Great Valley below, partially filling San Francisco Bay, and altering the volume of the tidal flow through the Golden Gate (entrance to San Francisco Bay). This, in turn, changed the configuration of the semicircular sand bar deposited on the seaward side of the Golden Gate by tidal erosion. Thus, the effects of surface mining by this method spread far beyond the immediate diggings just as they do in strip mining.

Ever since humans found valuable and useful metals in rocks, their extraction and refining (smelting) has drastically altered landscapes. Every ancient civilization turned landforms and rock outcrops into both useful mineral and discarded waste. Ancient mines dot the European landscape from the Near East and the Mediterranean to the British Isles and Scandinavia. (See archaelogist R. F. Tylecote's *The Early History of Metallurgy in Europe,* 1987, for a thorough and fascinating account of ancient mineral extraction and kindred arts.) The extensive system of ore extraction on Crete and Cyprus is described by McPhee (1993); the extraction of copper and precious metals required both vast amounts of ore and native wood for fueling the smelters. Cyprus cedar (*Cedrus libani* var. *brevifolia*) was nearly exterminated on the island by the demand for fuel.

The classic work on medieval metallurgy is Georgius Agricola's *De Re Metallica,* published in 1556, at the beginning of the Renaissance and soon after Gutenberg's revolutionary invention of printing. The English translation by Herbert C. and Lou C. Hoover (1950) is a massive work of over 600 pages; its text and the numerous original woodcuts reveal the magnitude of the many impacts of the mining industry on the environments of those early times. This magnum opus and the one by Tylecote (1987) cited above can open the doors to viewing the manifold nature of the environmental consequences of early and ancient mining and metallurgical activities. And from the Industrial Revolution onward, the impacts on terrains (landforms, rocks, and soil) from mining have vastly escalated. Rather than belabor the self-evident, I choose to examine two specific mining activities: the exploitation of ultramafic deposits for nickel, magnesium, and asbestos and then the extraction of ores for certain heavy metals. Both illustrate the heavy hand of human industry on landscapes worldwide.

Ultramafic Rock and Human Uses

With their occurrence worldwide (mostly at plate sutures), it is not surprising that the mineral-rich ultramafic rocks have been exploited for human uses. Serpentine and kindred metamorphic rocks have had a long history of use in art and architecture. In their review of the serpentine-group minerals, Faust and Fahey (1962) give a brief account of the aesthetic values of serpentinites. The rock has been used since ancient times in sculpture; it is said that Michelangelo used serpentine marble for some of his masterworks. For architects, various serpentine rocks that can be fashioned into slabs and polished have gained use as building stones. A host of serpentines and serpentine marbles (ophicalcites) are known in architecture as "verdes," each having a special place of origin, (e.g., verde de Genoa, verde de Prato). And Irish green marble (an ophicalcite) from Connemara, Ireland, is also well known. Such ornamental stone has been quarried in Europe (and probably elsewhere) for centuries, leaving telltale excavations that have altered landform and rock outcrops.

The mining of ultramafic areas as a source of nickel has made the greatest impact on both landscapes and vegetation. Open-pit nickel mining in New Caledonia has transformed the terrain on a massive scale (Jaffré, Latham, and Schmid 1977). Vegetation is destroyed, soil is removed, and catastrophic erosion results in this tropical environment of high rainfall. Inevitably, the rich and diverse endemic flora of New Caledonia suffers attrition on a massive scale. In other regions (e.g., western Canada), where the extraction of nickel-chromium ore is subterranean, the surface disturbance is less destructive of the vegetation.

The mineral chromite, the primary source of the industrial metal chromium, is nearly universally associated with ultramafic rocks. The mineral occurs either diffusely, banded, or massively, in ultramafic deposits, especially peridotite (Ramp 1961). It can be extracted from exposed (surface) outcrops or from underground mines. In North America, the states of Montana, California, and Oregon are the major producers. Chromite extraction is important in other countries, such as Greece, Russia, and Cuba. The major sources of chromite worldwide are South Africa and India (Bureau of Mines 1991). Because of the mineral's often diffuse deposition, much country rock must be removed to secure the desired ore. Hence chromite mines create major disturbances at the site, where soil and vegetation are often destroyed. Since the floras of ultramafic terrains are often unique and rich in endemics, the mining activities run the risk of destroying or decimating this botanical treasure. The account of chromite mining in southwestern Oregon (Ramp 1961) reveals the extent of environmental intrusions made on the land when mineral extraction goes on. The Oregon Chrome Mine, the largest in the state, yielded 31,918 long tons of chromite from 1917 to 1958.

At this writing (2001), a classic confrontation between conservation and mining interests is being witnessed in the Siskiyou National Forest of southwestern Oregon. A rich and diverse flora, including local endemics, occurs on ultramafic substrates in the 24,000 acre (9600 hectare) drainage of Rough and Ready Creek. The area has been proposed for preservation as a National Wild and Scenic River. Yet the U.S. Forest Service is being forced to yield access to a mining operation: a major strip mine to extract low-grade nickel and iron. The priority of mining over preservation derives from the 1872 Federal Mining Law, still in effect. The outcome of this controversy still hangs in the balance, despite the overwhelming national outcry against the mining project (Anon. 1999). Yet the following year (2000), the U.S. Forest Service denied the mining company the permits to go ahead with their project, stating that the mine operation was not economically feasible. So, for now, the Rough and Ready serpentine ecosystem is saved from major disturbance.

A unique intrusion on ultramafic landscapes has been the exploitation of geothermal power in northern California. For many years, the hot springs and steam vents of The Geysers area served only as the site of a health resort. But in the last decades of the twentieth century, the area on both sides of the Mayacamas mountain range became pockmarked with the massive plumbing apparatus required to tap this geothermal energy producing field, the world's largest. Most of the Mayacamas Range of Sonoma and Lake Counties, where the power works are located, are in extensive, though discontinuous, outcrops of serpentine rock. The characteristic serpentine chaparral vegetation (Kruckeberg 1985) harbors several herbaceous species known to be local serpentine endemics. Thus the development would have disturbed not only the serpentine community structure but also the populations of the narrow endemics. The potential for disturbance is substantial. Crane and Malloch (1985) noted that geothermal development has two major impacts on vegetation: habitat loss and operational effect. "The construction of well pads, steam lines, and roads by steam suppliers, and the construction of electrical generating facilities and roads by power plant operators results in habitat loss." California state law requires that any project for energy development must determine if rare plants will be impacted. Pursuant to this requirement, Pacific Gas and Electric's study (Crane and Malloch 1985) identified twenty-one species considered to be rare and thus in potential jeopardy from geothermal development. One of The Geyser's rarities, *Streptanthus brachiatus,* epitomizes the premium put on the taxonomic status of a rare and endangered species. Is *S. brachiatus* a "good" species and do its several variant populations in The Geysers area warrant taxonomic recognition? As named taxa, the variants in this section of the genus *Streptanthus* would have to be monitored for protection or mitigation. The federal government (Bureau of Land Management) supported contract research to make the necessary taxonomic determinations (Dolan and La Pré 1987; Dolan 1988). The politics of conservation thus provoked the scientific assessment of a rare species. That study supported the validity of the species and recognized at least two infraspecific variants of *S. brachiatus.* So geothermal power development will have to guard against extensive disturbance of these rarities.

Asbestos is found in many of the world's serpentine deposits. Major sources in New England, especially Vermont, have been mined for years. Kevin Dann (1988) recounts the fascinating story of the asbestos industry in Vermont. The Belvedere Mountain deposit produced 90 percent of the asbestos mined in the United States, and the side effects of the enterprise were substantial. Thirty million tons of waste rock (serpentine and dunite) created new mountains nearly barren of vegetation. And then, of course, asbestos fibers got into the environment, not only in natural ecosystems (mostly aquatic habitats) but in the bodies of those persons associated with the mining enterprise. Some species of asbestos fibers may cause pulmonary fibrosis, a form of lung cancer (asbestosis). In fact, Vermont's mining industry has had more than the asbestos health problem; quarrying of granite may induce another lung disease, silicosis. Worthy of note is the quality (species) of asbestos fiber as the source of the lung cancer. All forms of asbestos are dangerous to miners, but the form of asbestos found in the United States (chrysotile, or white asbestos) is not known to be dangerous (lethal) in nonoccupational exposures; blue asbestos (crocidolite) mined in South Africa and Australia and brown asbestos (actinolite and amosite) "are unquestionably dangerous" (Coleman 1985).

Ultramafic rock has also figured in human enterprises in other ways. As materials for art objects (jewelry, sculpture, etc.), nephrite and jadeite used to fabricate jade objects have been sought for ages by many cultures. Both minerals are associated with ultramafic outcrops (e.g., Holland 1961). An example of pre-Columbian (primitive) exploitation of yet another variant—steatite—is engagingly described in Dann's (1988) account of the natural history of serpentinites in eastern North America. Indians on the east coast happened upon a rock along the Conowingo River that had the fortuitous properties of soft malleability (hardness of 1, the lowest on the Mohs ten-point hardness scale) and high heat retention. Steatite, like soapstone, is an impure form of talc found in massive deposits invariably associated with serpentinites. In the West, the Klamath Indians also used this "soft rock" (which they called *asaxusas*) for domestic implements. Dann tells of the many steatite (serpentine) quarries on the eastern seaboard that were worked by Indians. The soft rock could easily be fashioned into thick, blocky bowls, many of which have been recovered by archeologists. Ancient quarry sites have been found mostly in the Pennsylvania and Maryland serpentine beds; even the site of present downtown Washington, D.C., had its steatite workings. All of these primitive quarryings of steatite and serpentines made just modest dents in the landscape. It was only later when Europeans occupied the land that major inroads were made on the outcrops. Early in the nineteenth century, chromite was mined in these same ultramafic sites that had yielded the Indians' steatite.

Ultramafic rocks are one of several sources of the valuable metallic element magnesium (Comstock 1963). An active mining operation yielding iron magnesium silicate is in the Twin Sisters Mountain area of western Washington, which contains the largest pure olivine (dunite) deposit in North America. The main mine, operated by Olivine Corporation, converts the dunite to fine sand for industrial use as refractory sand and for solid waste incineration. Since the dunite rock containing olivine is readily accessible at the surface, its extraction makes a significant impact on landforms and vegetation (Kruckeberg 1969b).

Other Human Impacts

Whatever the human enterprise may be, the living and the physical environments will be perturbed. While the major assaults on landforms and geologic deposits have been addressed above, there are yet other impacts. The development of population centers ranging from villages to vast cities (there is even a geology of cities, e.g., Legget 1973), the communication network of roads, power lines, and aqueducts, and the intermittent but devastating military inroads on landscapes have all disturbed or destroyed natural ecosystems. Not only are geologic settings transformed by these activities, but their biospheric skins are changed or destroyed. There were few times in human history when some attrition of natural resources did not occur. An apocryphal anecdote from the Roman era reveals the timelessness of human assaults on the environment. The Roman historian Livy tells of Hannibal's solution to an impasse in his trek through the Alps. His military column was held up by an impassable cliff. Seeing the futility of trying to detour around the cliff, he decided to attack it: "The soldiers were then set to work to construct a road across the cliff—their only possible way. Since they had to cut through the rock, they felled some huge trees that grew near at hand, and lopping off their branches, made an enormous pile of logs. This they set on fire, as soon as the wind blew fresh enough to make it burn, and pouring vinegar over the glowing rocks, caused them to crumble. After thus heating the crag with fire, they opened a way in it with iron tools, and relieved of the steepness of the slope with zigzags of an easy gradient, so that not only the baggage animals but even the elephants could be led down." Whether fact or fiction, the story exemplifies how trees and rock have succumbed at the hand of man.

What I have attempted here is a brief account of the environmental consequences of the human pursuit of natural resources. Some might call this environmental geology: the linkage between geology as a pure science and the human use of the Earth's resources. This orientation is exemplified in a series of Benchmark Papers in Geology, entitled *Environmental Geology* (Betz 1975). Yet in the same collection a telling remark is made that epitomizes any branch of natural science that takes on the cloak of the environment: "'Environmental geology' is a ridiculous term! *All* geology is environmental, and the most basic element in every man's environment is the geologic factor" (Oakeshott 1970, in Betz 1975, p. 4).

Preservation of Geoedaphic Resources

If one side of the geoedaphic resource coin is exploitation, then the other surely *should* be preservation. But exploitation and its accompanying transformation of Earth's crust have greatly outstripped the preservation of landforms and other geologic features having some novel attributes. The realms of conservation of resources and environmental protection are so vast and global—potentially, at least—that I am constrained to focus on one aspect of this universal theme. Given the bias of this book, it is fitting to ask: What has been the outcome of preservation of floras and vegetation types associated with particular and unique landforms and lithologies? Have preserves been established for the express purpose of saving a distinctive geology-flora linkage? It seems fair to say that the majority of preserves (national and regional parks, nature preserves, etc.) have been saved for perpetuity because of their geologic attributes. On any continent, the spectacular preserves commemorate some remarkable geologic phenomenon. In western North America, consider the major national parks. Nearly all exhibit a singular and mostly spectacular geology. Yosemite, Crater Lake, Grand Canyon, Yellowstone, Glacier, Banff, and Jasper national parks are all examples of the natural bias toward preservation of outstanding landforms or geologic features (Chronic 1984). Yet every one of these major preserves has a flora with some elements tied to a particular geoedaphic syndrome and merits preservation in its own right. Beyond North America, this same pattern of preservation—geology favored over flora—largely repeats itself. Polunin and Walters's *Guide to the Vegetation of Britain and Europe* (1985) lists the national parks and nature preserves of Europe (pp. 192–213). Distinctive geomorphology and lithology characterize most of these preserves. But, given the focus on vegetation, the Polunin and Walters roster emphasizes the rich and unique flora of many of Europe's preserves. For instance, wherever calcium carbonate rocks occur in European preserves (limestone, dolomite, karst, etc.), one can expect a distinctive flora to have been included. Landscapes of limestone, dolomite, and karst are frequently protected as preserves, for both their landform attributes and their unique floras. Two entries from the Polunin and Walters list give the flavor of the intimate ties between geology and flora. The first is for Cévennes, "Limestone plateau with deep gorges, with rich and interesting flora, including over 40 species of orchids. Limestone grasslands, forests of oak, sweet chestnut, and beech, and planted conifers. Also granite and schist peaks with distinctive and different flora" (p. 200). The second is for Mount Olympus, Greece: "Unique alpine rock and scree vegetation [on limestone] with about 1500 species of flowering plants and ferns and with 19 endemic species and many Balkan endemics" (p. 203).

The creation of national parks and nature preserves with unique geoedaphic features in other countries has an uneven record. A number of volcanoes, as spectacular displays of tectonic forces, have been given such status in the United States, Japan, New Zealand, Mexico, and Chile. Since the biota of volcanoes often includes distinctive floras and often endemic species, their protection is usually assured by the protection of the geologic phenomena. The linkage of geology with plant life is exhaustively documented in the valuable reference work, *World Directory of National Parks and Other Protected Areas,* published by the IUCN (United Nations Secretariat 1977, 1985). For each listing, a summary is given of physical features and the biota are noted. For instance, Vanoise National Park (France) is described as having "great floristic richness due to the variety of rock types."

The conservation status of ultramafic areas and their remarkable floras can serve as a fair example of the limited extent to which floras on geoedaphically unique sites are protected. A curious hiatus in the rich array of European preserves is the apparent absence of serpentine preserves. Countries like Sweden, Switzerland, Britain, Italy, Greece, and Yugoslavia all have substantial ultramafic exposures, usually supporting a singular plant life. No doubt, some serpentine habitats have been included in preserves if they are associated with some other unique landform or geologic formation. Incidental preservation of ultramafic habitats seems to be the rule worldwide. For instance, the Great Dyke of Zimbabwe, the huge ultramafic exposure, embraces one small preserve; its man-made lake was the object of preservation. Coincidentally (and fortunately!), the nearby serpentine flora is included in the preserve. In New Zealand, Dun Mountain with its unique ultramafic (dunite) vegetation is preserved only because the mountain is the watershed for the city of Nelson. For years, New Zealand conservationists agitated to secure protection of a segment of the Cascade-Olivine country on the South Island (Malloy 1983). Red Mountain and the Red Hills were the target areas for protected status. While no serpentine endemics are known for this ultramafic belt, its serpentine scrub vegetation type spectacularly contrasts with the adjoining mature *Nothofagus* forests (Lee 1980) (Fig. 8.7). The potential mineral resource of the high montane ultramafic barrens had delayed action on the proposal for protected wilderness status (Malloy 1983). Robert Brooks (pers. comm.) informs me that the area in question has now gained wilderness status.

Incidental protection of ultramafic floras in North America also has been the rule. In California, Mount Tamalpais State Park includes some striking and extensive serpentine habitats; yet Mount Tamalpais is a state park preserved for other amenity reasons. Other preserves in California are described in Kruckeberg (1985, chap. 10). In Canada, none of the western exposures of ultramafic areas are preserved (e.g., the Lilloet area, the Cassiar fault zone of British Columbia). Fortunately the world-famous Gaspé serpentines are included in the Gaspasian Provincial Park of Quebec. The park embraces Mount Albert, a serpentine-peridotite massif with its unique arctic-alpine flora (Scoggan 1950; Rune 1954; Sirois and Grandtner 1992).

New Caledonia, with the richest ultramafic flora in the world, has as yet no park or preserve that protects this rich natural resource. Ultramafic areas here will eventually succumb to mining of the vast ferronickel deposits. The 2nd Interna-

8.7 Once threatened by mining, this pristine ultramafic area has been accorded protection. Red Mountain, South Island, New Zealand. Photo by L. Homer.

tional Conference on Serpentine Ecology, held in New Caledonia in 1995 (Jaffré et al. 1997), offered a fine opportunity to make known to the authorities the need to preserve samples of the island's botanical treasures on ultramafic areas.

Preservation of serpentine outcrops and their floras in the United States is spotty and largely incidental. Some exceptional landmark or endangered endemic species may coexist on a serpentine site, and thus the habitat gains protected status. There are few examples where the serpentine syndrome (rock-soil-plants) has provoked, for its own sake, the establishment of a preserve.

In the eastern United States, where serpentines extend in a thin, discontinuous ribbon from Georgia to Vermont (and on to the Gaspé Peninsula and Newfoundland in Canada; Dann 1988), apparently *none* of the exposures are preserved. Although the Conowingo barrens (at the Pennsylvania-Maryland state line) and the larger outcrops at Newfane and Westfield in Vermont are well known to botanists (Dann 1988), the fates of most are uncertain.

In pre-Columbian times, over 50,000 acres (20,235 ha) of serpentine habitat existed in the state of Maryland (Mitchell 1998). Before European settlement there were two community types on these serpentines: bluestem dominated (*Andropogon scoparius*) grasslands and oak savanna (stunted *Quercus marilandica* and *Q. stellata*). "Serpentine oak savannah is considered the State's rarest plant community" (Mitchell). The two communities were maintained by periodic fire. But after European settlement, fire was controlled, and the communities came to be dominated by the invasive conifers. Mining and human settlement further diminished the habitats. Now only about 2500 acres (1010 ha) persist. Recent efforts to restore the remaining habitats include removal of the conifers and controlled burning. Not only are the habitats being restored, but two rare species, sandplain gerardia (*Agalinis acuta*) and serpentine aster (*Aster depauperatus*), can begin recovery (Mitchell 1998).

In the Pacific Coast states of California, Oregon, and Washington, the number and kind of serpentine preserves are barely adequate; the more spectacular exposures are either in jeopardy or inadvertently untouched—as yet. Since there is such an abundance of ultramafic exposures with exceptional topographic and botanical features, preservation of choice samples should have already occurred. While a few can be listed, their number is not sufficient to capture the "serpentine syndrome" in its diverse manifestations. In Washington State, preservation of serpentine habitats is minimal and has been both deliberate and incidental. Fortunately, the Alpine Lakes Wilderness captures some of the Wenatchee Mountain serpentines, especially some surrounding the Mount Stuart granodiorite massif. Similar coincidental preservation by the Mount Baker Wilderness captures a part of the subalpine dunite of the Twin Sisters Mountain. Intentional preservation is provided by two Research Natural Areas. One is on Twin Sisters dunite: the Olivine Bridge Natural Area, State Department of Natural Resources. The other is the recently established Eldorado Creek Research Natural Area, Wenatchee National Forest. Both preserve choice samples of ultramafic vegetation, including several endemics in the latter preserve (Kruckeberg 1969b).

Oregon's serpentines have gained modest protection, mostly coincidental with some other feature. In the Strawberry Range of east-central Oregon, Baldy Mountain would be a prime candidate for a natural area; it has extensive exposures of subalpine flora on peridotite, including the only populations of *Polystichum lemmonii* and *Aspidotis densa* (both ferns) east of the Siskiyou Mountains (Kruckeberg 1969b). Baldy Mountain is probably protected de facto, since it is within the Strawberry Mountain Wilderness. In southwestern Oregon's Klamath-Siskiyou mountain complex, ultramafic areas abound. A vast mountainous tract is preserved as the Kalmiopsis Wilderness (U.S. Forest Service), for the protection of the endemic ericaceous shrub *Kalmiopsis leachiana*. Within this wilderness, serpentine outcrops are abundant, interspersed with other lithologies (Whittaker 1960). Just east of the Kalmiopsis Wilderness is Eight Dollar Mountain, a floristically rich ultramafic body. Its preservation has been proposed, but mining interests and mixed ownership have prevented consummation of this potential preserve with its wealth of serpentine endemics and indicator species.

Preservation of serpentine habitats in California is spotty, inadequate, and largely coincidental. The vast majority of serpentine landscapes persist relatively untouched through "benign neglect." That is, they abide only because they have yet to be exploited for mineral deposits, timber, or grazing. Much of California's serpentines occur on federal and state lands. By law, public agencies must consider protection of rare and endangered species. Hence preserves that include such taxa have been (or will be) established. The only preserve established specifically for the full display of the serpentine syndrome (the physical environment of geology and soils coupled with its flora and vegetation) is the Frenzel Creek drainage in Colusa County. Although much serpentine land is managed by the U.S. Forest Service and the Bureau of Land Management, scarcely any samples have been singled out for special natural area designation. Only if rare or endangered species are on serpentines do the federal custodians contemplate protected status. A case in point is the vast ultramafic exposure at New Idria, San Benito County. The summit area (San Benito Mountain) has been accorded protection. The endangered species, *Camissonia benitensis* Raven, restricted to the New Idria area serpentines, has gained only partial protection, since some of its marginal populations lie outside the San Benito Mountain preserve boundary (Taylor 1990).

The Nature Conservancy has established a serpentine preserve on Ring Mountain, Tiburon Peninsula, Marin County. This serpentine grassland is the local habitat of two rare species, *Castilleja neglecta* and *Calochortus tiburonensis*. The nearby St. Hillary Church tract (Howell Wildflower Garden) that preserves the very local *Streptanthus niger* is outside the Ring Mountain Preserve; its protection appears to be voluntary on the part of Tiburon citizens. Another serpentine grassland in the thickly urbanized San Francisco Bay region is in the midst of a growing suburbia.

Edgewood Park, San Mateo County, surrounded by a densely populated suburban community, is generously endowed with serpentine rarities in its grassland community. Pressures by the human community nearly overwhelmed the natural wildflower amenity—a golf course was proposed to "replace" the serpentine grassland (Sommers 1983). Ultimately the local government sided with the conservationists to preserve this suburban treasure. The protection of one of the few remaining populations of the Bay Checkerspot Butterfly (*Euphydryas editha bayensis*) is a valued bonus in this successful outcome. Papers by the author (Kruckeberg 1985, 1986, 1992, 1993) provide a partial listing of serpentine preserves (de facto and realized) in California. The April 1984 issue of *Fremontia* has descriptions of several California serpentine landscapes (various authors, *Fremontia* 11[5]:11–30).

Two recent (2001) preserves for serpentine habitat in Napa and Lake counties, California, are heartening prospects. The Homestake Mining Company, now nearing the end of its gold extraction, will be deeding its large acreage—both managed and pristine habitat—to the University of California (Davis) as a permanent preserve for ecological research. In the same region, ecologists at Davis (S. Harrison, pers. comm.) have petitioned the Bureau of Land Management to manage as a preserve a much larger tract, mostly serpentine, in the Napa-Lake area.

Management of Geoedaphic Preserves

Most geoedaphic sites are not only azonal, but in that designation they may also harbor narrow endemics or possess other singular features. Practices necessary to manage a serpentine site may not serve for a karst landscape. The dynamics of azonal habitats fit well in the framework of successional theory: management techniques take cues from the successional status of the vegetation. Two very different serpentine habitats illustrate this approach. The serpentine talus habitat in the Dun Mountain Ophiolite Belt of the South Island, New Zealand, is considered unstable (Lee and Hewitt 1981). Seral communities on unstable talus coexist with stable ultramafic forest. To perpetuate the seral stages, measures to prevent the establishment of forest would need to be invoked. If the ultramafic forest is the desired habitat for preservation, then it should not require any overt management practice other than the prevention of fire. In contrast, the complex of contrasting lithologies, moisture gradients, and associated vegetation in the Klamath-Siskiyou country of Oregon was judged by Whittaker (1960) to be a mosaic of edaphic climaxes. Hence any proposed preserves on either serpentine or adjoining nonserpentine rocks could persist in perpetuity without human intervention.

There is good reason to suspect that "benign neglect" would work for other azonal sites, such as the limestone vegetation of the Balkans, dolomite vegetation of the White Mountains of eastern California, and so forth. But this generalization merits caution. Only by careful ecological study can the successional status of a particular azonal habitat be assessed.

Besides using climax or seral status as a guide to preserve management, there are other precautions called for to protect the community. Entry by humans, their vehicles, and their livestock needs to be controlled. Fire protection, as well as controlled burning, should be options open to the preserve manager, as should the removal of weedy alien plants.

Levels of protection for geoedaphic habitats will vary for a variety of reasons: the size of the site, as well as political and scientific biases. The hierarchy—and nomenclature—of preserves used by the federal agencies in the United States can illustrate the range of protection status (Table 8.1). For example, the U.S. Forest Service accords full and perpetual protection to Research Natural Areas and to Wilderness. The perpetuation of the natural state of the preserve is the key objective.

Other countries with policies (or intentions) for establishing preserves are likely to have similar categories of preservation. Enforcement of a given level of protection can be strict or lax, so that a given level of statutory protection is effective only if enforced.

Preservation of Geoedaphic Habitats for the Future

Throughout this book, I have singled out examples of a wide range of geoedaphically unique phenomena and their botanical attributes. Some are fortuitously or even intentionally protected; but most are not. Rather than simply deplore the low level of protection, we should be challenged by it. Scientists, naturalists, land managers, and private citizens worldwide can take up this challenge. Inventories of candidate preserves should be made first, and with these prioritized lists in hand, the conservationist will lobby the relevant agencies, public and private. As an initial stimulus for this conservation effort, I have tabulated the kinds of sites meriting protection, along with some representative examples (Table 8.2). It behooves conservation-minded readers to add to this list for their own region or country. The list emphasizes the unique botanical attributes of geologic phenomena; I leave it to geologists to identify for preservation those one-of-a-kind geologic areas where flora and vegetation are not considered critical or of special interest. For geologists, such a "wish list" apparently has not been compiled.

Preservation of exceptional geoedaphic phenomena as yet unprotected will be

Table 8.1 Designation of lands protecting natural diversity: federal (United States) system

Types of preserve	*Attributes, with examples*	*Level of protection*	*Federal agency*
National Parks	Largest preserves with a variety of geological features and ecosystems: Yosemite NP, Mount Rainier NP, Grand Canyon NP, Canyonlands NP, etc.	Adequate	NPS
National Monuments	Smaller areas featuring a unique natural feature: Cedar Breaks NM, Utah; Devils Postpile NM, California	Adequate	NPS
Wilderness Areas	Large, nearly pristine areas, usually forested. Can have several ecosystems and geological features; Alpine Lakes, Washington; Pasayten, Washington	Variable due to mining, grazing	USFS, BLM
Research Natural Areas	Local samples of ± pristine biota; usually a distinct vegetation type; may have geological features; Meeks Table RNA; Lake 22 RNA, Metolius RNA.	Adequate, in perpetuity	USFS, NPS, BLM, USFWS, DOE
Special Interest Areas	Local sites with geological, biological, or historical features; includes Botanical Areas: Tumwater Botanical Area, Washington	Potentially adequate	USFS
Wild and Scenic Rivers	Major rivers and their natural amenities: Salmon River, Idaho	Mostly inadequate	Various agencies
Outstanding Natural Areas	Emphasis on some unusual natural feature	"Potentially adequate"	BLM
Areas of Critical Environmental Concern	Local areas of unique natural features: Juniper Dunes, Washington	Mostly adequate	BLM
National Wildlife Refuges	Land and water areas with significant wildlife (mostly avian); Preserves wetlands as well: Nisqually Wildlife Refuge, Washington	Adequate	USFWS
National Estuarine Sanctuary Research Reserves	System of preserves that represents diversity of estuarine ecosystems	Mostly adequate	Joint state and federal
National Natural Landmarks	Encourages preservation of ecological and geological features	Inadequate; no assured protection: Mima Mounds, Washington (Mima State Natural Area Preserve)	NPS

NOTES: In Washington State various state agencies manage a variety of preserves (e.g., Biological Study Areas, Heritage Areas, Marine Biological Preserves, Natural Areas, Natural Forest Areas, Natural Area Preserves, Natural Resource Conservation Areas, Registered Natural Areas, Wildlife Areas). In the private sector, The Nature Conservancy establishes Natural Area Preserves.

NPS = National Park Service; USFS = US Forest Service; BLM = Bureau of Land Management; USFWS = US Fish and Wildlife Service; DOE = Department of Energy.

SOURCE: Washington State Department of Natural Resources, Natural Heritage Plan, 1993.

Table 8.2. Some geoedaphic features meriting protection

Types of site	*Geobotanical features*	*Locations, with examples*
Landforms		
Karst types		
Cockpit karst	Endemic flora	Jamaica
Limestone pavement	Arctic disjuncts	UK: Yorkshire and The Burren, Ireland
Tower karst	Potentially insular biota	Guilin, China; mogotes of Cuba (endemism)
Cone karst	Same as tower karst	Puerto Rico, Cuba
Dolines, poljes, sinks	Rim of basins with limestone flora	Northern Yugoslavia and Istria
Biokarst types		
Sedimentary landforms		
Tepuis	Rich endemism; isolated floras	Guyana Shield of Venezuela
Mesas, plateaus	Isolated floras	Western USA
Erosional types	Vegetational mosaics on alluvial fans	Western USA; on all continents
Mima-type relief	Patterned vegetation on microtopography	Washington State, USA; South Africa; Kenya
Vernal pools	Seasonal variation in flora; endemism	Great Valley and San Diego mesas, California
Natural landslides	Successional flora	Frequent in tropics (Thomas 1994)
Marine terraces	Edaphic endemics and unique vegetation	Pygmy conifer forests, California
Igneous landforms		
Granite Inselbergs	Insular isolation; endemism	Southeastern USA (granite balds); Western Australia (granite domes); kopjes of South Africa; granite boulders and domes in tropical Africa (Thomas 1994)
Lava flows and their erosional features	Successional floras; rimrock/coulee communities	Western USA: Columbia Basin; Hawaii
Volcanoes	Insularity of biota; endemism	East Africa; Central America
Lithology/Soils		
Limestone (other than karst)		
Cedar glades	Endemics in openings of hardwood forests	Southeastern USA: Kentucky, Tennessee
Limestone fynbos	Endemism; unique vegetation type	Cape Province, South Africa
Other limestone outcrops	Endemism, disjunctions, unique vegetation types	Europe: Alps, Balkans, etc. USA: White Mountains, California
Shale barrens	Endemics and range disjunctions; unique communities	Southeastern USA: Brallier shales, Virginia
Hydrothermal sites	Insularity: islands of pines in sagebrush ecosystem on altered andesite; geothermal vents and geysers with unique floras	USA: Western Nevada (Billings 1950); Mayacamas Mountains, California
Gypsum deposits	Rich endemism and unique floras	Southwestern USA and Northern Mexico

Table 8.2. *(continued)*

Types of site	*Geobotanical features*	*Locations, with examples*
Duricrusts (laterites and bauxites; etchplains)	Unique vegetation types, often locally xeric	Many topical areas (Thomas 1994)
Ultramafic rocks		
Dunite	Unique plant associations; some endemism	Dun Mountain, New Zealand; Twin Sisters Mountain, Washington
Serpentinite		
Serpentine barrens	Endemism, unique vegetation	Western USA: California, Oregon, Washington
Serpentine chaparral	Endemism, unique scrub vegetation	USA: Napa and Lake Counties, California
Serpentine mixed woodland	Endemics, open, sparse forest type	USA: Oregon, California; Balkans; Cuba
Serpentine maqui	Endemism, unique scrub-type vegetation	New Caledonia; Balkan Peninsula; Cuba

a long-term, uphill battle. The visual appeal of spectacular formations makes them the most likely candidates. And if unique flora clothes such a landform, preservation is doubly rewarding. Making a case for preservation of distinctive lithological or pedological sites will be harder. A serpentine barren or a sere limestone exposure is not likely to capture the eye nor call forth broad support for preservation. If the lithological anomaly has fostered the evolution of narrow endemic plants, the case for preservation can be strengthened. In the United States, endangered species programs, both federal and state, focus on rare organisms that are endangered. The conflict for conserving such sites then often becomes a contest pitting resource extraction (usually minerals) against saving rare biota.

Gaining support for—and ultimate preservation of—unusual geoedaphic landscapes calls for a multiplicity of stratagies. The first step is to identify the unique physical and biological attributes of the projected preserve. Good inventory data, including photographs, form an invaluable basis for gathering support. Citizen support is essential. Working with local and national conservation groups like native plant societies, natural history societies, and the like, gains the grass-roots support necessary to bring a project to the attention of private and public land managers. These independent lay organizations can campaign for the project, use their newsletters or periodicals to publicize it, and speak to authorities on behalf of their constituencies.

Some examples, from very different parts of the world, show how effective this grass-roots approach has been. The grand landforms, geological diversity, and rich biota of the North Cascades of western Washington State came into preserve sta-

8.9 *Erica occulta*, a strange and rare heather, endemic to the Hagelkraal limestone. Photo by C. Willis.

tus only through the sustained efforts of conservation organizations, especially the North Cascades Conservation Council and the Alpine Lakes Preservation Society. The protection of this vast montane wilderness is assured through National Park and Forest Service Wilderness status. On a more local level, one entire drainage in the Wenatchee National Forest of Washington State recently became a Research Natural Area to preserve the endemic flora and unique plant communities on serpentine soils. The process of securing preserve status for the Eldorado Creek Research Natural Area took several years; mining claims had to be voided (vacated) and the U.S. Forest Service faced other, more pressing issues. The impetus for getting this splendid example of a cool temperate serpentine habitat preserved was the effort of one person—your author!

Two natural history societies in the Southern Hemisphere exemplify the efforts of the "independent sector" of society to achieve conservation goals. In New Zealand, the Federated Mountain Clubs of New Zealand and kindred groups succeeded in getting the Red Mountain and Red Hills area of the South Island's Ophiolite Belt preserved; they are the most spectacular and remote of the ultramafic exposures in New Zealand (Malloy 1983). The Botanical Society of South Africa champions conservation causes on behalf of

8.8 Limestone fynbos of South Africa is rich in endemic species. Karstic cliffs at Hagelkraal, Cape Province. Photo by C. Willis.

8.10 *Acacia cyclops* (Rooikrans), an alien species massively invading limestone fynbos near Cape Agulhas, South Africa. Photo by C. Willis.

its members through its colorful and informative periodical *Veld and Flora*. In 1991 it took on the rescue of a pristine fynbos region, the Kogelberg, which was threatened by a projected dam. This spectacular montane wilderness landscape with riverine canyons and steep cliff walls supports a fine example of that unique edaphically induced flora, the fynbos. Another unique variant of fynbos habitat occurs on limestone in the extreme southern border of the Cape province, South Africa (Figs. 8.8 and 8.9). The several disjunct outcrops of limestone support a rich flora: 110 species are limestone endemics. Efforts to preserve several of the limestone outcrops are being sustained by the Flora Conservation Committee of the Botanical Society of South Africa (Heydenrych, Willis, and Burgers 1994). This endemic flora is as yet "underconserved"; the most critical threat to the limestone fynbos is the invasive alien vegetation, notably the shrubby Australian, *Acacia cyclops* (Fig. 8.10). So we see that places of exceptional geological wonder and exceptional floras do have their champions.

Epilogue

The biospheric fabric of our small planet is exorbitantly rich and diverse. A spherical blob of matter and energy has been transformed over billions of years into a teeming ark of living things and their inanimate settings—probably unparalleled in our solar system. The shaping of the Earth's crust into countless landforms and lithologies in turn contributes to the drama of biological diversity through evolution. The myriad processes and products of the geology-life interface have nurtured much of this diversity.

"Geology as a life force" has been the major emphasis in this book. The phrase is the mirror image of *Life as a Geological Force,* the title of Westbroek's (1991) fascinating little book. Geology, the fabric of the inanimate environment, influences the manner in which life populates our planet. Life processes in turn extract matter and energy ultimately from inanimate sources, only to leave the organically transformed end products behind to alter the face of the earth. Throughout the book, I have been mindful of the vast range of synergisms between the inanimate and animate spheres. The reader may feel that in my enthusiasm for the linkage between geology and plant life, I have slighted the reticulate nature of the life/geology paradigm. So I must repeat an earlier-cited self-evident "profundity." The planet and its cosmic setting are holocoenotic—a web of interconnectedness.

Some answers to the quintessential question of ecology "Why is what where?" will have come from the scrutiny of what I have termed the geoedaphic component of our planet's vast heterogeneity. I cannot leave the reader without hoping that the new word geoedaphics is not just another neologism. It is meant to be (and can become) shorthand for a worldwide dynamic in the structure and function of our biosphere. So I commend its use. Yet let the user be aware of its shortcomings. Geoedaphics not only stimulates synthesis—it *is* a synthesis. To flesh out its meanings and relevance for the biosphere, I have drawn upon major fields of natural science. Botany, ecology, geology, geomorphology, climatology, and soil science all contribute to the geoedaphic synthesis. And their contributions are examined in detail in this volume. Note that I have omitted microbiology and zoology from this roster of contributory disciplines. While both fields do make use of the geoedaphic synthesis, I have not felt competent to enlarge upon the linkages. The microbial world is a geologic force (Westbroek 1991), and in turn geology shapes the microbial habitat. Similarly, animals have taken environmental cues for the evolution of their structure and function from the geoedaphic environments. The connections are rarely voiced by zoologists, in part because they are some steps removed from geologic influences. The sequence of connection is via the food chain: carnivores to herbivores to plants and then to the geoedaphic milieu. Landforms and other geologic features may also more directly fashion the animal and its evolved adaptations. Would there be any mountain goats or pikas if we did not have mountains, talus, rocky outcrops, and so forth? I leave it to other authors (e.g., Butler 1995) to cement the heterotrophic connection of the animal world to geology.

Finally, a word about the future. Will the geoedaphic approach to the living world have a life beyond this book? The answer is yes, but with an unexpected twist. First off, geoedaphics is already flourishing, though not under that banner. There is an active cadre of scientists, mostly ecologists and geomorphologists, who contribute to our knowledge of the plant-geology connection. Just two contemporary examples: First, the exceptional ultramafic and other heavy-metal substrates continue to attract scientific inquiry, mostly from geobotanically inclined plant ecologists. Two reviews illustrate this area of research (Shaw 1989; Baker, Proctor, and Reeves 1992). Second, the duality of climates begetting landforms and landforms begetting climate is the active domain of geomorphologists. To this duality we need to add the organism component: How do plants and animals fit into the landform ⟷ climate equation? In early chapters, I pointed out the singular lack of attention given to the living skein as a part of geomorphologic context. I hope I have compensated for this hiatus by suggestive inferences that may spur further attention.

Study of geoedaphic systems for their own intrinsic merit is worthwhile. We need to know more about their manifold connections. But geoedaphics provides a heuristic context for seeking answers to fundamental biological questions. Parochial answers to the ecologist's question "Why is what where" are all well and good—but not enough. The "why" question in biology goes deeper. It seeks to understand the evolutionary origins of organisms and biological systems; it probes the adaptive value of a biological end product. The further question "How come?" initiates searches for the evolutionary history of living things. It is my contention that both the "Why?" and the "How come?" questions, so central in evolutionary thought, can draw upon geoedaphic case histories as experimental systems. For example, those who work with metallophytes (plants adapted to serpentines and other heavy metal substrates) see the physiological tolerance and endemic status of this flora as prime tools for study of the cellular and molecular bases of the adaptations. Here is a fundamental question in ecology, "Why do organisms occur where they do?" Baker (1987) says it well: "There is no doubt . . . that metal tolerance in plants will continue to intrigue all those plant scientists attempting to understand the nature and scale of plant adaptation to the environment and will remain an evolutionary paradigm *par excellence.*" Others have alluded to the research opportunities provided by metallophytes (Antonovics 1975; Kruckeberg and Kruckeberg 1989; and Kruckeberg 1992). I quote from the last reference:

> A major premise underpinning much biological research is the concept that study of contrasting members of a natural system can get at a given fundamental process. Recognizing the striking contrast of serpentine with nonserpentine ecologies, a host of fundamental biological problems could be tackled. In the realm of evolution and systematics, the serpentine milieu can serve heuristically to shed light on such questions as modes of speciation, gene flow across edaphic boundaries, the genetic basis of edaphic (serpentine) races, and molecular phylogeny of species groups with taxa on and off serpentine.

At the molecular level, Kruckeberg and Kruckeberg (1989) pose some intriguing approaches to basic biological questions, using, for instance, nickel-tolerant or copper-tolerant plants. Beginnings are on the horizon. Studies of Macnair and Cumbes (1989) illustrate how copper tolerance evokes protocols for the study of the genetic architecture of a tolerant species.

Will there be a place on our crowded planet for protected preserves with exceptional geoedaphic attributes? In chapter 8, I made a case for their protection. Unusual landforms are more likely to be set aside as preserves, but fortunately the potential salvation of places with exceptional parent materials and their floras is abetted by frequent occurrence of edaphic endemic species as well as vegetation types of exceptional physiognomies or species composition. At least in the United States such coincidental protection is fostered by the Endangered Species Act. But in other countries, edaphic endemics and their habitats have not fared well at all. For instance, there appears to be no protection as yet for the richest edaphic flora in the world—the highly endemic serpentine flora of New Caledonia. During the 1991 International Conference on Serpentine Ecology (Baker, Proctor, and Reeves 1992), serpentine floras were deemed imperiled enough to inspire a resolution on conservation of serpentine ecosystems (Kruckeberg 1993); see below. Those scientists with knowledge of such special places are best able to promote their protection.

TEXT OF RESOLUTION

This Resolution was endorsed by members and the organizing committee of the First International Conference on Serpentine Ecology. The resolution supports the conservation of a unique biological resource—the vegetation and flora of serpentine areas, found worldwide:

(1) whereas, in many parts of the world, there occurs a special geologic formation, ultramafic rocks (serpentine, peridotite, dunite and other iron-magnesium rocks) and these rocks form soils of unique chemical composition (high in magnesium, iron and nickel, and low in calcium and other nutrients);

(2) whereas, plants growing in habitats created by ultramafic (serpentine) rocks, are unique species (often narrowly restricted—endemic—to ultramafic soils), have singular ecological properties and can form distinctive vegetation types (communities and associations);

(3) whereas, the plant life of many ultramafic (serpentine) formations of the world are substantially disturbed by human activities (mining, forestry, grazing, etc.) and serpentine species are threatened by decimation or extinction;

be it resolved, that governments, public and private agencies, and private industry, be made aware of the threat to ultramafic (serpentine) biota, and that steps be taken by the aforementioned bodies, to preserve this rich and unique biodiversity.

Respectfully submitted by,
The Organizing Committee

Dr. A. J. M. Baker   Dr. A. R. Kruckeberg   Dr. S. N. Martens
Dr. R. D. Reeves   Dr. Lin Wu

on behalf of the 70 delegates coming from 18 countries.

Does the subject matter of this book have any relevance to current concerns over global climatic change? I believe it does and in the following contexts: First, global warming (or cooling) would change the composition of edaphically or geomorphically unique floras. Given their azonal natures and high fidelity to local habitats, such floras probably could not migrate in step with the changing climate.

Extinction is their probable fate. There is also the matter of landform influence—even control—over some regional climates. In an earlier chapter, I make a strong case for the ability of landform heterogeneity (especially mountains) to shape regional climates. With global climatic change, the landform constraints on climate could have significant effects: they could either modulate or ameliorate the climate change, or could accentuate it. Two major reviews barely touch upon this likely (or inevitable) influence of landform on global climate change, yet both provide the basis in landform heterogeneity to draw the inference I have made here (Bull 1991; Mooney, Fuentes, and Kronberg 1992).

I end this book with some pithy wisdom from other—wiser—writers. It may be unconventional to end a book with epigraphs; yet I will take that liberty. As some unknown wit has said, "Metaphors and other pithy sayings can serve as pogo sticks for crossing the terrain of knowledge." I am not unmindful of a possible criticism of this book and its focus. Fellow ecologists will say: "This is nothing new. We knew all along that the physical world with all its diversity determines much of the nature of life's diversity." That criticism may be valid. But on the matter of belaboring the self-evident, J. B. S. Haldane had this to say: "It is, in my opinion, worthwhile devoting some energy to proving the obvious."

The geology-plant connection has had many other protagonists—beginning with Dioscorides and Alexander von Humboldt on through Franz Unger, Anton Kerner, Braun-Blanquet, Harold Lutz, to Dwight Billings and Robert Brooks. These notables have been recognized contributors to the field throughout this text. But my most quotable "geoedaphologist" is soil scientist Hans Jenny in his landmark book of 1980: "The description of parent materials and their configurations as the initial state of landscape evolution is a benchmark like the birth record of a child. The infant develops, but the registry remains, and is basic to growth studies" (p. 247). And in the same lithological vein, he says: "Within a given climatic region, the growth of vegetation is mainly determined by the character of the parent material, whether limestone, igneous rock, sand deposit or clayey shale" (Jenny 1941). And finally, there is my own paraphrasing of Will Durant's epigram ("Civilization exists by geologic consent, subject to change without notice"), which I transform to read, "Plant life exists by geologic consent—subject to change without notice." Yes, cast in a time-bounded context, the plant world is subject to geologic change without notice—all the way from sudden geologic events to those occurring in shallow (Pleistocene or Quaternary) time, and on to those changes in deep, geologic time.

From transient equilibrium to change through time, *geology,* in all its manifestations, dominates among the inanimate forces that foster diversity of plant life. And with this assertion, patient reader, I rest my case.

Glossary

This glossary contains definitions of technical terms from botany, geology, geomorphology, and soil science. Sources include Akin (1991), Hunt (1972), Jenny (1980), and Skinner and Porter (1992). Several dictionaries of geological terms were consulted as well.

Accordant summits (alpine summit accordance). Nearly level array of high summits. (Thompson 1990) See *Gipfelflur*

Adret. The side of a mountain receiving direct sunlight.

Allopatry. Populations or species occupying mutually exclusive (but often adjacent) geographical areas. (Mayr 1970) See *Sympatry*

Allopolyploidy. Doubling of chromosome number in a sterile interspecific hybrid. (Stebbins 1950)

Azonal. Soil or vegetation under the influence of local, unique geomorphic or lithologic sets of conditions. Contrasts with *Zonal* (which see), where regional climate influences the quality of soils/vegetation. (Hunt 1972)

Biocrust. Surface veneer ("varnish") on soils or rocks, caused mostly by microorganisms. (Viles 1988)

Biokarst. Surface heterogeneity (usually microtopographic) caused by organisms; usually occurs on limestones and gypsum. (Viles 1988)

Bodenstet. Species with narrow edaphic tolerance, often limited to a single soil (parent material). (Unger 1836; Kruckeberg 1969b)

Bodenvag. Species of wide edaphic tolerance, occurring on more than one soil (parent material); literally, "soil wanderer." (Unger 1836; Kruckeberg 1969b)

Calcrete. Hard layer of carbonate forming over soft limestone. (Ollier 1984)

Catastrophic selection. Survival of exceptional genotypes following rapid extinction of other members of a population. (Lewis 1962)

Catena. Topographically induced variations of soil types, usually on the same parent material and in the same locality. Same as soil toposequence. (Ollier 1984)

Clastic sediments. Loose fragmented debris produced by the mechanical breakdown of older rocks. Also called detritus or detrital sediments. (Skinner and Porter 1992)

Desert varnish. Thin coating of metallic oxides (manganese and iron) on rocks and rock outcrop surfaces in desert areas after long exposure. (Skinner and Porter 1992)

Diagenesis (-etic). Chemical, physical, and biological processes that affect sediment after its initial deposition and during and after its slow transformation into sedimentary rock.

Drift (in genetics). Random fluctuations in gene frequencies in small populations.

Drift (in geology). Transported regolith; see also *Glacial drift.*

Dysgeogenous. Describing rock not easily weathered, producing only a small amount of detritus.

Ecotype. Within a species, the genotypic response of a population to a local habitat (e.g., climatic ecotype or edaphic ecotype).

Edaphic. Pertaining to soil.

Genecology. The study of the genetical structure of plant populations in relation to their habitats.

Geobotany. Variously used: (1) the linkage of landforms and lithology to flora and vegetation; (2) plant geography or the geographic distribution of plants.

Geoecology. The interactions between topography, climate, soils, and vegetation. (Gerrard 1990; Huggett 1995)

Geoedaphic. The reciprocal interactions of topography (landforms), lithology, and soils with floras and vegetation. (Kruckeberg 1986)

Geomorphology. Descriptive and analytical (causational) study of landforms.

Geosyncline. A great trough that has received thick deposits of sediment during its slow subsidence through long geological periods. (Skinner and Porter 1992)

Gipfelflur (plural *-en*). Accordant summits; array of mountain summit peaks of nearly the same elevation; literally, "summit plain" in German. (Thompson 1990)

Glacial drift. Sediment deposited directly by glaciers or indirectly by glacial meltwater; also called drift. (Skinner and Porter 1992)

Gypsophily. Vegetation and flora tolerant to, or restricted to, gypsum deposits.

Holocoenotic. The components of the environment do not act separately or independently, but mutually and have a concerted effect on organisms.

Hyperaccumulation. Uptake of metal elements (e.g., nickle) by plants in excess of 1000 ppm (0.1% in dried tissue). (Brooks 1987, 1998; Shaw 1989)

Karren. Small-scale solutional sculpturing of carbonate rocks; associated with karst terrain. (White 1988)

Karst. A large family of named landforms usually associated with differential solution of limestones.

Krummholz. Literally, "crooked wood" (German); stunted, often deformed trees occurring at upper timberline.

Lithification. The process that converts a sediment into a sedimentary rock. (Skinner and Porter 1992)

Lithology. Descriptive study of rocks, their mineral content and texture. (Skinner and Porter 1992)

Lithosequence. A sequence of different rock types in a local area. (Jenny 1980)

Mafic. Referring to rock types high in iron-magnesium silicate minerals.

Massenerhebung effect. Influence of large mountainous plateaus in decreasing temperature change with altitude (lapse rate). (Arno and Hammerly 1984)

Mesic. Moderate, equitable moisture conditions for vegetation.

Mogote. Tower-like conical karst in the Caribbean; inclined conical karst on Cuban limestone. (Borhidi 1991)

Nebka. Sand traps formed by vegetation. (Pitty 1971)

Ophiolite. Elongate areas of obducted oceanic crust; assemblages of rocks associated with plate tectonics, including ultramafic rocks. (Coleman and Jove 1992)

Orographic uplift. Rise of moisture-laden air along mountain ranges lying across the path of prevailing winds. (Akin 1991)

Palsa (or *Palsen*). Mounds of earth pushed up by glacial action.

Parent material. Rock or other materials that weather to form soils; usually the C horizon of a soil profile.

Peaked plain. See *Gipfelflur*

Phytogeomorphology. The interface between landforms and vegetation.

Phytokarst. Biogenic origin of karstic (limestone) microrelief; action of plants (lichens, algae, etc.) on carbonate rocks. (Viles 1988)

Plagiotropy. Oblique or horizontal growth form.

Preadaptation. Fortuitous prior existence of genes in populations that may have high fitness for a new environment.

Rain shadow. A dry area downwind from a mountain range. (Akin 1991)

Regolith. Loose noncemented rock particles; can include soil, talus (colluvial materials).

Rendzina. Humus-carbonate (lime) soil. (Jenny 1980)

Saltation. Rapid evolutionary change of major magnitude; for example, "instantaneous" speciation. (Lewis 1962)

Sclerophylly. Having hard, stiff, and usually small leaves. Nearly the same as xerophylly.

Serpentine. Loosely applied to minerals, rocks, soils, vegetation, and floras associated with ultramafic (ferromagnesian) substrates. (Kruckeberg 1969b; Brooks 1987)

Serpentine syndrome. The manifold causes and effects of ultramafic (serpentine) environments as they influence plant life. (Jenny 1980)

Serpentinite. Rock composed of serpentine minerals (e.g., chrysotile and antigorite).

Shale. A fine-grained clastic sedimentary rock. (Skinner and Porter 1992)

Sial. A layer of rocks rich in silica and alumina.

Sima. A layer of rocks rich in silica and magnesium, found beneath ocean floors and the sial of continents.

Slate. A low-grade metamorphic rock with a pronounced slaty cleavage. (Skinner and Porter 1992); metamorphosed shale.

Sympatry. The occurrence of two or more populations (or species) in the same area, potentially enabling them to interbreed. See *Allopatry*

Temperature inversion. Reversal of normal temperature gradient, high to low, with altitude.

Tepui (tepuy). "Archipelago" of massive, vertical-sided sedimentary plateaus, surrounded by lowland tropical forest in Venezuela.

Till. Nonsorted sediments of glacial origin.

Timberline. Uppermost montane limit of tree growth.

Toposequence. Succession of landforms in a local area.

Travertine. Solution-deposited calcite and/or aragonite (calcium carbonate minerals), often in caves; calcareous tufa in seeps, springs.

Trimline. Border between glaciated (deforested) and nonglaciated (forested) areas associated with montane glaciers.

Ubac. Opposite side of a mountain from the one receiving direct sunlight (adret side).

Ultrabasic. Older term for ultramafic (ferromagnesian) lithology.

Ultramafic. Referring to ferromagnesian silicate rocks.

Vadose water. Water above the water table (shallow water).

Xeric. Referring to dry environments; dry vegetation types.

Zonal. Areas of soils and vegetations largely controlled by the prevailing regional climate. Here climate, not exceptional landforms or rock types, determines the nature of soils and vegetation. See *Azonal*

Bibliography

Agricola, G. 1556. *De Re Metallica.* English translation (1912) by H. C. and L. C. Hoover. New York: Dover PI.

Akin, W. E. 1991. *Global Patterns: Climate, Vegetation and Soils.* Norman: University of Oklahoma Press.

Alexander, E. B., W. E. Wildman, and W. C. Lynn. 1985. No. 12, Ultramafic (serpentinitic) mineralogy class. In *Mineral Classification of Soils.* Soil Sci. Soc. Amer. Spec. Pub. 16:135–146.

Allee, W. C., and T. Park. 1939. Concerning ecological principles. *Science* 89:166–169.

Allen, J. E., and M. Burns. 1986. *Cataclysms on the Columbia.* Portland, Ore.: Timber Press.

Allmon, W. D., and R. M. Ross. 1990. Specifying causal factors in evolution: the paleontological contribution. In *Causes of Evolution,* ed. R. M. Ross and W. D. Allmon, pp. 1–17. Chicago: University of Chicago Press.

Anderson, E. 1936. The species problem in *Iris. Annals of the Missouri Botanical Garden.* 23:457–509.

Anderson, R. C., J. S. Fralish, and J. M. Baskin, eds. 1999. *Savannas, Barrens, and Rock Outcrop Plant Communities of North America.* New York: Cambridge University Press.

Anon. 1999. *Voice of the Wild Siskiyou.* Winter 1999. Newsletter of the Siskiyou Regional Educational Project, PO Box 220, Cave Junction, OR 97523.

Antonovics, J. 1975. Metal tolerance in plants: perfecting an evolutionary paradigm. In *Symposium Proceedings, International Conference on Heavy Metals in the Environment,* pp. 169–186. Toronto, Canada.

Antonovics, J., A. D. Bradshaw, and R. G. Turner. 1971. Heavy metal tolerance in plants. *Advances in Ecological Research* 7:1–85.

Antos, J. A., and D. B. Zobel. 1985a. Plant form, developmental plasticity, and survival following burial by volcanic tephra. *Canadian Journal of Botany* 63:2083–2090.

———. 1985b. Upward movement of underground plant parts into deposits of tephra from Mount St. Helens. *Canadian Journal of Botany* 63:2091–2096.

———. 1985c. Recovery of forest understories buried by tephra from Mount St. Helens. *American Journal of Botany* 73:495–499.

Arno, S. F., and R. P. Hammerly. 1984. *Timberline: Mountain and Arctic Forest Frontiers.* Seattle: The Mountaineers.

Arthur, M. A. 1992. Vegetation. In *Biogeochemistry of a Subalpine Watershed: Loch Vale Watershed,* ed. J. Baron, pp. 76–92. New York: Springer-Verlag.

Asprey, G. F., and R. G. Robbins. 1953. The vegetation of Jamaica. *Ecological Monographs* 22:359–412.

Atwater, B. 1992. Geologic evidence for earthquakes during the past 2000 years along the Copalis River, southern coastal Washington. *Journal of Geophysical Research* 97: 1901–1919.

Avias, J. 1972. Karst of France. In *Karst: Important Karst Regions of the Northern Hemisphere,* ed. H. Herak and V. T. Stringfield, pp. 129–188. Amsterdam: Elsevier.

Baars, D. L. 1994. *The Canyon Revisited: A Rephotography of the Grand Canyon, 1923/1991.* Salt Lake City: University of Utah Press.

Bach, R. 1950. Die Standorte jurassischer Buchenwaldgesellschaften mit besonderer Berüksichtigung der Böden (Humuskarbonatböden und Rendzinen). *Schweizerisch Botanische Gesellschaft Berichte* 60:51–152.

Baker, A. J. M. 1987. Metal tolerance. *New Phytologist* 106 (Suppl.):93–111.

Baker, A. J. M., J. Proctor, and R. D. Reeves, eds. 1992. *The Vegetation of Ultramafic (Serpentine) Soils.* Andover, UK: Intercept.

Baker, H. G. 1965. Characteristics and modes of origin of weeds. In *Genetics of Colonizing Species,* ed. H. G. Baker and G. L. Stebbins, pp. 147–168. New York: Academic Press.

———. 1995. Aspects of the genecology of weeds. In *Genecology and Ecogeographic Races,* eds. A. R. Kruckeberg, R. B. Walker, and A. Leviton, pp. 189–224. Washington, D.C.: American Association for the Advancement of Science.

Ball, I. R. 1983. Planarians, plurality and biogeographical explanations. In *Evolution, Time and Space: The Emergence of the Biosphere,* ed. R. W. Simsims, J. H. Price, and P. E. S. Whalley, pp. 409–430. London: Academic Press.

Barbour, M., and A. Johnson. 1988. Beach and Dune. In *Terrestrial Vegetation of California,* ed. M. Barbour and J. Major, pp. 223–261. New York: John Wiley and Sons.

Baron, J., ed. 1992. *Biogeochemistry of a Subalpine Watershed.* New York: Springer-Verlag.

Barry, R. G. 1992. *Mountain Weather and Climate.* 2d ed. London: Routledge.

Baskin, J. M, and C. C. Baskin. 1986. Distribution and geographical/evolutionary relationships of cedar glade endemics in southeastern United States. *Bulletin of the Association of Southeastern Biologists* 33:138–194.

———. 1988. Endemism in rock outcrop communities of unglaciated eastern United States: an evaluation of the roles of the edaphic, genetic and light factors. *Journal of Biogeography* 15:829–840.

———. 1989. Cedar glade endemics in Tennessee and a review of their autecology. *Vegetation and Flora of Tennessee: Journal of the Tennessee Academy of Science* 68(3):63–74.

Bateman, R. M., and W. A. DiMichele. 1994. Saltational evolution of form in vascular plants: a neoGoldschmidtian synthesis. In *Shape and Form in Plants and Fungi,* ed. D. S. Ingram and A. Hudson, pp. 61–160. Symposium Series, no. 16. Linnaean Society of London. New York: Academic Press.

Bean, A. 1991. Kogelberg—is its fate to be under water? *Veld and Flora* 77:80.

Becking, R. W. 1986. *Hastingsia atropurpurea* (Liliaceae: Asphodeleae), a new species from southwestern Oregon. *Madroño* 33:175–181.

Behrensmeyer, A. K. et al., eds. 1992. *Terrestrial Ecosystems Through Time: Evolutionary Paleoecology of Terrestrial Plants and Animals.* Chicago: University of Chicago Press.

Belloni, S., B. Martinis, and G. Orombelli. 1972. Karst of Italy. In *Karst: Important Karst Regions of the Northern Hemisphere,* ed. H. Herak and V. T. Stringfield, pp. 83–128. Amsterdam: Elsevier.

Betz, F., Jr., ed. 1975. *Environmental Geology.* Benchmark Papers in Geology, no. 25. Stroudsburg, Penn.: Dowden, Hutchinson and Ross.

Bilderback, D. E., ed. 1987. *Mount St. Helens 1980: Botanical Consequences of the Explosive Eruptions.* Berkeley: University of California Press.

Billings, W. D. 1950. Vegetation and plant growth as affected by chemically altered rocks in the western Great Basin. *Ecology* 31:62–74.

———. 1952. The environmental complex in relation to plant growth and distribution. *Quarterly Review of Biology* 27:251–265.

———. 1954. Temperature inversions in the pinyon-juniper zone of a Nevada mountain range. *Butler University Botanical Studies* 12:112–117.

———. 1978. Alpine phytogeography across the Great Basin. *Great Basin Naturalist Memoirs* 2:105–117.

Billings, W. D., and H. A. Mooney. 1968. The ecology of arctic and alpine plants. *Biological Review* 43:481–529.

Birkeland, P. W. 1974. *Pedology, Weathering, and Geomorphological Research.* London: Oxford University Press.

Birrell, K. S., and A. C. S. Wright. 1945. A serpentine soil in New Caledonia. *New Zealand Journal Science and Technology* 27A:72–76.

Borhidi, A. 1991. *Phytogeography and Vegetation Ecology of Cuba*. Budapest: Akadémiai Kiadó.

Borhidi, A., A. J. M. Baker, M. Fernandez Zequeira, and R. Oviedo Prieto. 1992. Preliminary studies on possible Ni-accumulator plants of Cuba. *Acta Botanica Hungarica* 37:279–286.

Bowen, F., R. Hess, H. Planeto, V. Stansell, and D. Werschkul. 1982. *Hunter Creek and Springs, Bogs*. Wedderburn, Ore.: Kalmiopsis Audubon Society.

Bradshaw, A. D. 1976. Pollution and evolution. In *Effects of Air Pollutants on Plants*, ed. T. Mansfield, pp. 35–159. Cambridge: Cambridge University Press.

Braun-Blanquet, J. 1923. L'origine et le développement des flores dans le Massif Central de France. Paris and Zurich: n.p.

———. 1932. *Plant Sociology: The Study of Plant Communities*. New York: McGraw-Hill.

Braun-Blanquet, J., and H. Jenny. 1926. Vegetationsentwickelung und Bodenbildung in der alpinen Stufe der Zentralalpen. *Neue Denkschrift Schweizer Naturforschung Gesellschaft* 63:175–349.

Bretz, J. H. 1959. The Lake Missoula floods and the channelled scabland of Washington: new data and interpretations. *Journal of Geology* 77:503–543.

Brewer, W. H. 1949. *Up and Down California in 1860–1864*. Berkeley: University of California Press.

Briggs, J. C. 1987. *Biogeography and Plate Tectonics*. Developments in Paleontology and Stratigraphy, vol. 10. Amsterdam: Elsevier.

Brooks, R. R. 1972. *Geobotany and Biogeochemistry in Mineral Exploration*. New York: Harper and Row.

———. 1987. *Serpentine and Its Vegetation: A Multidisciplinary Approach*. Portland, Ore.: Dioscorides Press.

Brooks, R. R., ed. 1998. *Plants that Hyperaccumulate Heavy Metals*. New York: CAB International.

Brooks, R. R., and D. Johannes. 1990. *Phytoarchaeology*. Portland, Ore.: Dioscorides Press.

Brooks, R. R., and F. Malaisse. 1985. *The Heavy Metal–Tolerant Flora of Southcentral Africa*. Rotterdam: A. A. Balkema.

Brooks, R. R., and C. C. Radford. 1978. Nickel accumulation by European species of the genus *Alyssum*. *Proceedings, Royal Society of London* Section B 200:217–224.

Brooks, R. R., R. D. Reeves, A. J. M. Baker, J. A. Rizzo, and H. D. Ferreria. 1990. The

Brazilian serpentine plant expedition (BRASPEX), 1988. *National Geographic Research* 6:205–219.

Brown, J. H. 1988. Species diversity. In *Analytical Biogeography,* ed. A. A. Myers and P. S. Giller, pp. 57–89. London: Chapman and Hall.

———. 1995. *Macroecology.* Chicago: University of Chicago Press.

Brown, J. H., and A. C. Gibson. 1983. *Biogeography.* St. Louis: C. V. Mosby Co.

Brunt, M. A., and J. E. Davies. 1994. *The Cayman Islands: Natural History and Biogeography.* Dordrecht: Kluwer Academic Press.

Buck, L. J. 1949. Association of plants and minerals. *Journal of the New York Botanical Garden* 50:265–269.

Bull, W. B. 1991. *Geomorphic Responses to Climatic Change.* Oxford: Oxford University Press.

Burbanck, M. P., and R. B. Platt. 1964. Granite outcrop communities of the Piedmont Plateau in Georgia. *Ecology* 45:292–306.

Bureau of Mines. 1991. *Mineral Industries of Asia and Pacific.* 1991 International Review. U.S. Dept. of Interior Minerals Yearbook 111. Washington, D.C.: Government Printing Office.

Burgess, I. C., and M. Mitchell. 1994. Origin of limestone pavements. *Cumberland Geological Society Proceedings* 5:405–412.

Butler, D. 1995. *Zoogeomorphology: Animals and Geomorphic Agents.* Cambridge: Cambridge University Press.

Butler, J., H. Goetz, and J. L. Richardson. 1986. Vegetation and soil-landscape relationships in the North Dakota Badlands. *American Midland Naturalist* 116:378–386.

Cain, S. A. 1944. *Foundations of Plant Geography.* New York: Harper and Brothers.

Cannon, H. L. 1960. Botanical prospecting for ore deposits. *Science* 132:591–598.

———. 1971. The use of plant indicators in ground water surveys, geologic mapping, and mineral prospecting. *Taxon* 20:227–256.

Carlquist, S. 1970. *Hawaii: A Natural History.* Garden City, N.Y.: Natural History Press.

Carroll, D. 1970. *Rock Weathering.* New York: Plenum Press.

Chapin, D. M., and L. C. Bliss. 1988. Soil-plant water relations of two subalpine herbs from Mount St. Helens. *Canadian Journal of Botany* 66:809–818.

Christiansen, N. I. 1971. Fabric, seismic anisotrophy, and tectonic history of the Twin Sisters dunite, Washington. *Geological Sciences of America Bulletin* 82: 1681–1694.

Chronic, H. 1984. *Pages of Stone: Geology of Western National Parks and Monuments.* Vol. 1, *Rocky Mountains and Western Great Plains.* Vol. 2, *Sierra Nevada, Cascades and Pacific Coast.* Vol. 3, *The Desert Southwest.* Seattle: The Mountaineers.

Cittadino, E. 1990. *Nature as the Laboratory: Darwinian Plant Ecology in the German Empire, 1880–1900.* Cambridge: Cambridge University Press.

Clausen, J., D. Keck, and W. M. Hiesey. 1940. *Experimental Studies on the Nature of Species. I. Effect of Varied Environments on Western North American Plants.* Carnegie Institution of Washington Publications, no. 520. 452 pp.

Clements, F. E. 1928. *Plant Succession and Indicators.* New York: H. W. Wilson Co.

Coleman, R. G. 1967. Low-temperature reaction zones and alpine ultramafic rocks of California, Oregon and Washington. *United States Geological Survey Bulletin* 1247:1–49.

Coleman, R. 1985. Not all forms of asbestos are dangerous. *Stanford Observer,* October.

Coleman, R. G., and C. Jove. 1992. Geological origin of serpentinites. In *The Vegetation of Ultramafic (Serpentine) Soils,* ed. A. J. M. Baker, J. Proctor, and R. D. Reeves, pp. 1–17. Andover, UK: Intercept.

Coleman, R. G., and A. R. Kruckeberg. 1999. Geology and plant life of the Klamath-Siskiyou bioregion. *Natural Areas Journal* 19:320–340.

Comstock, H. B. 1963. *Magnesium and Magnesium Compounds.* Bureau of Mines Information Circular IC 8201. Washington, D.C.: Bureau of Mines, U.S. Department of the Interior.

Conard, H. S. 1951. *The Background of Plant Ecology.* (Translation of *The Plant Life of the Danube Basin* by Anton Kerner, 1863.) Ames: Iowa State College Press.

Cook, R. J., J. C. Barron, R. I. Papendick, and G. J. Williams, III. 1980. Impact on agriculture of the Mount St. Helens eruptions. *Science* 211:16–22.

Cooke, S. S. 1994. The edaphic ecology of two western North American composite species. Ph.D. dissertation, University of Washington.

Cowles, H. C. 1901a. The influence of underlying rocks on the character of vegetation. *Bulletin of the American Bureau of Geography* 2:1–26.

———. 1901b. The physiographic ecology of Chicago and vicinity: a study of the origin, development and classification of plant societies. *Botanical Gazette* 31:73–108; 145–182.

Cox, G. W. 1984. Mounds of mystery. *Natural History* 93:36–45.

Cox, G. W., and C. G. Gakahu. 1985. Mima mound microtopography and vegetation pattern in Kenyan savannas. *Journal of Tropical Ecology* 1:23–36.

Crane, N. L., and B. S. Malloch. 1985. A study of rare plants for the Geysers-Calistoga known geothermal resource area. Report 417-84.53, Department of Engineering Research, Pacific Gas and Electric Co., San Francisco, California.

Croizat, L., G. J. Nelson, and D. E. Rosen. 1974. Centers of origin and related concepts. *Systematic Zoology* 23:71–90.

Cronquist, A. 1981. *An Integrated System of Classification of Flowering Plants.* New York: Columbia University Press.

Cronquist, A., A. H. Holmgren, N. H. Holmgren, and J. L. Reveal. 1972. *Intermountain Flora.* Vascular Plants of the Intermountain West, U.S.A., vol. 1, vol. 5 (1984), vol. 6 (1977). New York: New York Botanical Garden/Hafner Publishing Co.

Crowther, J. 1987. Ecological observations in tropical karst terrain, West Malaysia. III. Dynamics of the vegetation-soil bedrock system. *Journal of Biogeography* 14:157–164.

Cushman, M. J. 1981. The influence of recurrent snow avalanches on vegetation patterns in the Cascades of Washington. Ph.D dissertation, University of Washington.

Cymerman, C. 1988. The effects of serpentine soils on the morphology of two Wenatchee Mountain flowering plants. M.S. thesis, University of Washington.

Dahlquest, W. W., and V. B. Scheffer. 1942. The origin of the Mima mounds of western Washington. *Journal of Geology* 50:68–84.

Dale, V. H. 1991. Revegetation of Mount St. Helens debris avalanche 10 years posteruption. *National Geographic Research and Exploration* 7:328–341.

Dann, K. T. 1988. *Traces on the Appalachians: A Natural History of Serpentine in Eastern North America.* New Brunswick, N.J.: Rutgers University Press.

Dansereau, P. 1957. *Biogeography: An Ecological Perspective.* New York: Ronald Press.

Daoxian, Y. 1993. Environmental change and human impact on karst in southern China. In *Karst Terrains: Environmental Changes and Human Impact,* ed. P. W. Williams, pp. 99–107. Cremlingen-Destedt, Germany: Catena-Verlag.

Darwin, C. 1839. *Journal of Researches into the Geology and Natural History of the Various Countries Visited by H. M. S. Beagle Under the Command of Captain Fitzroy, R. N. from 1832 to 1838.* London.

———. 1887 [1868]. *The Variation of Animals and Plants under Domestication.* 2d ed. New York: D. Appleton and Co.

Daubenmire, R. F. 1947. *Plants and Environment: A Textbook of Plant Autecology.* New York: John Wiley and Sons.

———. 1970. Steppe vegetation of Washington. Exp. Sta. Tech. Bull. 62. Pullman: College of Agriculture, Washington State University.

Davidse, G., ed. 1983. Biogeographical relationships between temperate eastern Asia and temperate eastern North America. The Twenty-ninth Annual Systematics Symposium. *Annals of the Missouri Botanical Garden* 70(3,4).

Davis, W. E., and H. E. LeGrand. 1972. Karst of the United States. In *Karst: Important Karst Regions of the Northern Hemisphere,* ed. H. Herak and V. T. Stringfield, pp. 467–506. Amsterdam: Elsevier.

Delucia, E. H., W. H. Schlesinger, and W. D. Billings. 1989. Edaphic limitations to growth and photosynthesis in Sierran and Great Basin vegetation. *Oecologia* 78:184–198.

Derbyshire, E., ed. 1976. *Geomorphology and Climate.* London and New York: John Wiley and Sons.

De Sequeira, E. M., and A. R. Pinto Da Silva. 1992. The ecology of serpentinized areas of north-east Portugal. In *The Ecology of Areas with Serpentinized Rocks,* ed. B. A. Roberts and J. Proctor, pp. 169–197. Dordrecht: Kluwer Academic Publishers.

Dice, J. C. 1997. Algodones Dunes. In *California's Wild Gardens,* ed. P. M. Faber, pp. 216–217. Sacramento: California Native Plant Society.

Dierschke, H. 1975. *Vegetation und Substrat.* Berichte der Intern. Symp. der Internationale Vereinigung für Vegetationskunde, ed. Reinhold Tuxen. Vaduz, Liechtenstein: J. Cramer.

Dietrich, R. V., and B. J. Skinner. 1979. *Rocks and Rock Minerals.* New York: John Wiley and Sons.

DiMichele, W. A., T. L. Phillips, and R. G. Olmstead. 1987. Opportunistic evolution: abiotic environmental stress and the fossil record of plants. *Review of Paleobotany and Palynology* 50:151–178.

Dobzhansky, T., F. J. Ayala, G. L. Stebbins, and J. W. Valentine. 1977. *Evolution.* San Francisco: W. H. Freeman and Co.

Dodd, J. R., and R. J. Stanton, Jr. 1990. *Paleoecology: Concepts and Applications.* 2d ed. New York: John Wiley and Sons.

Dolan, R. W. 1988. A re-examination of the serpentine endemic *Streptanthus morrisonii* F. W. Hoffman complex. *American Journal of Botany* 75:169–170. (Abstract 457.)

Dolan, R. W., and L. F. La Pré. 1987. *Streptanthus morrisonii* complex. Final report. Riverside, Calif.: Tierra Madre Consultants.

———. 1989. Taxonomy of *Streptanthus* Sect. Biennes, the *Streptanthus morrisonii* complex (Brassicaceae). *Madroño* 36:33–40.

Drew, D. 1985. *Karst Processes and Landforms.* Aspects of Geography Series. London: Macmillan Education Ltd.

Dunham, K. C. 1969. Geologic map of the British Islands. Geologic Survey of Great Britain.

Egger, B. 1994. Végétation et stations alpines sur serpentine pres de Davos. Zurich: Geobotanisches Institute der ETH Stiftung Rubel.

Ehrlich, A. H., and P. R. Ehrlich. 1987. *Earth.* New York: Franklin Watts.

Ehrlich, P., and A. Ehrlich. 1981. *Extinction: The Causes and Consequences of the Disappearance of Species.* New York: Random House.

Eldredge, N., and S. J. Gould. 1972. Punctuated equilibria: an alternative to phyletic gradualism. In *Models in Paleobiology,* ed. T. J. M. Schopf, pp. 82–115. San Francisco: Freeman, Cooper and Co.

Ellenberg, H. 1988. *Vegetation Ecology of Central Europe.* 4th ed. English translation. Cambridge: Cambridge University Press.

Erickson, R. O. 1945. The *Clematis fremontii* var. *riehlii* population in the Ozarks. *Annals of the Missouri Botanical Garden* 32:413–460.

Etherington, J. R. 1982. *Environment and Plant Ecology.* 2d ed. New York: John Wiley and Sons.

Faber, P., ed. 1984. Serpentine flora: notes on prominent sites in California. *Fremontia* 11(5):11–30.

Faber, P. M., ed. 1997. *California's Wild Gardens: A Living Legacy.* Sacramento: California Native Plant Society.

Falk, D., and K. Holsinger. 1991. *Genetics and Conservation of Rare Plants.* New York: Oxford University Press.

Faust, G. T., and J. J. Fahey. 1962. The serpentine-group minerals. Geological Survey, Paper 384A. Washington, D.C.: U.S. Department of the Interior.

Fernald, M. L. 1907. The soil preferences of certain alpine and subalpine plants. *Rhodora* 9:149–193.

Fiedler, P. L. 1985. Heavy metal accumulation and the nature of edaphic endemism in the genus *Calochortus* (Liliaceae). *American Journal of Botany* 72:1712–1718.

Franklin, J. F., and C. T. Dyrness. 1973. *Natural Vegetation of Oregon and Washington.* USDA For. Serv. Gen. Tech. Rep. PNW-8. Corvallis: Oregon State University Press, 1988 (reprint).

Franklin, J. F., J. A. MacMahon, F. J. Swanson, and J. R. Sedell. 1985. Ecosystem responses to the eruption of Mount St. Helens. *National Geographic Research* 1:199–216.

Fredriksen, R. L. 1970. Erosion and sedimentation following road construction and timber harvest on unstable soils in three small western Oregon watersheds. USDA For. Serv. Res. Pap. PNW-104.

Fridriksson, S. 1975. *Surtsey: Evolution of Life on a Volcanic Island.* London: Butterworth.

Friederichs, K. 1927. Grundsätzliches über die Lebenseinheiten höherer Ordnung und die ökologischen Einheitsfactor. *Die Naturwissenschaften* 15:153–157, 182–186.

Futuyma, D. J. 1979. *Evolutionary Biology.* Sunderland, Mass.: Sinaur Associates, Inc.

Gankin, R., and J. Major. 1964. *Arctostaphylos myrtifolia,* its biology and relationship to the problem of endemism. *Ecology* 45:792–808.

Garner, H. F. 1974. *The Origin of Landscapes: A Synthesis of Geomorphology.* New York: Oxford University Press.

Gartside, D., and T. McNeilly. 1974. The potential for evolution of heavy metal tolerance in plants. II. Copper tolerance in normal populations of different plant species. *Heredity* 32:335–348.

George, U. 1989. Venezuela's islands in time. *National Geographic* 175 (May): 526–561.

Gerrard, A. J. 1988. *Rocks and Landforms.* London: Unwin Hyman.

———. 1990. *Mountain Environments: An Examination of the Physical Geography of Mountains.* London: Belhaven Press.

Gieseking, J. E. 1975. *Soil Components.* Vol. 2, *Inorganic Components.* New York: Springer-Verlag.

Gigon, A. 1971. Vergleich alpiner Rasen auf Silikat und auf Karbonatboden: Konkurrenz- und Stickstofformenversuche sowie standortskundliche Untersuchungenen im Nardetum und im Seslerietum bei Davos. Geobotanisches Instituts der ETH Stiftung Rubel, no. 48.

Giles, L. J. 1969. The ecology of the mounds on the Mima Prairie, with special reference to Douglas Fir invasion. M.S. thesis, University of Washington.

Gillispie, C. C., ed. 1970. *Dictionary of Scientific Biography.* New York: Scribners.

Gleason, H. A. 1939. The individualistic concept of the plant association. *American Midland Naturalist* 21:92–108.

Gleason, H. A., and M. T. Cook. 1927. *Scientific Survey of Porto Rico and the Virgin Islands.* Vol. 7, pt. 1, *Plant Ecology of Porto Rico.* New York: New York Academy of Sciences.

Gleason, H. A., and A. Cronquist. 1964. *The Natural Geography of Plants.* New York: Columbia University Press.

Goldschmidt, R. B. 1940. *The Material Basis of Evolution.* New Haven: Yale University Press.

Good, R. D'O. 1931. A theory of plant geography. *New Phytologist* 30:149–171.

———. 1974. *The Geography of Flowering Plants.* 3d ed. Harlow, UK: Longman.

Gorham, E. 1954. An early view of the relation between plant distribution and environmental factors. *Ecology* 85:97–98.

Goudie, A. S. 1988. The geomorphological role of termites and earthworms in the tropics. In *Biogeomorphology,* ed. H. A. Viles, pp. 166–192. New York: Basil Blackwell.

Gray, A. 1846. Analogy between the flora of Japan and that of the United States. *American Journal of Science Arts* 2:135–136.

———. 1860. Botany of Japan and its relations to that of Central and Northern Asia, Europe and North America. *Proceedings of the American Academy of Arts* 4:130–135.

Greene, E. L. 1904. Certain west American Cruciferae. *Leaflets of Botanical Observation and Criticism* 1:81–90.

Greenland, D. J., and M. H. B. Hayes. 1978. *The Chemistry of Soil Constituents*. Chichester, N.Y.: John Wiley and Sons.

Griffin, J. R. 1965. Digger pine seedling response to serpentine and non-serpentine soil. *Ecology* 46:801–807.

Grimes, J. P., and J. G. Hodgson. 1969. An investigation of the ecological significance of lime-chlorosis by means of large-scale comparative experiments. In *Ecological Aspects of the Mineral Nutrition of Plants,* ed. L. H. Rorison, pp. 67–99. Oxford: Blackwell Scientific Publications.

Habeck, J. R., and E. Hartley. 1968. *A Glossary of Alpine Terminology.* Missoula: University of Montana.

Hall, C. A., Jr., and V. D. Jones. 1988. Plant biology of eastern California. In *Natural History of the White-Inyo Range Symposium,* vol. 2. (The Mary DeDecker Symposium.) Los Angeles: University of California White Mountain Research Station.

Hallam, A. 1994. *An Outline of Phanerozoic Biogeography.* Oxford: Oxford University Press.

Hamilton, W. 1983. Cretaceous and Cenozoic history of the northern continents. *Annals of the Missouri Botanical Garden* 70:440–458.

Hanes, T. L. 1977. California chaparral. In *Terrestrial Vegetation of California,* ed. M. G. Barbour and J. Major, pp. 417–469. New York: John Wiley and Sons.

Hardham, C. B. 1962. The Santa Lucia *Cupressus sargentii* groves and their associated northern hydrophilous and endemic species. *Madroño* 16:173–178.

Harper, J. L. 1967. A Darwinian approach to plant ecology. *Journal of Ecology* 55:247–270.

———. 1977. *Population Biology of Plants.* London and New York: Academic Press.

———. 1981. The meanings of rarity. In *The Biological Aspects of Rare Plant Conservation,* ed. H. Synge, pp. 189–203. New York: John Wiley and Sons.

Harper, K. T., L. L. St. Clair, K. H. Thorne, and W. M. Hess, eds. 1994. *Natural History of the Colorado Plateau and Great Basin.* Niwot: University Press of Colorado.

Harris, S. 1980. *Fire and Ice: The Cascade Volcanoes.* Seattle: The Mountaineers.

Harrison, A. E. 1954. Fluctuations of the Nisqually Glacier Mt. Rainier, Washington, during the last two centuries. *Association Internationale d'Hydrologie* 39:506–510.

Harrison, S. 1997. How natural habitat patchiness affects the distribution of diversity in California serpentine chaparral. *Ecology* 78:1898–1906.

———. 1999a. Local and regional diversity in a patchy landscape: native, alien, and endemic herbs on serpentine. *Ecology* 80:70–80.

———. 1999b. Native and alien species diversity at the local and regional scales in a grazed grassland. *Oecologia* 121:99–106.

———. 1999c. Population persistence and community diversity in a naturally patchy landscape: plants on serpentine soils. In *The Biology of Biodiversity,* ed. M. Kato. Tokyo and New York: Springer-Verlag.

Harshberger, J. W. 1911. *Phytogeographic Survey of North America.* Leipzig: W. Engelman; 1958 (reprint). New York: Hafner.

Hedberg, O. 1970. Evolution of the Afroalpine flora. *Biotropica* 2:16–23.

Henderson, L. J. 1913. *The Fitness of the Environment.* Boston: Beacon Press.

Herak, H., and V. T. Stringfield. 1972. *Karst: Important Karst Regions of the Northern Hemisphere.* Amsterdam: Elsevier.

Heydenrych, B., C. Willis, and C. Burgers. 1994. Limestone fynbos: unique, useful and under-conserved. *Veld and Flora* 80:68–72.

Hickman, J. C., ed. 1993. *The Jepson Manual: Higher Plants of California.* Berkeley: University of California Press.

Holland, R. F., and S. K. Jain. 1981. Insular biogeography of vernal pools in the central valley of California. *American Naturalist* 117:24–37.

Holland, S. S. 1961. Jade in British Columbia. Annual Report, Minister of Mines and Petroleum Resources, Province of British Columbia, pp. 118–126.

Hotz, P. E. 1964. Nickeliferous laterites in southwestern Oregon and northwestern California. *Economic Geology* 59:354–397.

Howard, J. A., and C. W. Mitchell. 1985. *Phytogeomorphology.* New York: John Wiley and Sons.

Howard, R. A. 1955. The vegetation of Beata and Alta Vela islands, Hispaniola. *Journal of Arnold Arboretum* 36:209–240.

Howard, R. A., and W. R. Briggs. 1953. The vegetation on coastal dogtooth limestone in southern Cuba. *Journal of Arnold Arboretum* 34:88–96.

Howard, R. A., and G. R. Proctor. 1957. The vegetation on bauxitic soils in Jamaica. *Journal of Arnold Arboretum* 38:1–50, 151–169.

Huber, O., ed. 1992. *El Macizo del Chimantá: Escudo de Guyana.* Caracas: Oscar Todtmann Editores.

Huggett, R. J. 1995. *Geoecology: An Evolutionary Approach.* London and New York: Routledge.

Humboldt, A. von. 1849. *Cosmos: A Sketch of a Physical Description of the Universe,* vol. 2. Trans. E. C. Otté. London: H. G. Bohn.

Humboldt, A. von, and A. Bonpland. 1807. *Ideen zu einer Geographie der Pflanzen.* Leipzig: Akademische Verlag, 1960 (reprint).

Hunt, C. B. 1972. *Geology of Soils and Their Evolution.* San Francisco: W. H. Freeman and Company.

Huntoon, P. W., G. H. Billingsley, Jr., and W. J. Breed. 1982. Geologic Map of Canyonlands National Park and Vicinity, Utah. Moab, Utah: Canyonlands Natural History Association.

Huxley, A., and W. Taylor. 1977. *Flowers of Greece and the Aegean.* London: Chatto and Windus.

International Union for the Conservation of Nature (IUCN). 1977–85. *World Directory of National Parks and Other Protected Areas,* vols. 1 and 2. Morges, Switzerland: IUCN, United Nations Secretariat.

Ives, J. D., and R. G. Barry, eds. 1974. *Arctic and Alpine Environments.* London: Methuen.

Jaffré, T. 1980. *Etude écologique du peuplement végétal des sols dérivées de roches Ultrabasiques en Nouvelle Calédonie.* Paris: ORSTOM.

Jaffré, T., M. Latham, and M. Schmid. 1977. Aspects de l'influence de l'extraction du mineral de nickel sur la végétation et les sols en Nouvelle Calédonie. *Cah. ORSTOM, ser. Biologique* 12:307–321.

Jaffré, T., P. Morat, J. M. Veillon, and H. S. MacKee. 1987. Changements dans la végétation de la Nouvelle Calédonie ou cours du Tertiaire: la végétation et la flora des roches ultrabasiques. *Adansonia* 4:365–391.

Jaffré, T., R. D. Reeves, and T. Becquer, eds. 1997. *The Ecology of Ultramafic and Metalliferous Areas.* Documents Scientifiques et Techniques, no. III2. Noumea, New Caledonia: Centre ORSTOM de Noumea.

Jeffrey, D. W. 1987. *Soil-Plant Relationships: An Ecological Approach.* Portland, Ore.: Timber Press.

Jennings, J. N. 1985. *Karst Geomorphologg.* Oxford: Basil Blackwell.

Jenny, H. 1941. *Factors of Soil Formation.* New York: McGraw-Hill.

———. 1980. *The Soil Resource: Origin and Behavior.* Ecological Studies, no. 37. New York: Springer-Verlag.

Jenny, H., R. J. Arkley, and A. M. Schultz. 1969. The pygmy forest-podsol ecosystem and its dune associates of the Mendocino coast. *Madroño* 20:64–74.

Johnson, A. W., and J. G. Packer. 1965. Polyploidy and environment in arctic Alaska. *Science* 148:237–239.

Johnston, I. M. 1941. Gypsophily among Mexican desert plants. *Journal of Arnold Arboretum* 22:145–170.

Jokerst, J. D. 1991. A revision of *Acanthomintha obovata* (Lamiaceae) and a key to the taxa of *Acanthomintha. Madroño* 38:278–286.

Jones, G. N. 1936. *A Botanical Survey of the Olympic Peninsula, Washington.* University of Washington Publications in Biology, vol. 5.

Jordan, C. F. 1988. The tropical rain forest landscape. In *Biogeomorphology,* ed. H. A. Viles, pp. 145–165. New York: Basil Blackwell.

Just, T. 1947. Geology and plant distribution. *Ecological Monographs* 17:129–137.

Kabata-Pendias, A., and H. Pendias. 1984. *Trace Elements in Soils and Plants.* Boca Raton, Fla.: CRC Press.

Keener, C. S. 1983. Distribution and biohistory of the endemic flora of the mid-Appalachian shale barrens. *Botanical Review* 49:65–115.

Kerner von Marilaun, A., and F. W. Oliver. 1902–1903 [1888–1891]. *The Natural History of Plants,* vols. 1 and 2. London: Blackie and Son, Ltd.

Kilmer, V. J. 1982. *Handbook of Soils and Climate in Agriculture.* Boca Raton, Fla.: CRC Press.

Kinzel, H. 1982. *Pflanzenökologie und Mineralstoffwechsel.* Stuttgart: Eugen Ulmer.

———. 1983. Influence of limestone, silicates and soil pH on vegetation. In *Physiological Plant Ecology,* vol. 3, ed. O. L. Lange, P. S. Nobel, C. B. Osmond, and H. Zeigler, pp. 201–244. Berlin: Springer-Verlag.

Klemmendson, J. O., and H. Jenny. 1966. Nitrogen availability in California soils in relation to precipitation and parent material. *Soil Science* 102:215–222.

Knopf, A. 1906. An alteration of Coast Range serpentine. *University of California Publications in Geology* 4(18):425–430.

Krause, W. 1958. Andere Bodenspecialisten. In *Handbuch der Pflanzenphysiologie,* vol. 4, ed. G. Michael, pp. 758–806. Berlin: Springer-Verlag.

Kruckeberg, A. R. 1951. Intraspecific variability in response of certain native plant species to serpentine soil. *American Journal of Botany* 38:408–419.

———. 1954. The ecology of serpentine soils: A symposium. III. Plant species in relation to serpentine soils. *Ecology* 35:267–274.

———. 1957. Variation in fertility of hybrids between isolated populations of the serpentine species, *Streptanthus glandulosus* Hook. *Evolution* 11:185–211.

———. 1958. The taxonomy of the species complex, *Streptanthus glandulosus* Hook. *Madroño* 14:217–227.

———. 1959. Ecological and genetic aspects of metallic ion uptake by plants and their possible relation to wood preservation. In *Marine Boring and Fouling Organisms,* ed. D. L. Ray, pp. 526–536. Seattle: University of Washington Press.

———. 1964. Ferns associated with ultramafic rocks in the Pacific Northwest. *American Fern Journal* 54:113–126.

———. 1967. Ecotypic response to ultramafic soils by some plant species of northwestern United States. *Brittonia* 19:133–151.

———. 1969a. Soil diversity and the distribution of plants with examples from western North America. *Madroño* 20:129–154.

———. 1969b. Plant life on serpentinite and other ferromagnesian rocks in northwestern North America. *Syesis* 2:15–114.

———. 1985. *California Serpentines: Flora, Vegetation, Geology, Soils, and Management Problems.* University of California Publications in Botany, vol. 78. Berkeley: University of California Press.

———. 1986. An essay: The stimulus of unusual geologies for plant speciation. *Systematic Botany* 11:455–463.

———. 1987a. Plant life on Mount St. Helens before 1980. In *Mount St. Helens 1980,* ed. D. Bilderback, pp. 3–23. Berkeley: University of California Press.

———. 1987b. Serpentine endemism and rarity. In *Conservation and Management of Rare and Endangered Plants,* ed. T. S. Elias, pp. 121–128. Sacramento: California Native Plant Society.

———. 1991a. *Natural History of Puget Sound Country.* Seattle: University of Washington Press.

———. 1991b. An essay: Geoedaphics and island biogeography for vascular plants. *Aliso* 13:225–238.

———. 1992. Plant life of western North American ultramafics. In *The Ecology of Areas with Serpentinized Rocks: A World View,* ed. B. A. Roberts and J. Proctor, pp. 31–73. Dordrecht: Kluwer Academic Publishers.

———. 1993. Serpentine biota of western North America. In *The Vegetation of Ultramafic (Serpentine) Soils,* ed. A. J. M. Baker, J. Proctor, and R. D. Reeves, pp. 19–33. Andover, UK: Intercept.

———. 1995. Ecotypic variation in response to serpentine soils. In *Genecology and Ecogeographic Races,* ed. A. K. Kruckeberg, R. B. Walker, and A. Leviton, pp. 57–66. Washington, D.C.: American Association for the Advancement of Science.

———. 1997. Serpentine and its plant life in California. In *California's Wild Garden: A Living Legacy,* ed. P. M. Faber, pp. 4, 5. Sacramento: California Native Plant Society.

———. 1999. Serpentine barrens of western North America. In *Savannas, Barrens, and Rock Outcrop Plant Communities of North America,* ed. R. C. Anderson, J. S. Fralish, and J. M. Baskin, pp. 309–321. Cambridge: Cambridge University Press.

Kruckeberg, A. R., N. Adiguzel, and R. Reeves. 1999. Glimpses of the flora and ecology of Turkish (Anatolian) serpentines. *Karaca Arboretum Magazine* 5:67–86.

Kruckeberg, A. R., and A. L. Kruckeberg. 1989. Endemic metallophytes: their taxonomic, genetic and evolutionary attributes. In *Heavy Metal Tolerance in Plants: Evolutionary Aspects,* ed. A. J. Shaw, pp. 301–312. Boca Raton, Fla.: CRC Press.

Kruckeberg, A. R., and J. L. Morrison. 1983. New *Streptanthus* taxa (Cruciferae) from California. *Madroño* 30:230–244.

Kruckeberg, A. R., P. J. Peterson, and Y. Samiullah. 1993. Hyperaccumulation of nickel by *Arenaria rubella* (Caryophyllaceae) from Washington State. *Madroño* 40:25–30.

Kruckeberg, A. R., and D. Rabinowitz. 1985. Biological aspects of rarity in higher plants. *Annual Reviews of Ecology and Systematics* 16:447–479.

Kruckeberg, A. R., and R. D. Reeves. 1995. Nickel accumulation by serpentine species of *Streptanthus* (Brassicaceae): field and greenhouse studies. *Madroño* 42:458–469.

Kruckeberg, A. R., and W. Tanaka. 1983. Goat Rocks (Annual High Country Trip). *Douglasia* 7(4):2.

Kruckeberg, A. R., R. B. Walker, and A. Leviton, eds. 1995. *Ecogeographic Races—Turesson to the Present—A Symposium.* San Francisco: Pacific Division, American Association for the Advancement of Science.

Landolt, E. 1960. *Unsere Alpenflora.* Zurich: Zollikon-Zurich Schweizer Alpen Club.

Lang, F. A., and P. F. Zika. 1997. A nomenclatural note on *Hastingsia bracteosa* and *H. atropurpurea* (Liliaceae). *Madroño* 44:189–192.

Lee, W. G. 1980. Ultramafic plant ecology of the South Island, New Zealand. Ph.D. dissertation, University of Otago.

———. 1992. The serpentine areas of New Zealand, their structure and ecology. In *The Ecology of Areas with Serpentinized Rocks: A World View,* ed. B. A. Roberts and J. Proctor, 375–417. Dordrecht: Kluwer Academic Publishers.

Lee, W. G., and A. E. Hewitt. 1981. Soil changes associated with development of vegetation on an ultramafic scree, northwest Otago, New Zealand. *Journal of the Royal Society of New Zealand* 12:229–242.

Lee, W. G., A. F. Mark, and J. B. Wilson. 1983. Ecotypic differentiation in the ultramafic flora of the South Island, New Zealand. *New Zealand Journal of Botany* 21:141–156.

Legget, R. F. 1973. *Cities and Geology.* New York: McGraw Hill.

Lewis, H. 1962. Catastrophic selection as a factor in speciation. *Evolution* 16: 257–271.

Lind, E. M., and M. E. S. Morrison. 1974. *East African Vegetation.* London: Longman.

Link, H. F. 1789. Florae Gottingensis specimen, sistens vegetabilia saxo calcareo propria. In *Usteri's Delectus Opusculorum Botanicorum,* vol. 1, pp. 299–336.

Lloyd, R. M., and R. S. Mitchell. 1973. *A Flora of the White Mountains, California and Nevada.* Berkeley: University of California Press.

Lousley, J. E. 1950. *Wild Flowers of Chalk and Limestone.* London: Collins.

Lutz, H. J. 1958. Geology and soil in relation to forest vegetation. In *First North American Forest Soils Conference Proceedings,* pp. 75–85. East Lansing: Agricultural Experiment Station, Michigan State University.

Mabberley, D. J. 1987. *The Plant-book.* Cambridge: Cambridge University Press.

MacArthur, R. H., and E. O. Wilson. 1967. *The Theory of Island Biogeography.* Princeton: Princeton University Press.

Mack, R. N. 1981. Initial effects of ashfall from Mount St. Helens on vegetation in eastern Washington and adjacent Idaho. *Science* 213:537–539.

Macnair, M. R. 1983. The genetic control of copper tolerance in the yellow monkey flower *Mimulus guttatus. Heredity* 50:283–293.

———. 1989. A new species of *Mimulus* endemic to copper mines in California. *Botanical Journal of the Linnaean Society* 100:1–14.

———. 1993. Tansley Review No. 49: The genetics of metal tolerance in vascular plants. *New Phytologist* 124:541–559.

Macnair, M. R., and Q. J. Cumbes. 1989. The genetic architecture of interspecific variation in *Mimulus. Genetics* 122:211–222.

Major, J. 1951. A functional, factorial approach to plant ecology. *Ecology* 32:392–412.

———. 1988. Endemism: a botanical perspective. In *Analytical Biogeography,* ed. A. A. Myers and P. S. Giller, pp. 117–146. London: Chapman and Hall.

Major, J., and S. A. Bamberg. 1963. Some cordilleran plant species new for the Sierra Nevada of California. *Madroño* 17:93–109.

Malloy, L. 1983. How much longer before Red Mountain is protected? *Forest and Bird* 14(6): 17–24.

Malpas, J. 1992. Serpentine and the geology of serpentinized rocks. In *The Ecology of Areas with Serpentinized Rocks,* ed. B. A. Roberts and J. Proctor, pp. 11–30. Dordrecht: Kluwer Academic Publishers.

Mani, M. S. 1978. *Ecology and Phytogeography of High Altitude Plants of the Northeast Himalaya: Introduction to High Altitude Botany.* London: Chapman and Hall.

Marbut, C. F. 1928. A scheme for soil classification. Proceedings and Papers of the First International Congress of Soil Science, Washington, D.C., 1927.

Martin, W. E., J. Vlamis, and N. W. Stice. 1953. Field correction of calcium deficiency on a serpentine soil. *Agronomy Journal* 45:204–208.

Martin, W. H., S. G. Boyce, and A. C. Echternacht, eds. 1993. *Biodiversity of the Southeastern United States*. Vol. 2, *Upland Terrestrial Communities*. New York: John Wiley and Sons.

Mason, H. L. 1936. The principles of geographic distribution as applied to floral analysis. *Madroño* 3:181–190.

———. 1946a. The edaphic factor in narrow endemism. I. The nature of environmental influences. *Madroño* 8:209–226.

———. 1946b. The edaphic factor in narrow endemism. II. The geographic occurrence of plants of highly restricted patterns of distribution. *Madroño* 8:241–257.

Mast, M. A. 1992. Geochemical characteristics. In *Biogeochemistry of a Subalpine Watershed: Loch Vale Watershed,* ed. J. Baron, pp. 93–107. New York: Springer-Verlag.

Matthews, J. A. 1992. *The Ecology of Recently-Deglaciated Terrain.* Cambridge: Cambridge University Press.

Mayr, E. 1970. *Populations, Species, and Evolution.* Cambridge, Mass.: Belknap Press.

McClelland, L., T. Simkin, M. Summers, E. Nielsen, and T. Stein. 1989. *Global Volcanism, 1975–1985.* Englewood Cliffs, N.J.: Prentice-Hall.

McDougall, W. B. 1949. *Plant Ecology.* Philadelphia: Lea and Febiger.

McKenna, M. C. 1983. Holarctic landmass rearrangement, cosmic events and Cenozoic terrestrial organisms. *Annals of the Missouri Botanical Garden* 70:459–489.

McMillan, C. 1956. The edaphic restriction of *Cupressus* and *Pinus* in the Coast Ranges of central California. *Ecological Monographs* 26:177–212.

McNaughton, S. J., T. C. Folsom, T. Lee, F. Park, C. Prive, D. Roeder, J. Schmits, and C. Stockwell. 1974. Heavy metal tolerance in *Typha latifolia* without the evolution of tolerant races. *Ecology* 55:1163–1165.

McPhee, J. 1993. *Assembling California.* New York: Farrar, Straus and Giroux.

Meloy, M. 1986. Islands on the prairie: the mountain ranges of eastern Montana. Montana Geographic series, no. 13. Helena: *Montana Magazine.* 102 pp.

Menard, H. W. 1974. *Geology, Resources, and Society.* San Francisco: W. H. Freeman and Co.

Merriam, C. H. 1895. Laws of temperature control of the geographical distribution of terrestrial plants and animals. *National Geographic Magazine* 6:229.

———. 1898. Life zones and crop zones of the United States. U.S. Department of Agriculture Division. *Biological Survey Bulletin* 10:1–79.

Miklos, G. L. G. 1993. Emergence of organizational complexities during metazoan evolution: perspectives from molecular biology, paleontology and neo-Darwinism. *Memoirs of the Association of Australasian Paleontologists* 15:7–41.

Mitchell, L. 1998. Partners restore rare serpentine ecosystem. *Endangered Species Bulletin* 23:14–15. Washington, D.C.: Fish and Wildlife Service, U.S. Department of the Interior.

Mizuno, N., and S. Nosaka. 1992. The distribution and extent of serpentinized areas in Japan. In *The Ecology of Areas with Serpentinized Rocks,* ed. B. A. Roberts and J. Proctor, pp. 271–311. Dordrecht: Kluwer Academic Publishers.

Mooney, H. A. 1966. Influence of soil type on the distribution of two closely related species of *Erigeron. Ecology* 47:950–958.

Mooney, H. A., E. Fuentes, and B. I. Kronberg, eds. 1992. *Earth System Responses to Global Change: Contrasts Between North and South America.* New York: Academic Press.

Moore, D. 1983. *Flora of Tierra del Fuego.* London: Anthony Nelson.

Moral, R. del. 1972. Diversity patterns in forest vegetation of the Wenatchee Mountains, Washington. *Bulletin of the Torrey Botanical Club* 99:57–64.

———. 1974. Species patterns in the upper North Fork Teanaway River drainage, Wenatchee Mountains, Washington. *Syesis* 7:13–30.

———. 1982. Control of vegetation on contrasting substrates: herb patterns on serpentine and sandstone. *American Journal of Botany* 69:227–238.

Moral, R. del, and D. Deardorff. 1976. Vegetation of the Mima Mounds, Washington State. *Ecology* 57:520–530.

Moral, R. del, and A. F. Watson. 1978. Gradient structure of forest vegetation in the central Washington Cascades. *Vegetatio* 38:29–48.

Moral, R. del, A. F. Watson, and R. S. Fleming. 1976. Vegetation structure in the Alpine Lakes region of Washington State: classification of vegetation on granitic rocks. *Syesis* 9:291–316.

Moral, R. del, and D. M. Wood. 1988. Dynamics of herbaceous vegetation recovery on Mount St. Helens, Washington, USA, after a volcanic eruption. *Vegetatio* 74:11–27.

Morony, J. 1999. Andean volcanism as a biogeographic determinant in South America and a new paradigm for vicariance. Manuscript.

Munz, P., and D. R. Keck. 1959. *A California Flora.* Berkeley: University of California Press.

Murdy, W. H. 1968. Plant speciation associated with granite outcrop communities of the southeastern Piedmont. *Rhodora* 70:394–407.

Myers, A. A., and P. S. Giller, eds. 1988. *Analytical Biogeography*. London: Chapman and Hall.

Nelson, E. C. 1991. *The Burren: A Companion to the Wildflowers of a Limestone Wilderness*. Kilkenny, Ireland: Boethius.

Neustruev, S. S. 1927. *Genesis of Soils*. Russian Pedological Investigations, vol. 3. Leningrad: Academy of Sciences, USSR.

Nicholls, M. K., and T. McNeilly. 1985. The performance of *Agrostis capillaris* L. genotypes differing in copper tolerance, in ryegrass swards on normal soil. *New Phytologist* 101:207–217.

Nordenskiold, E. 1935. *The History of Biology*. New York: Tudor Publishing Company.

Norris, R. M. 1995. Sand dunes of the California desert. In *The California Desert: An Introduction to Natural Resources and Man's Impact*, eds. J. Latting and P. G. Rowlands, pp. 15–25. Riverside, CA: June Latting Books.

Ollier, C. 1984. *Weathering*. 2d ed. London: Longman.

Orians, G. H., G. M. Brown, Jr., W. E. Kunin, and J. E. Swierzbinski. 1990. *The Preservation and Valuation of Biological Resources*. Seattle: University of Washington Press.

Ornduff, R. 1987. Islands on islands: plant life on the granite outcrops of Western Australia. *Harold L. Lyon Arboretum Lecture*, 15.

Parsons, R. B., and C. A. Balster. 1966. Morphology and genesis of six "red hill" soils in the Oregon Coast Range. *Soil Science Society of America Proceedings* 30:90–93.

Parsons, R. B., and R. C. Harriman. 1975. A lithosequence in the mountains of southwestern Oregon. *Soil Science Society of America Proceedings* 39:943–948.

Parsons, R. F. 1976. Gypsophily in plants: A review. *American Midland Naturalist* 96:1–20.

Pearse, R. O. 1978. *Mountain Splendour: Wild Flowers of the Drakensberg*. Cape Town: Howard Timmins.

Petersen, K. L. 1994. Modern and pleistocene climate patterns in the West. In *Natural History of the Colorado Plateau and Great Basin*, ed. K. T. Harper, L. L. St. Clair, K. H. Thorne, and W. M. Hess. Niwot: University Press of Colorado.

Pichi-Sermolli, R. 1948. Flora e vegetazione delle serpentine e delle altre ofioliti dell'alta valle del Trevere (Toscana). *Webbia*. 6:1–380.

Pielou, E. C. 1979. *Biogeography*. New York: John Wiley and Sons.

Pitty, A. F. 1971. *Introduction to Geomorphology*. London: Methuen.

Platt, R. B. 1951. An ecological study of the mid-Appalachian shale barrens and of the plants endemic to them. *Ecological Monographs* 21:269–300.

Poldini, L. 1989. *La vegetazione del Carso Isontino e Triestino*. Trieste: Edizioni Lint.

Polunin, O., and M. Walters. 1985. *A Guide to the Vegetation of Britain and Europe.* Oxford: Oxford University Press.

Porembski, S., and W. Barthlott, eds. 2000. *Inselbergs: Biotic Diversity of Isolated Rock Outcrops in Tropical and Temperate Regions.* Ecological Studies, vol. 146. Berlin and New York: Springer.

Porter, S. C. 1981. Lichenometric studies in the Cascade Range of Washington: establishment of *Rhizocarpon geographicum* growth curves at Mt. Rainier. *Arctic and Alpine Research* 13:11–23.

Price, L. W. 1981. *Mountains and Man.* Berkeley: University of California Press.

Price, M. F. 1995. *Mountain Research in Europe.* Man and The Biosphere Series, vol. 14. UNESCO, Paris, and Carnforth, UK: Parthenon Publishing Group Ltd.

Proctor, G. R. 1984. *Flora of the Cayman Islands.* Kew Bull. Addit. Series XI.

———. 1986. Cockpit country and its vegetation. In *Forests of Jamaica,* ed. D. A. Thompson, P. K. Bretting, and M. Humphries, pp. 43–47, 140–143. Kingston: Jamaican Society of Scientists and Technologists.

Proctor, J. 1992a. Chemical and ecological studies on the vegetation of ultramafic sites in Britain. In *The Ecology of Areas with Serpentinized Rocks: A World View,* ed. B. A. Roberts and J. Proctor, pp. 135–168. Dordrecht: Kluwer Academic Publishers.

———. 1992b. The vegetation over ultramafic rocks in the tropical Far East. In *The Ecology of Areas with Serpentinized Rocks: A World View,* ed. B. A. Roberts and J. Proctor, pp. 249–270. Dordrecht: Kluwer Academic Publishers.

Proctor, J., and M. M. Cole. 1992. The ecology of ultramafic areas in Zimbabwe. In *The Ecology of Areas with Serpentinized Rocks: A World View,* ed. B. A. Roberts and J. Proctor, pp. 313–331. Dordrecht: Kluwer Academic Publishers.

Proctor, J., and S. R. J. Woodell. 1975. The ecology of serpentine soils. *Advances in Ecologic Research* 9:255–365.

Pye, K., and H. Tsoar. 1990. *Aeolian Sand and Sand Dunes.* London and Boston: Unwin Hyman.

Quarterman, E., M. P. Burbanck, and J. S. Fralish. 1993. Rock outcrop communities: limestone, sandstone and granite. In *Biodiversity of the Southeastern United States, Upland Terrestrial Communities,* vol. 2, ed. W. H. Martin, S. G. Boyce, and A. C. Echternacht, pp. 35–86. New York: John Wiley and Sons.

Rambler, M. B., L. Margulis, and R. Fester. 1989. *Global Ecology: Towards a Science of the Biosphere.* Boston: Academic Press/Harcourt Brace Jovanovich.

Ramp, L. 1961. Chromite in southwestern Oregon. Portland: Bulletin 52, Department of Geology and Mineral Industries, State of Oregon.

Raunkiaer, C. 1934. *The Life Forms of Plants and Statistical Plant Geography.* Oxford: Clarendon Press.

Raup, D. 1986. *The Nemesis Affair.* New York: Norton.

Raven, P. H. 1964. Catastrophic selection and edaphic endemism. *Evolution* 18:336–338.

———. 1972. An introduction to continental drift. *Australian Natural History,* December 1972, pp. 245–248.

Raven, P. H., and D. I. Axelrod. 1972. Plate tectonics and Australasian paleobiogeography. *Science* 176:1379–1386.

———. 1974. Angiosperm biogeography and past continental movements. *Annals of the Missouri Botanical Garden* 61:539–673.

———. 1975. History of the flora and fauna of Latin America. *American Scientist* 63:420–429.

———. 1978. Origin and relationships of the California flora. *University of California Publications in Botany* 72:1–134.

Reeves, R. D., A. J. M Baker, A. Borhidi, and R. Berazain. 1996. Nickel-accumulating plants from the ancient serpentine soils of Cuba. *New Phytologist* 113:217–224.

Reeves, R. D, A. J. M. Baker, A. Borhidi, and R. Berazain. 1999. Nickel hyperaccumulation in the serpentine flora of Cuba. *Annals of Botany* 83:29–38.

Reeves, R. D., R. R. Brooks, and T. R. Dudley. 1983. Uptake of nickel by species of *Alyssum, Bornmuellera,* and other genera of Old World Tribus Alyssae. *Taxon* 32:184–192.

Reeves, R. D., R. R. Brooks, and R. M. MacFarlane. 1981. Nickel uptake by Californian *Streptanthus* and *Caulanthus* with particular reference to the hyperaccumulator, *S. polygaloides* Gray (Brassicaceae). *American Journal of Botany* 68: 708–712.

Reeves, R. D., A. R. Kruckeberg, N. Adiguzel, and U. Kramer. 2001. Studies on the flora of serpentine and other metalliferous areas of western Turkey. *African Journal of Science* (in press).

Reiners, W. A., I. A. Worley, and D. B. Lawrence. 1971. Plant diversity in a chronosequence at Glacier Bay, Alaska. *Ecology* 52:55–69.

Ritter-Studnicka, H. 1968. Die Serpentinomorphosen der Flora Bosniens. *Botanische Jahrbuch* 88:443–465.

Roberts, B. A. 1992. Ecology of serpentinized areas, Newfoundland, Canada. In *The Ecology of Areas with Serpentinized Rocks: A World View,* ed. B. A. Roberts and J. Proctor, pp. 75–113. Dordrecht: Kluwer Academic Publishers.

Roberts, B. A., and J. Proctor, eds. 1992. *The Ecology of Areas with Serpentinized Rocks: A World View.* Dordrecht: Kluwer Academic Publishers.

Ross, R. M., and W. D. Allmon, eds. 1990. *Causes of Evolution: A Paleontological Perspective.* Chicago: University of Chicago Press.

Rune, O. 1953. Plant life on serpentines and related rocks in the north of Sweden. *Acta Phytogeographica Suecica* 31:1–139.

———. 1954. Notes on the flora of the Gaspé Peninsula. *Svensk Botanisk Tidskrift* 48:117–138.

Sauer, J. 1988. *Plant Migration: The Dynamics of Geographic Patterning in Seed Plant Species.* Berkeley: University of California Press.

Schimper, A. F. W. 1903. *Plant-Geography upon a Physiological Basis.* Oxford: Clarendon Press.

Schuster, R. M. 1976. Plate tectonics and its bearing on the geological origin and dispersal of angiosperms. In *Origin and Early Evolution of Angiosperms,* ed. C. B. Beck, pp. 48–136. New York: Columbia University Press.

Scoggan, H. J. 1950. The flora of Bic and The Gaspé Peninsula, Quebec. National Museum of Canada Bulletin, no. 115 (Biological Series, no. 39). Ottawa: Edmond Cloutier.

Sellers, P. 1991. Modelling and observing land-surface-atmosphere interactions on large scales. In *Land Surface–Atmosphere Interactions for Climate Modelling,* ed. E. F. Wood, pp. 85–114. Dordrecht: Kluwer Academic Publishers.

Sharsmith, H. 1961. The genus *Hesperolinon* (Linaceae). *University of California Publications in Botany* 32:235–314.

Shaw, A. J., ed. 1989. *Heavy Metal Tolerance in Plants: Evolutionary Aspects.* Boca Raton, Fla.: CRC Press.

Shimizu, T. 1962–1963. Studies on the limestone flora of Japan and Taiwan. Part I, Part II. *Journal of the Faculty of Textile Science and Technology,* Shinshu University, 30 (Series A, Biology) 11:1–105; 12:1–88.

Shiro, T., J. H. Titus, and R. del Moral. 1997. Seedling establishment patterns on the Pumice Plain, Mount St. Helens, Washington. *Journal of Vegetation Science* 8:727–734.

Sigafoos, R. S., and E. L. Hendricks. 1969. The time interval between stabilization of alpine glacial deposits and establishment of tree seedlings. *U.S. Geological Survey Professional Paper* 650-B:B89–B93.

Simkin, T., and R. S. Fiske. 1983. *Krakatau 1883: The Volcanic Eruption and Its Effects.* Washington, D.C.: Smithsonian Institution Press.

Simpson, B. 1966. *Rocks and Minerals.* Oxford: Pergamon Press.

Singer, C. 1950. *A History of Biology.* Rev. ed. New York: Henry Schuman.

Sirois, L., and M. M. Grandtner. 1992. A phyto-ecological investigation of the Mount Albert serpentinized plateau. In *The Ecology of Areas with Serpentinized Rocks,* ed. B. A. Roberts and J. Proctor, pp. 115–133. Dordrecht: Kluwer Academic Publishers.

Skinner, B. J., and S. C. Porter. 1992. *The Dynamic Earth: An Introduction to Physical Geology.* 2d ed. New York: John Wiley and Sons.

Smathers, G. A., and D. Mueller-Dombois. 1974. *Invasion and Recovery of Vegetation after a Volcanic Eruption in Hawaii.* National Park Service Scientific Monograph 5. Washington, D.C.: Superintendent of Documents, U.S. Government Printing Office.

Snaydon, R. W. 1962. The growth and competitive ability of contrasting natural populations of *Trifolium repens* L. on calcareous and acid soils. *Journal of Ecology* 50:439–447.

Soil Survey Staff. 1960. *Soil Classification, A Comprehensive System,* 7th approx. Soil Conservation Service, U.S. Dept. of Agriculture. Washington, D.C.: Government Printing Office.

———. 1975. *Soil Taxonomy.* USDA Soil Conservation Service Agricultural Handbook 436. Washington, D.C.: Superintendent of Documents.

———. 1978. *Soil Survey of Napa County, California.* USDA Soil Conservation Service. Washington, D.C.: Superintendent of Documents, U.S. Government Printing Office.

Sommers, S. 1983: Edgewood Park background information. Unpublished report. Edgewood Park Committee, Santa Clara Valley Chapter, California Native Plant Society, San Mateo, California.

Sparks, B. W. 1971. *Rocks and Relief.* London: Longman Group Ltd.

Spellenberg, R. W. 1968. *Biosystematic studies in Panicum Group Lanuginosa, from the Pacific Northwest.* Ph.D. dissertation, University of Washington.

Stålfelt, M. G. 1960. *Plant Ecology.* English translation by P. Jarvis and M. Jarvis. New York: Halsted Press, John Wiley and Sons.

Starkel, L. 1976. The role of extreme (catastrophic) meterological events in contemporary evolution of slopes. In *Geomorphology and Climate,* ed. E. Derbyshire, pp. 203–246. London and New York: John Wiley and Sons.

Stebbins, G. L., Jr. 1950. *Variation and Evolution in Plants.* New York: Columbia University Press.

———. 1971. *Chromosomal Evolution in Higher Plants.* Reading, Mass.: Addison-Wesley.

———. 1982. *Darwin to DNA, Molecules to Humanity.* San Francisco: W. H. Freeman and Co.

Stebbins, G. L., Jr., and J. Major. 1965. Endemism and speciation in the California flora. *Ecological Monographs* 35:1–35.

Stork, A. 1963. Plant immigration in front of retreating glaciers with examples from the Kebnekajse area, northern Sweden. *Geografiska Annaler* 45:1–22.

Strain, B. R., and W. D. Billings, eds. 1974. *Vegetation and Environment.* The Hague: Dr. W. Junk b.v. Publ.

Strid, A. 1980. *Wildflowers of Mount Olympus.* Kifissia, Greece: Goulandris Natural History Museum.

———. 1989. Endemism and speciation in the Greek flora. In *The Davis and Hedge Festschrift,* ed. K. Tan, pp. 27–44. Edinburgh: Edinburgh University Press.

———. 1993. Phytogeographical aspects of the Greek mountain flora. *Fragmenta Floristica et Geobotanica Supplementum* 2:411–433.

Tansley, A. G. 1917. On competition between *Galium saxatile* L. (*G. hercynicum* Weig.) and *Galium sylvestre* Poll. (*G. asperum* Schreb.) on different types of soil. *Journal of Ecology* 5:173–179.

Tansley, A. G., and T. F. Chipp, eds. 1926. *Aims and Methods in the Study of Vegetation.* London: The British Empire Vegetation Committee and The Crown Agents for the Colonies.

Tatic, B., and V. Veljovic. 1992. Distribution of serpentinized massives on the Balkan peninsulas and their ecology. In *The Ecology of Areas with Serpentinized Rocks,* ed. B. A. Roberts and J. Proctor, pp. 199–215. Dordrecht: Kluwer Academic Publishers.

Taylor, D. W. 1990. Ecology and life history of the San Benito evening primrose (*Camissonia benitensis*). Sacramento: Bureau of Land Management; Santa Cruz: Biosystems Analysis, Inc.

———. 1993. Ecology and life history of the San Benito evening primrose (*Camissonia benitensis*). 1992 Final Supplemental Report. Santa Cruz: Biosystems Analysis, Inc.

Theophrastus. *De Causis Plantarum.* 3 vols. (1976–1990). Cambridge: Harvard University Press.

Thomas, D. S. 1988. The biogeomorphology of arid and semi-arid environments. In *Biogeomorphology,* ed. H. A. Viles, pp. 222–254. New York: Basil Blackwell.

Thomas, M. F. 1994. *Geomorphology in the Tropics: A Study of Weathering and Denudation in Low Latitudes.* Chichester, UK: John Wiley and Sons, Ltd.

Thompson, D. A., F. K. Bretting, and M. Humphreys. 1986. *Forests of Jamaica.* Kingston: Jamaican Society of Scientists and Technologists.

Thompson, W. F. 1962. Cascade alp slopes and Gipfelfluren as clima-geomorphic phenomena. *Erdkunde* 16(2).

———. 1964. How and why to distinguish between mountains and hills. *Professional Geographer* 16:6–8.

———. 1990. Climate-related landscapes in world mountains. *Annals of Geomorphology* 78 (Suppl.):1–92.

Thrower, N. J. W., and D. E. Bradbury, eds. 1977. *Chile-California Mediterranean Scrub Atlas: A Comparative Analysis.* Stroudsburg, Penn.: Dowden, Hutchinson and Ross.

Thurmann, J. 1849. *Essai de phytostatique appliquée à la Châine du Jura.* Berne, Switzerland.

Tiffney, B. H., ed. 1985. *Geological Factors and the Evolution of Plants.* New Haven: Yale University Press.

Tiffney, B. H., and K. J. Niklas. 1990. Continental area, dispersion, latitudinal distribution, and topographic variety: a test of correlation with terrestrial plant diversity. In *Causes of Evolution,* ed. R. M. Ross and W. D. Allmon, pp. 76–102. Chicago: University of Chicago Press.

Titus, J., and R. del Moral. 1998. Seedling establishment in different microsites on Mount St. Helens, Washington, USA. *Plant Ecology* 134:13–26.

Tricart, J. 1965. *Principes et méthodes de la géomorphologie.* Paris: Masson.

Trimble, S. 1989. *Sagebrush Ocean: The Natural History of the Great Basin.* Reno: University of Nevada Press.

Turesson, G. 1914. Slope exposure as a factor in the distribution of *Pseudotsuga taxifolia* in arid parts of Washington. *Bulletin of the Torrey Botanical Club* 41:337–345.

Turner, B. L. 1973. *Machaeranthera restiformis* (Asteraceae): a bizarre new gypsophile from northcentral Mexico. *American Journal of Botany* 60:836–838.

Turner, B. L., and A. M. Powell. 1979. Deserts, gypsum and endemism. In *Arid Land Plant Resources: Proceedings of the International Arid Lands Conference on Plant Resources,* ed. J. R. Goodin and D. K. Northington, pp. 96–116. Lubbock: International Center for Arid and Semi-arid Land Studies, Texas Tech University.

Turrill, W. B. 1929. *The Plant Life of the Balkan Peninsula: A Phytogeographical Study.* Oxford: Clarendon Press.

Tutin, T. G., et al., eds. 1980. *Flora Europaea.* Vols. 1–5. Cambridge: Cambridge University Press.

Tylecote, R. F. 1987. *The Early History of Metallurgy in Europe.* London: Longman.

Tyndall, R. W. 1994. Preface. Contributions of the Barrens Symposium, 1993. *Castanea* 59:182–183.

Tyrrell, G. W. 1929. *The Principles of Petrology.* New York: E. P. Dutton.

Unger, F. 1836. *Über den Einfluss des Bodens auf die Verteilung der Gewächse.* Vienna: Rohrmann und Schweigerd.

Vernano-Gambi, O. 1992. The distribution and ecology of the vegetation of ultramafic soils in Italy. In *The Ecology of Areas with Serpentinized Rocks,* ed. B. A. Roberts and J. Proctor, pp. 217–247. Dordrecht: Kluwer Academic Press.

Viles, H. A., ed. 1988. *Biogeomorphology.* New York: Basil Blackwell.

Vlamis, J. 1949. Growth of lettuce and barley as influenced by degree of Ca saturation of soil. *Soil Science* 67:453–466.

Vlamis, J., and H. Jenny. 1948. Calcium deficiency in serpentine soils as revealed by absorbent technique. *Science* 107:549–551.

Wagner, D. 1979. Systematics of *Polystichum* in Western North America north of Mexico. Pteridologia No. 1. American Fern Society Publication.

Wahlenberg, G. 1814. *Flora Carpatorum.* (Cited in Gigon 1971.)

Walker, R. B. 1948. Molybdenum deficiency in serpentine barren soils. *Science* 108:473–475.

———. 1954. Factors affecting plant growth on serpentine soils. *Ecology* 35:258–266.

Walker, R. B., H. W. Walker, and P. R. Ashworth. 1955. Calcium-magnesium nutrition with special reference to serpentine soils. *Plant Physiology* 30:214–221.

Wallace, D. R. 1983. *The Klamath Knot.* San Francisco: Sierra Club Books.

Walley, K. A., M. S. I. Khan, and A. D. Bradshaw. 1974. The potential for evolution of heavy metal tolerance in plants. I. Copper and zinc tolerance in *Agrostis tenuis. Heredity* 32:309–319.

Walter, H. 1979. *Vegetation of the Earth and Ecological Systems of the Biosphere.* 2d ed. Berlin: Springer-Verlag.

Walters, T. W., and R. Wyatt. 1982: The vascular flora of granite outcrops in the central mineral region of Texas. *Bulletin of the Torrey Botanical Club* 109: 344–364.

Wardle, P. 1973. New Zealand timberlines. *Arctic and Alpine Research* 5:A127–A135.

Warming, E. 1906. *Oecology of Plants: An Introduction to the Study of Plant Communities.* English translation by P. Groom and I. B. Balfour. London: Oxford University Press.

Washburn, A. L. 1980. *Geocryology.* New York: Halsted Press, John Wiley and Sons.

Waterfall, U. T. 1946. Observations on the desert gypsum flora of southwestern Texas and adjacent New Mexico. *American Midland Naturalist* 36:456–466.

Watson, H. C. 1833. Observations on the affinities between plants and subjacent rocks. *Magazine of Natural History* 6:424–427.

Weaver, J. E., and F. E. Clements. 1938. *Plant Ecology.* 2d ed. New York: McGraw-Hill.

Webb, D. A., and M. J. P. Scannell. 1983. *Flora of Connemara and The Burren.* Cambridge: Royal Dublin Society and Cambridge University Press.

Weinmann, F., M. Boule, K. Brunner, J. Malek, and V. Yoshino. 1984. Wetland plants of the Pacific Northwest. Seattle: U.S. Army Corps of Engineers.

Welch, T. G., and J. O. Klemmendson. 1973. Influence of the biotic factor and parent material on distribution of nitrogen and carbon in ponderosa pine ecosystems. In *Forest Soils and Forest Land Management,* ed. B. Bernier and C. H. Winget, pp. 159–178. Quebec: Laval University Press.

Werger, M. J. A. 1978. *Biogeography and Ecology of Southern Africa.* Monographiae Biologicae, vol. 31. Netherlands: Junk.

West, N. E. 1988. Intermountain deserts, shrub steppes, and woodlands. In *North American Terrestrial Vegetation,* ed. M. Barbour and D. E. Billings, pp. 209–230. New York: Cambridge University Press.

Westbroek, P. 1991. *Life as a Geologic Force.* New York: W. W. Norton.

Westman, W. E. 1979. Californian coastal forest heathlands. In *Heathlands and Related Shrublands of the World: A Descriptive Study,* ed. R. L. Secht, pp. 465–470. Amsterdam: Elsevier.

Wharton, C. H. 1989. *The Natural Environments of Georgia.* Bulletin 114. Atlanta Geologic Survey Branch, Georgia Department of Natural Resources.

White, P. S., R. I. Miller, and G. S. Ramseur. 1984. The species-area relationship of the southern Appalachian high peaks: vascular plant richness and rare plant distributions. *Castanea* 49:47–61.

White, W. B. 1988. *Geomorphology and Hydrology of Karst Terrains.* New York: Oxford University Press.

Whitney, J. D. 1865. Geology. In *Geological Survey of California,* vol. 1. Sacramento: Legislature of California.

Whittaker, R. H. 1954a. Plant populations and the basis of plant indication. *Angewandte Pflanzensoziologie, Festschrift Aichinger* 1:183–206.

———. 1954b. The vegetational response to serpentine soils. *Ecology* 35:275–288.

———. 1960. Vegetation of the Siskiyou Mountains, Oregon and California. *Ecological Monographs* 30:279–338.

———. 1975. *Communities and Ecosystems.* 2d ed. New York: Macmillan.

Whittaker, R. H., and W. A. Niering. 1968. Vegetation of the Santa Catalina Mountains, Arizona. IV. Limestone and acid soils. *Journal of Ecology* 56:523–544.

Wild, H. 1975. Termites and the serpentines of the Great Dyke of Rhodesia. *Transactions of the Rhodesia Scientific Association* 57:1–11.

Wild, H., and A. D. Bradshaw. 1977. The evolutionary effects of metalliferous and other anomalous soils in south central Africa. *Evolution* 31:282–293.

Wildman, W. F., M. L. Jackson, and L. D. Whiting. 1968. Iron-rich montmorillonite formation in soils derived from serpentinite. *Soil Science Society of America Proceedings* 32:787–794.

Williams, P. W., ed. 1993. *Karst Terrains: Environmental Changes and Human Impact.* Cremlingen-Destedt, Germany: Catena-Verlag.

Williams, R. B. G. 1988. The biogeomorphology of periglacial environments. In *Biogeomorphology,* ed. H. A. Viles, pp. 222–252. New York: Basil Blackwell.

Wilson, E. O., ed. 1988. *Biodiversity.* Washington, D.C.: National Academy Press.

Wolf, A., P. A. Brodmann, and S. Harrison. 1999. Distribution of the rare serpentine sunflower, *Helianthus exilis* (Asteraceae): the roles of habitat availability, dispersal limitation and species interactions. *Oikos* 84:69–76.

Wood, C. E., Jr. 1971. Some floristic relationships between the southern Appalachians and western North America. In *The Distributional History of the Biota of the Southern Appalachians. Part II. Flora,* ed. P. C. Holt, pp. 331–404. Blacksburg: Virginia Polytechnic Institute.

Wood, E. F., ed. 1991. *Land Surface–Atmosphere Interactions for Climate Modeling.* Dordrecht: Kluwer Academic Publishers.

Wright, R. D., and H. A. Mooney. 1965. Substrate-oriented distribution of bristlecone pine in the White Mountains of California. *American Midland Naturalist* 73:257–284.

Wu, L. 1989. Colonization and establishment of plants in contaminated sites. In *Heavy Metal Tolerance in Plants: Evolutionary Aspects,* ed. A. J. Shaw, pp. 269–284. Boca Raton, Fla.: CRC Press.

Wulff, E. V. 1943. *An Introduction to Historical Plant Geography.* Waltham, Mass.: Chronica Botanica.

Wyatt, R., and N. Fowler. 1977. The vascular flora and vegetation of North Carolina granite outcrops. *Bulletin of the Torrey Botanical Club* 104:245–253.

Zedler, P. H. 1987. The ecology of Southern California vernal pools: a community profile. Biological Report 85 (7.11). Washington, D.C.: Fish and Wildlife Service, U.S. Department of the Interior.

Zobel, D. B., and J. A. Antos. 1986. Survival of prolonged burial by subalpine forest understory plants. *American Midland Naturalist* 115:282–287.

———. 1987. Survival of *Veratrum viride,* a robust herbaceous perennial plant, when buried by volcanic tephra. *Northwest Science* 61:20–22.

Zwinger, A. H., and B. E. Willard. 1972. *Land Above the Trees: A Guide to American Alpine Tundra.* New York: Harper and Row.

Index

Boldface numerals indicate illustrations.

ARTHUR R. KRUCKEBERG, professor emeritus of botany at the University of Washington, is the author of *The Natural History of Puget Sound Country* and *Gardening with Native Plants of the Pacific Northwest*. Photo © by Mary Randlett.